ADVANCED NANO MATERIALS FOR INDUSTRIAL APPLICATIONS

ADVANCED NANO MATERIALS FOR INDUSTRIAL APPLICATIONS

Editors

V. Rajendran
P. Manivasakan
K.E. Geckeler

BLOOMSBURY
London • New Delhi • New York • Sydney

First Published, 2013

BLOOMSBURY PUBLISHING INDIA PVT. LTD.
London New Delhi New York Sydney

ISBN: 978-93-82563-34-1

Published by Bloomsbury Publishing India Pvt. Ltd.
VISHRUT Building, DDA Complex, Building No. 3, Pocket C-6 & 7
Vasant Kunj, New Delhi 110 070

Laser Typeset by FORTUNE GRAPHICS
WZ-911/2, Shankarlal Street, Naraina, New Delhi

Printed at Anvi Composers
1 DDA Market, Block A-1B, Pachim Vihar, New Delhi 110 063

Defence Institute of Advanced Technology

(DEEMED UNIVERSITY)

Established u/s 3 of the UGC act 1956. ISO 9001 Certified Institute

(An Autonomous Organization, Department of Defence Research & Development, Ministry of Defence, Govt. of India)

Foreword

I am glad to congratulate the initiative taken by Centre for Nano Science and Technology of K. S. Rangasamy College of Technology (KSRCT), KSR Group of Institutions, Tamil Nadu, and India in organising the MAM-12, an International Symposium on Macro- and Supramolecular Architectures and Materials during November 21-25, 2012. I hope that this is a grand fest in the specialised field of nanosystems and applications through sharing knowledge among people worldwide at one place.

Being a private organisation, it is an achievement by the Centre for Nano Science and Technology (CNST) in its R & D activity in association with World class University, South Korea. It is my pleasure to say about the initiation of the Director, Dr. V. Rajendren for the department who developed a good infrastructure and research facilities to the international standards, making people excel in the field of Nanoscience and technology within a short period of time. Centre made research collaborations and Memorandum of Understanding (MoU) with several government Organisations, universities and laboratories both in India and abroad.

MAM-12 will be the centre for international scientists and research engineers to distribute innovative concepts, technologies and important landmarks in the areas of nanotechnology. I understand that 36 invited speakers and more than 350 delegates from different research institutes and universities around the world are participating in the MAM-12. I am sure that this will be a great opportunity to the participants to utilise this knowledge fest.

I am very happy to give the foreword message for the book released under the title of Advanced Nanomaterials for Industrial Applications which comprises of 65 research papers identifies the industrial importance of nanomaterials and product developments.

I ensure that the participants will have effective deliberations through this conference. I wish to congratulate the management and CNST team members to make this conference a grand success.

Date : 25th October 2012 **(Dr. Prahlada)**

Phone: +91-020-24384021, Fax: 020 24389509 DIAT (DU) Girinagar, Pune 411025.

EDITORS

Dr. V. Rajendran FUSI, FASI, FInstP, is Director, Research & Development, K S R Group of Institutions and Centre for Nano Science and Technology, K.S Rangasamy College of Technology, Tamil Nadu, India. Under his able guidance, 11 scholars have completed and 14 scholars are pursuing their Ph.D. degrees. He has published more than 150 research papers in reputed international and national journals, 50 papers in conference proceedings, 28 refereed books, 4 R&D books and 11 patents. He has won many awards including the Prof. K. Arumugam National Award in 2011, Best faculty award in 2010, Raman-Chandra Sekhar silver medal (2010), Tamil Nadu Scientist Award, NDT Man of Year 2004, DAAD from Germany (2002), INSA, TNSCST Young Scientist, DAE/BRNS Visiting Scientist, Best paper award from MRSI, ASI, ASA and USI and Outstanding Organiser Award for the 7th National Symposium on Ultrasonics, 1996.

Dr. P. Manivasakan is an Assistant Professor, Centre for Nano Science and Technology (CNST), K.S. Rangasamy College of Technology. He obtained his Master of Philosophy degree in chemistry from Madurai Kamaraj University, Madurai. He has completed his Ph.D. in April 2012 from Anna University, under the guidance of Dr. V. Rajendran. He has worked as a Junior/Senior Research Fellow for three years under a DST sponsored research project. He has carried out extensive research work on synthesis and characterization of Al_2O_3, ZrO_2 and TiO_2 nanoparticles and their industrial applications. He has published 12 research papers in the peer reviewed international journals and five papers in international conference proceedings and presented his research work in five international conferences.

Dr. K.E. Geckeler is the Founding Chair of the Department of Nanobio Materials and Electronics, World Class University (WCU), Gwangju, South Korea. He is also a Professor in the Department of Materials Science and Engineering and the Department of Medical System Engineering at the Gwangju Institute of Science and Technology (GIST), South Korea and serves as Vice Director of the Gruenberg Center for Magnetic Nanomaterials (GCMN). He received his Ph.D. and M.D. degrees from the University of Tuebingen, Germany (both degrees: "magna cum laude") and spent sabbatical leaves at Harvard University, University of Montana, Clemson University (USA) and at the University of Montpellier (France). He received a series of awards such as the "Fonds der Chemischen Industrie", the "Fritz-Ter-Meer Award", the "Science Price of the President of Korea" and the "Outstanding Research Award of GIST". He has to his credit initiation and organizing the biannual MAM Symposia series. He has published more than 300 research articles and short communications, 12 book chapters, over 10 books, and over 80 patents.

PREFACE

The International Symposium on Macro- and Supramolecular architectures and materials (MAM-12) with a special topic on Nano systems and Applications is organised by Centre for Nano Science and Technology (CNST) of K. S. Rangasamy College of Technology. This conference is jointly organised with World Class University, Gwangju Institute of Science and Technology (GIST), South Korea.

More than 525 abstracts have been received from India and abroad. Out of 400 full papers received, 260 have been selected and peer reviewed by the expert committee for the publications in conference books. All the received books are classified under four headings namely Nano Material Synthesis and Characterisation, Advanced Nanomaterial for industrial applications, Nano biomaterials and Applications of Nanomaterials: Electronics, Energy, & Environment.

A total of 260 accepted papers 60 of contributed papers for the inclusion in the book entitled Advanced Nanomaterial for industrial applications after peer review. This book is assembled from the contributed papers containing the mass production of advanced nanomaterials for various industrial applications such as refractory electronic and automobiles. The most effort has been taken in getting the review of all the contributed papers. All the contributed authors are extended by our sincere thanks for their timely submission and cooperation in carrying out suggestions by the referees.

The CNST has well established the start–of–art of experimental facilities and looking for industrial collaboration and partners for the technology transfer in thrust areas of nanotechnology. We are sure that the outcome of MAM-12 open up new avenue between the researchers and industrialists for the development of nano products.

We owe our special thanks to members of technical committee for peer review of contributed papers. The various government funding agencies, private organizations and industries are thankful for their munificent support and sponsor for the successful conduct the conference. The various committee chairs and members are highly acknowledged for making this event a grant success.

The support extended by Mr. Suresh Gopal and. Mrs. Jyoti Mehrotra of Bloomsbury Publishing India Pvt. Ltd in bringing out this book on time is highly appreciated.

V. Rajendran
P. Manivasakan
Kurt E. Geckeler

CONTENTS

STUDIES ON HEAT TRANSFER BEHAVIOUR OF NANOFLUIDS

P.P. Kulkarni, P. Mandke[1], A.K. Nayak and P.K. Vijayan

Reactor Engineering Division, Bhabha Atomic Research Centre, Trombay, Mumbai
[1]Walchand College of Engineering, Sangli
E-mail: parimalk@barc.gov.in

ABSTRACT

Nanofluids are stable colloidal suspensions of nanoparticles, nanofibers, or nanocomposites in common base fluids, such as water, oil, ethylene-glycol mixtures (antifreeze), polymer solutions, etc. Nanoparticles are very small, nanometer-sized particles with dimensions usually less than 100 nm (nanometers). The smallest nanoparticles, only a few nanometers in diameter, contain only a few thousand atoms. These nanoparticles can possess properties that are substantially different from their parent materials. Study on nanofluids heat transfer is rather a very recent subject. However, it has attracted many researchers because of their promising heat transfer characteristics vis-à-vis water for applications in future nuclear and process industries.

In the similar framework, the authors have investigated the pool boiling heat transfer characteristics of Al_2O_3 Nanofluids. Experiments were carried out to measure the heat transfer coefficient of water and Alumina nanofluid prepared by dispersing nanoparticles inside water by ultrasonication. It was observed that the heat transfer coefficient of nanofluid was found to be higher in single phase convection region whereas in nucleate boiling region, the heat transfer coefficient is found to higher in water.

Further, experiments were carried out to determine the critical heat flux of Alumina Nanofluids for concentrations ranging from 0.001% to 0.5% by weight. In all the experiments, enhancement in CHF was observed as compared to water. This property of nanofluids certainly highlights the potential application of nanofluids in nuclear industries and need for further investigation

---

V. Rajendran, P. Manivasakan and K.E. Geckeler (eds.)
Advanced Nano Materials for Industrial Applications, pp. 1-6 (2012).

INTRODUCTION

Boiling phenomenon is of immense importance in nuclear reactor systems both as a means of achieving high heat-transfer rates from fuel to coolant and for generating steam in a heat exchanger. Till now many efforts have been done towards increasing the boiling heat transfer coefficients of conventional fluids by making use many additives in the base fluid. One of such method is adding suspended solid particles to increase the effective thermal conductivity of the base fluid. Although, this method has been used since long, traditionally, microscaled particles were used as dispersants. However, they were of little interest from practical point of view due to problems such as settling, erosion, fouling and increased pressure drop of the flow channel. The advent of nanotechnology gave rise to an alltogether different kind of fluids called nanofluids which overcame all the problems faced by traditinal suspensions. With very low settling rates and high stability, nanofluids have become topic of interest owing to their enhanced heat transfer characteristics as compared to base fluids[1-3]. In the present work, the authors have investigated the pool boiling heat transfer characteristics of Al_2O_3 Nanofluids. Experiments were carried out to measure the heat transfer coefficient of water and Alumina nanofluid prepared by dispersing nanoparticles inside water by ultrasonication.

In boiling reactor systems, occurrence of critical heat flux (CHF) is undesirable. Hence, for same operating conditions, increase in CHF can provide higher thermal margins. Use of nanofluids could be one such way to enhance the CHF [4-6]. To explore this, experiments were carried out to determine the critical heat flux of Alumina Nanofluids for concentrations ranging from 0.001% to 1% by weight.

POOL BOILING HEAT TRANSFER CHARACTERISTICS OF NANOFLUIDS

Experiments were carried out to measure the two phase heat transfer coefficient of water as well as nanofluid.

Experimental Setup

The schematic of the experimental test facilit is given in Fig. 1.

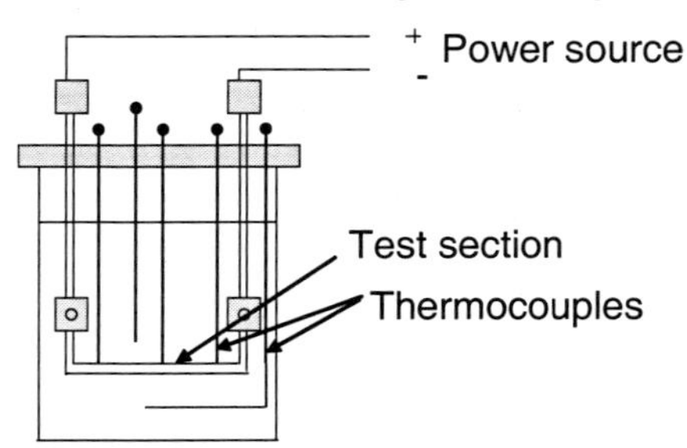

Fig. 1: Schematic of Test Section

The experiemntal setup consists of a 10 lit capacity glass beaker in which about 8 litres of nanfluid was poured. An SS 304 tube with diameter 6 mm and thickness 0.6 mm was used as a heater. The tube was directly heated with a DC source of 400 A, 50 V power supply. Three 0.5 mm 'K' type Thermocouples were brazed on the surface of the tube to measure the surface temperature. Besides, the water bulk and film temperature was measured using another 2 thermocouples.

Nanofluid Preparation

Nano-fluids were prepared by the two-step method, dispersing dry nanoparticles into the base liquid followed by uktrasonication. Distilled water was used as a base liquid, and Al_2O_3 (40 – 60 nm, 99.5 % purity Alfa Aesar make) nanoparticles were used. Some additives (dispersants or surfactants) could have been used to stabilize the nanoparticle suspensions, but the additives could exert a significant influence on the rheological behavior of the fluids and the boiling heat transfer. Therefore, in this work, no additive was used and ultrasonication was performed for 4-5 h just before pool boiling experiments. Nanoparticles dispersed without other additives do not change the surface tension of the base fluid. Concentration of 0.5 % by weight was used.

Experimental Procedure

Experiment was first carried out with pure water. Initially sub cooled water was put in the beaker. Power was applied to the test section. The power was raised in small steps and it was kept constant for a period of 5 mins for each power level. Transient history of temperatures on the surface, in the film just next to the heater and in bulk of the fluid was recorded. Experiment was continued till nucleate boiling was substantial. Then, similar experiment was repeated with same power steps for nanofluid.

Results

Fugure 2 shows the measured ΔT variation of nanofluid and water.

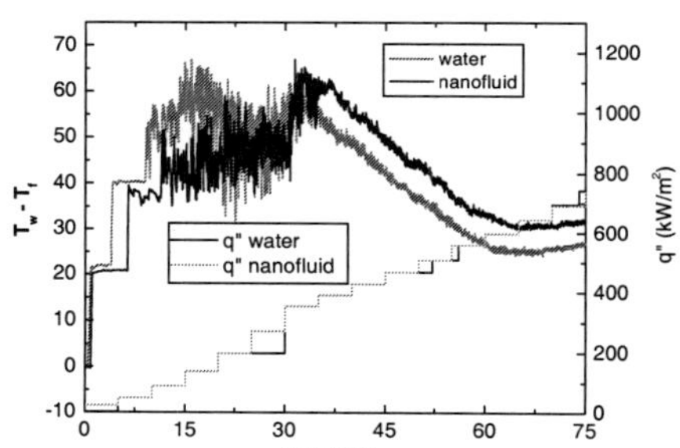

Fig. 2: ΔT Variation of Nanofluid and Water

It can be seen that, initially, the ΔT for water is higher than that of the nanofluid indicating that, the surface temperature is higher than that of fluid. But after boiling, it was seen that the ΔT of water reduced as compared to nanofluids. Figures 3 and 4 show the calculated heat transfer coefficients in single phase convective and two phase nucleate boiling region.

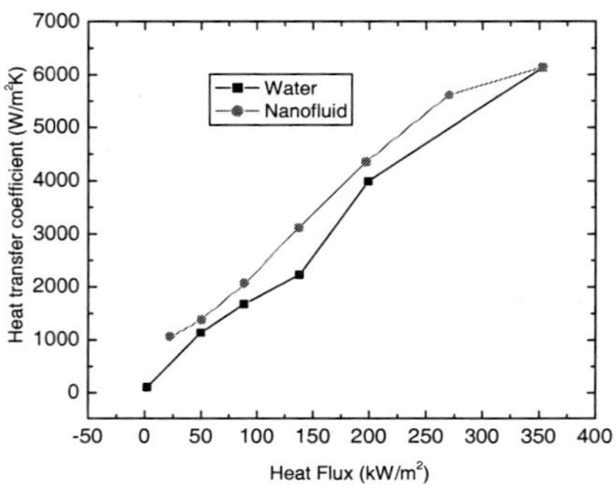

Fig. 3: Heat Transfer Coefficient in Single Phase Region

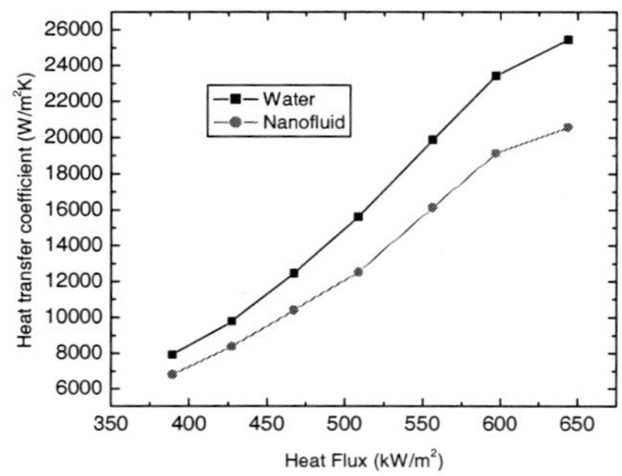

Fig. 4: Heat Transfer Coefficient in Nucleate Boiling Region

It is observed that, Initially, the heat transfer coefficient of nanofluid was found to be higher in single phase convection region whereas in nucleate boiling region, the heat transfer coefficient is found to higher in water.

Discussions

By addition of nanoparticles in water, the thermal conductivity is found to be enhanced as reported by [1]. As a result of increase in thermal conductivity, the single phase convective heat transfer coefficient is found to be enhanced. On the contrary, when boiling occurs, there is a microlayer deposition of nanoparticles [7] on the heater surface which causes deterioration of heat transfer coefficient in nucleate boiling condition.

CHF CHARACTERISTICS OF NANOFLUIDS

Experiments were carried out to determine the critical heat flux of Alumina Nanofluids for concentrations ranging from 0.001% to 1% by weight

Experimental setup

The experimental setup was slightly modified for the CHF experiements. The experiments were performed with a 0.2 m diameter NiChrome wire instead of SS tube. A 3 litre beaker was used in this case. The modified setup is given in Fig. 5.

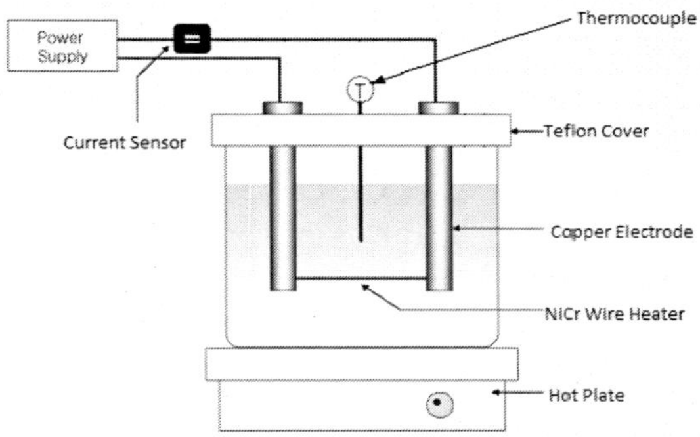

Fig. 5: Experimental Setup for CHF Experiments

Experimental Procedure

Initially, the experiment was performed with pure water. The water was heated using the heater and the temperature was brought to near saturation (95 °C). Then power was applied to the Nichrome wire. Power was raised in very small steps. At a point when CHF occurs, a bright red spot is observed in the Nichrome wire and it gets burned which trips the electirc current. This gives accurate indication of CHF. Similar experiments were repeated for nanofluids of concentrations ranging from 0.001% to 1% by weight of Alumina. For each concentration, experiment was repeated 3-4 times to get accurate results.

Results

Figure 6 shows the variation of CHF with increase in nanofluid concentration. It is seen that, even low concentration like 0.001 % enhances the CHF by 45 %. CHF first increases with increase in concentration and then it is found to be decreased. But as compared to water, all the concentrations tested had enhancement in CHF.

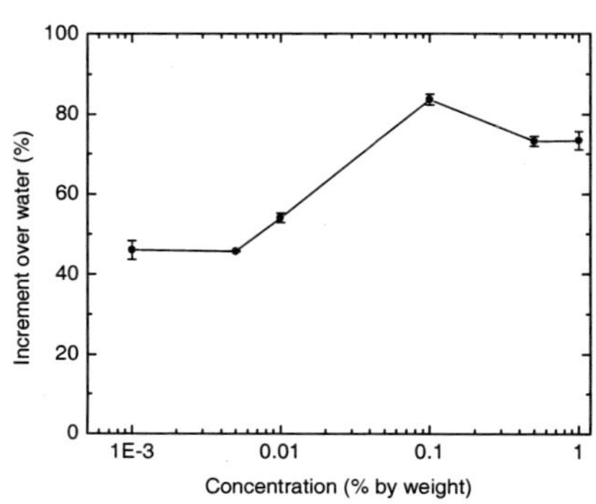

Fig. 6: Enhancement of CHF over Water in Case of Nanofluids

Discussion

It has been observed from Fig. 6 that, there is a particular trend in enhancement of CHF with increase in nanoparticle concentration. The main increase in CHF is because of change in surface properties of wire. In case of low concentration of nanoparticles, there is a fine deposition of nanoparticles on the wire which increases the surface roughness increasing the nucleation sites. This causes enhancement in the CHF. But as the concentration of nanoparticles increases, the deposition layer thickness increases, hence now the number of nucleation sites are less as comparted to those in case of nanoparticles with low concentration but more than that of bare wire. As a result, the CHF decreases but is more than that in case of pure water.

CONCLUSION

In this paper, we have studied the heat transfer bahaviour of nanofluids. The heat transfer coefficient of nanofluids was determined experimentally. It was observed that, single phase convection heat transfer of nanofluids was found to be higher than that of water whereas, the nucleate boiling heat trasnfer coefficient is found to be higher in water. Critical heat flux of nanofluids was also measure experimentally. It was observed that, at even low concentration of 0.001% by weight also, there was substantial enhancement in CHF (45 %). In case of nanofluieds, CHF first increased with increase in weight percentage and then decreased, but remained always higher than that of water. The reason for enhancement in CHF can be attributed to change in surface properties of wire.

REFERENCES

[1] Jing Fan and Liqiu Wang, *Transactions of the ASME*, 133, 040801-14, 2011.

[2] In Cheol Bang, Soon Heung Chang, Int. *J. Heat Mass Tr.*, 48, 2407–2419, 2005.

[3] Sadik Kakaç and Anchasa Pramuanjaroenkij, *Int. J.Heat Mass Tr.*, 52, 3187–3196, 2009.

[4] Ramakrishna Hegde, Srikanth S. Rao and R.P. Reddy, *Heat Transfer—Asian Research*, 39 (5), 323-331, 2010.

[5] Fujio Inasaka and Hideki *Nariai, Fusion Engineering and Design* 19, 329-337, 1992.

[6] M.N. Golubovic, H.D. Madhawa Hettiarachchi, W.M. Worek and W.J. Minkowycz, *Applied Thermal Engineering*, 29, 1281–1288, 2009.

[7] Dongsheng Wen, Michael Corr, Xiao Hua and Guiping Lin, *Int. J. of Therm. Sci.* 50, 480-485, 2011.

DIELECTRIC STUDY ON MECHANOCHEMICAL SYNTHESIS OF $CaTiO_3$ NANOPARTICLES

Subhanarayan Sahoo, S.K.S. Parashar[1] and S.M. Ali

School of Electrical Engineering, KIIT University, Bhubaneswar, Odisha
[1]School of Applied Sciences, KIIT University, Bhubaneswar, Odisha
E-mail: narayan_240@yahoo.com,subhanarayannanotech@gmail.com

ABSTRACT

The mechanochemical synthesis of calcium titanate ($CaTiO_3$) was carried out in a planetary high energy ball mill. The size and crystal structure of $CaTiO_3$ mixture were affected by the process parameters of the planetary mill. The crystallite size of the particles decreases with increase in milling time. The dielectric, impedance and ac conductivity were measured in a frequency range of 100Hz-1MHz at temperature 30°C to 500°C. It is interesting to observe that the incipient property was observed in a wider range of temperature for nanocrystalline $CaTiO_3$.

INTRODUCTION

Oxides with the perovskite structure are well known for their ability to produce high dielectric constants with ferroelectric property leading to many important applications[1]. Titanates as members of the perovskite structure family are widely used as ferroelectric materials with superior dielectric properties[2]. The properties of these materials are largely depend on the method of preparation, thermal treatments and usually modified by formation of solid solution Calcium titanates possess perovskite type structure with interesting electrical properties for technological applications, like thermisters, multilayer capacitors and DRAM memories[3], photoelectric and photochemical devices .$CaTiO_3$ is a promising material in the production of communication equipment ,resonators and filters[4,5,6].Also ,this material can be employed as a thermally sensitive resistor element due to negative temperature coefficient .Such unique

V. Rajendran, P. Manivasakan and K.E. Geckeler (eds.)
Advanced Nano Materials for Industrial Applications, pp. 7-12 (2012).

properties gave this material much attention, and many investigators regarding its many uses have been carried out in recent years[7]. High-energy ball milling has been used for many years in producing ultra fine powders in the range of sub-micron to a nanometer. Aside from size reduction, this process causes severe and intense mechanical action on solid surface, which was known to lead to physical and chemical changes in the near surface region where the solids come in contact under mechanical forces[8].

EXPERIMENTAL

The raw powders calcium oxide (CaO) and titanium dioxide (TiO_2) were weighed stoichiometrically for preparation of $CaTiO_3$ by high energy ball milling. The ball to powder ratio was maintained to be 20:1. The powders were milled at 300 rpm different milling time. The structural analysis of the milled powder was done by X-ray diffractometer (PAN anlytical, Japan). The powders were pressed into disc shape by applying 4 ton load for 180 sec by using hydraulic press. The pellets were sintered at 900°C for 2 hrs in a programmable furnace. The dielectric data were taken using computer interfaced LCR meter (HIOKI 3532-50 LCR HiTester) fitted with a furnace. The dielectric data were taken at a heating rate of 1°C/min.

RESULTS AND DISCUSSION

The structural changes and microstructure characterization of nano sample analyzed using X-ray defraction. Figure 1 shows the XRD pattern of the starting and milled powders. The diffraction patterns of milled sample have lower intensity and broader peak base compared to the starting powder, mainly due to the process of disordering of CaO and TiO_2 crystal structure from intensive milling. The XRD pattern indicated that powder ball milled for 15 hour in 20:1 BPR had single phase according to the JCPDS Card No22-0153. The average particle size of $CaTiO_3$ nanoparticles with approximately 90 nm.

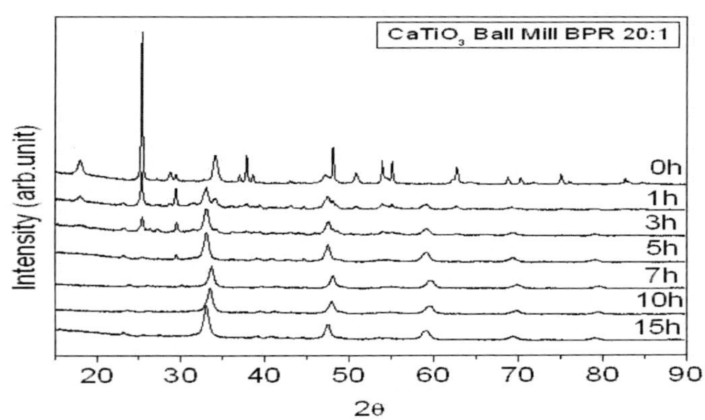

Fig. 1: XRD Patterns of $CaTiO_3$

Figure 2 shows the variation of dielectric constant with respect to frequency at various temperature. The nature of the dielectric permittivity for free dipoles oscillating in an alternating field may be described in the following way. At very low frequencies ($\omega < 1/\tau$, τ = relaxation time), dipoles follow the field and we have $\varepsilon' = \varepsilon_s$ (value of the dielectric constant at quasistatic fields). As the frequency increases (with $\omega < 1/\tau$), dipoles begin to lag behind the field and σ slightly decreases. When frequency reaches the characteristic frequency ($\omega = 1/\tau$), the dielectric constant drops (relaxation process). At very high frequencies ($\omega > 1/\tau$), dipoles can no longer follow the field and $\varepsilon' \approx \varepsilon_{\infty''}$ (high-frequency value of ε'). Qualitatively this behavior has been observed in Figure 3(a). The dielectric constant at low frequency is rather high, and is found to decrease with frequency at first and then becomes more or less stabilized. The high value of ε' at frequencies lower than 1 kHz, which increases with decreasing frequency and increasing temperature, correspond to bulk effect of the system.

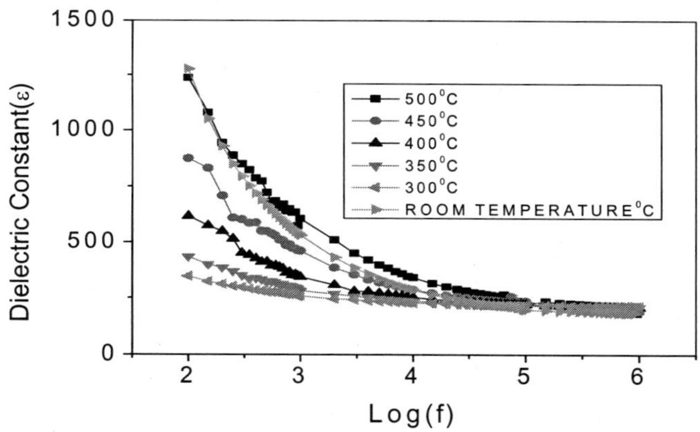

Fig. 2: Frequency vs Dielectric Constant at Different Temperature

Figure 3 shows the variation of the real part of impedance (Z') with frequency at various temperatures. It is observed that the magnitude of Z' decreases with the increase in both frequency as well as temperature, indicate an increase in *ac* conductivity with the rise in temperature and frequency. The Z' values for all temperatures merge above 1 kHz. This may be due to the release of space charges as a result of reduction in the barrier properties of material with the rise in temperature and may be a responsible factor for the enhancement of *ac* conductivity of material with temperature at higher frequencies. Further, at low frequencies the Z' values decrease with rise in temperature shows negative temperature coefficient of resistance (NTCR) type behavior like that of semiconductors. The coincidence of the impedance (Z') at higher frequencies at all the temperatures indicates a possible release of space charge as a reduction in barrier properties of the material with rise in temperature.

Figure 4 shows the variation of the imaginary part of impedance (Z'') with frequency at different temperatures. The curves show that the Z'' values reach a maxima peak (Z''_{max}) and the value of Z''_{max} shifts to higher frequencies with increasing temperature. A typical peak broadening which is slightly asymmetrical in nature can be observed with the rise in temperature. The broadening of peaks in frequency explicit plots of Z'' suggests that there is a spread of relaxation times, i.e. the existence of a temperature dependent electrical relaxation phenomenon in the material.

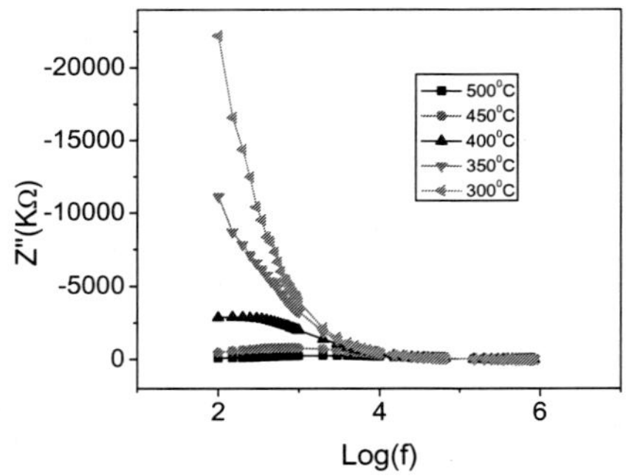

Fig. 3: Frequency vs *Z'*

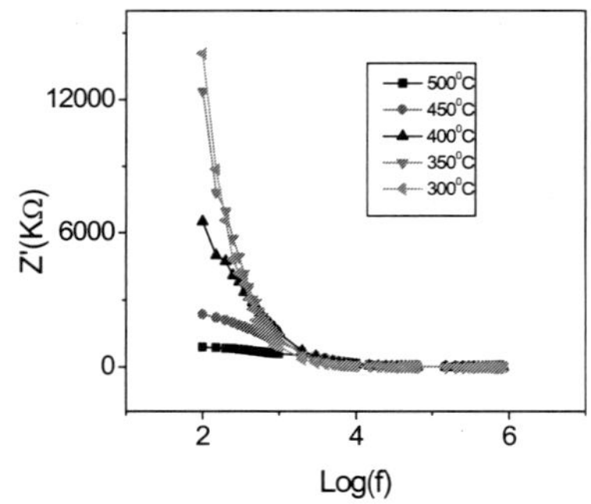

Fig 4. Frequency vs *Z'*

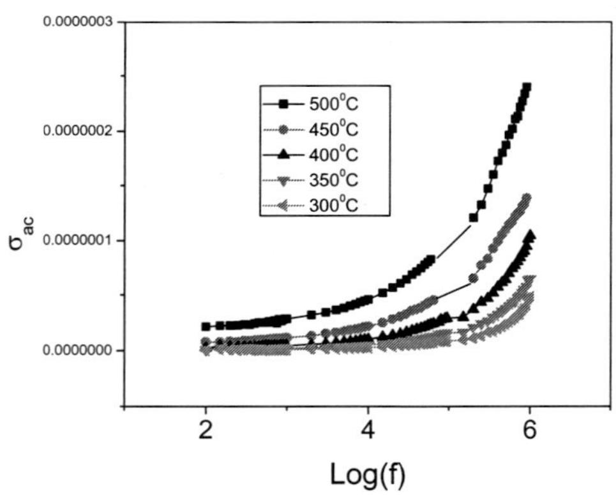

Fig. 5: Frequency vs AC Conductivity

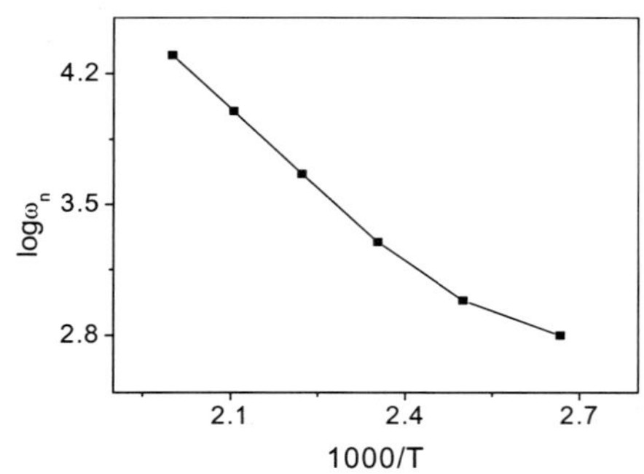

Fig. 6: Activation Energy Graph

Figure 5 shows the plot of *ac* conductivity with frequency at different temperatures. The frequency dependence of conductivity of the material at room temperature exhibits both low and high frequency dispersion phenomena. This obeys the Jonscher's power law equation given by

$$\sigma(\omega) = \sigma_{dc} + A\omega^n$$

where n is the frequency exponent in the range of $0 < n < 1$, A *and* n are thermally activated quantities, hence electrical conduction is thermally activated process. According to Jonscher [9], the origin of the frequency dependence of conductivity lies in the relaxation phenomena arising due to mobile charge carriers. When a mobile charge carriers hop to a new site from its original position, it remains in a state of displacement between two potential energy minima, which includes contributions from other mobile defects.

The frequency ω_m (corresponding to Z''_{max}) gives the most probable relaxation time τ_m from the condition $\omega_m \tau_m = 1$. The most probable relaxation time follows the Arrhenius law, given by

$$\omega_m = \omega_0 \, exp\left(\frac{-E_a}{k_B T}\right)$$

where ω_o is the pre-exponential factor and E_a is the activation energy. Figure 6 shows a plot of the *log* ω_m versus $1/T$, where the dots are the experimental data and the solid line is the least-squares straight-line fit. The activation energy E_a calculated from the least-squares fit to the points. From Figure 6 it can be seen that the activation energy from Arrhenius relation for relaxation of Z''.

CONCLUSIONS

The $CaTiO_3$ ceramics has been prepared by high energy ball milling at different milling time at room temperature with BPR 20:1. The X-ray analysis shows that the ceramic ball milled for 15 hrs presents a single phase. The particle size of the ceramic ball milled for 1 hr is found to be 90nm. The electrical parameters such as the real part of impedance (Z'), the imaginary part of impedance (Z) and AC conductivity as a function of both frequency and temperature have been studied through impedance spectroscopy. The electrical relaxation process occurring in the material has been found to be temperature-dependent. The AC conductivity spectrum is found to obey 'Jonscher's universal power law.

ACKNOWLEDGEMENT

The authors are grateful to Center for Nano Technology & Supporting Staffs, KIIT University.

REFERENCES

[1] G. Mi, Shindo FY., D Murakami, F. Saito, *Powder Technol*, 75-79, 104,1999.

[2] M. Sh. Khalil and F.F. Hammad, Egypt' Influence of Cu Addition on the Structure and Dielectric Properties of $CaTiO_3$'. *J. Sol., Vol.* (25), No. (2), 2002.

[3] National Cheng Kung Univ., Dept. Elect. Eng., Tainan, Taiwan, *Proc. Natl. Sci. Counc.* Part 3, 10, 229 ,1977.

[4] G. Pfaff, *Chem. Mater.* 6,1994.

[5] L.S. Cavalante,V.S. Marques, J.C. Scznacoski, M.T. Escote, M.R. Joya, J.A. Varela, M.R.M.C. Santos, P.S Pizani, E. Longo, Chem. Eng. J. 143, 2008.

[6] L.S. Cavalante, A.Z. Simoes, L.P.S. Santos, *Mater. Res. Bull.*, 1203, 2003.

[7] I.R. Evans, J.A.K Howard, T. Sreckovic, M.M. Ristic, *Mater. Res. Bull.*, 1203, 2008.

[8] K.S. Venkataraman, K.S. Narayan, *Powder Technol*, 190-201, 1998.

[9] J. Zhu, X.B. Chen, J.H. He and J.C. Shen, *Phys. Lett.* A 362-471,2007.

DEFECT DETECTION AND STRAIN ANALYSIS ON THE SURFACE USING CARBON NANOTUBE BASED ELECTRO MECHANICAL SYSTEMS

A. Deepak, V. Karthick[1], Vaidehi Ganesan[1], Varrla Eswaraiah[2], Sundara Ramaprabhu[2] and P. Shankar

Saveetha School of Engineering, Saveetha University, Chennai, Tamil Nadu
[1]Indira Gandhi Centre for Atomic Research, Kalpakkam, Tamil Nadu
[2]Indian Institute of Technology Madras, Chennai, Tamil Nadu
E-mail: madhurilakhane@gmail.com

ABSTRACT

Industrial Material Testing and evaluation is a process of finesse and accuracy. The use of Nanosensors in this application would take the right step towards its improvisation. This work proposes a novel approach in non-destructive evaluation technique of industrial materials using Multiwalled Carbon Nanotube (MWCNT) based NanoElecroMechanical system (NEMS) for defect identification and strain analysis of engineering components. The MWCNTs are blended with the PVDF composites at different weight percentages to form the sensor film. Different weight percentages of MWCNT/PVDF films were characterised using various techniques like Scanning Electron Microscope (SEM) and Raman analysis. Surface defects are measured by pasting the film directly over the defective metal plate and resistance variation are noted for the corresponding change in defect size. To measure strain, this sensor film is attached to the specimen with a non-conductive adhesive and the test load is applied using the Universal Testing Machine (UTM). Silver epoxy metal contacts are made on both ends of the film. The readings are obtained based on the changes in the resistance between these contacts. Current source is connected to one end of the two probe setup. The calibration curves are plotted based on the relation between

V. Rajendran, P. Manivasakan and K.E. Geckeler (eds.)
Advanced Nano Materials for Industrial Applications, pp. 13-18 (2012).

the applied load and voltage across the sensor. The results show that this method gives better accuracy than the conventional methods.

INTRODUCTION

Accuracy, robustness and sensitivity are very important in the advanced field of Engineering. Compromising over these factors in designing any electromechanical system would lead to an incompetent product. Nano-electro-mechanical systems can be used in various engineering applications due to their novel properties. In this work the main focus is to use a nanosensor to detect the surface defects and to find the surface strain of any engineering component made of materials like stainless steel, ferrite steel, carbon steel, Zr alloy etc. Non-destructive testing (NDT) is a wide group of analysis techniques used in industry to evaluate the properties of a material, component or system without causing physical damage to it. There are many non-destructive techniques available like ultrasonic, eddy current, thermal imaging, radiography, phased array, XRD residual stress etc., for failure analysis, structural integrity assessments, health monitoring and in-service inspection[1][2]. This paper proposes a new simple Non-destructive evaluation (NDE) technique using a Carbon Nanotube (CNT) based nano-electro-mechanical system for defect identification and strain measurements on engineering components[3].

EXPERIMENTAL PROCEDURE

The strain sensor was synthesized using polyvinylidene fluoride (PVDF) as the polymer material and Multi-walled carbon nanotubes (MWCNT) as filler material. The multi-walled carbon nanotubes were functionalized and dispersed over the PVDF polymer so as to form conductive networks. This helps in improving the conductivity as well as the piezoresistive property of the polymer material[4].

The multi-walled carbon nanotubes are purchased from Shriram Enterprises Pvt Ltd. 45 mg of multi-walled carbon nanotubes are dispersed in 3:1 ratio of sulphuric acid and nitric acid mixture. The solution was heated at 80°C for 6 hours and then cooled down to room temperature in order to impart hydroxyl and carboxyl functional groups over the side walls. Vacuum filtration is carried out to remove the moisture content. The f-MWCNT'S are separated by centrifugation at 400 rpm for 15 minutes[5]. The same procedure was repeated for 4 times. Distilled water was used for washing multiwalled carbon nanotubes. f-MWCNT'S are separated and dried at 80°C for 8 hours. The required amount of f-MWCNT'S (40 mg) was dispersed in DMF (7 ml) and 960 mg of PVDF in 7 ml of DMF separately. Then the two solutions were mixed by sonication for 1 hour and now the composite solution was transformed to a melt mixture and stirred for 2 hours to obtain homogeneous dispersion and it is heated at 80°C for half an hour. Homogeneous solution obtained was poured on glass substrate and dried in oven at 80°C for 8 hours. Solvent free films are obtained [6].

EXPERIMENTAL SETUP

For testing purpose, Zirconium Alloy, which is used in Pressurised Heavy Water Reactor (PHWR) as structural material and SS316LN (structural material of Proto type Fast Breeder Reactor (PFBR)),9 CR 1 MOD(steam generator material of PFBR),12 CR 1 MOD (turbine blade material of PHWR) and carbon steel which is used in many commercial applications are used. These materials are fabricated into dog bone shaped specimens according to ASTM standards as shown in Fig. 1. A two-point probe is attached to the carbon nanotube film by using silver epoxy. Tensile load is applied by connecting the material to Universal Testing Machine[7]. The ASTM material is subjected to tension and current is passed through one of the two probes and a resistor is connected in series to other end of the probe[8].

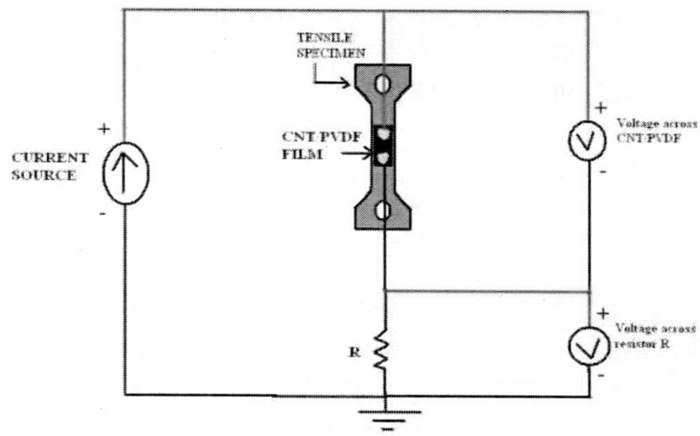

Fig. 1: Tensile Specimen With Two Probe Setup

RESULTS AND DISCUSSION

Characterization of the film was done using different techniques like SEM and Raman and some of the results are presented below.

Fig. 2: SEM Micrograph of CNT/PVDF Film

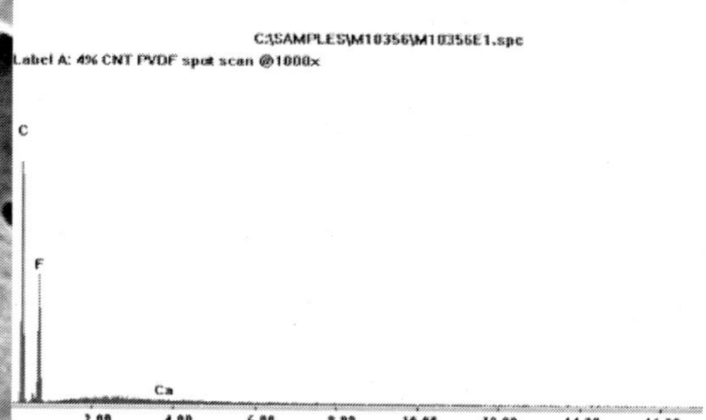

Fig. 3: EDAX of (wt. 4% CNT)

In the Fig. 2, the bright regions are attributed to MWCNTs as a result of their high electrical conductivity. The SEM micrographs show that the MWCNTs are uniformly dispersed over the PVDF matrix. Energy-dispersive X-ray spectroscopy (EDAX) analysis was conducted by means of SEM instrument. It determines the chemical composition of a specimen. Figure 3 shows the percentage of carbon present in the film is more compared to other elements and fluorine is the next element found with higher composition.

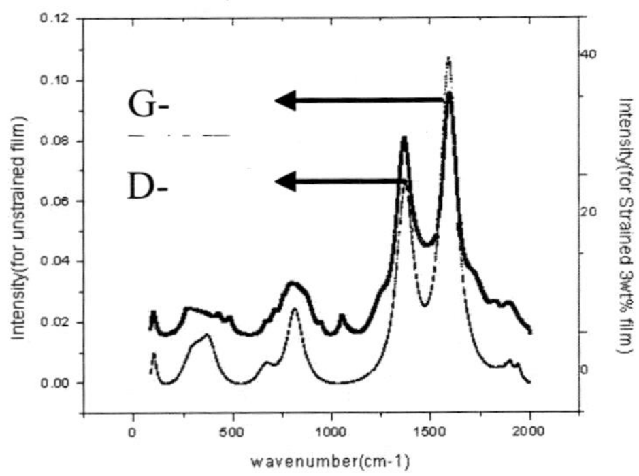

Fig. 4: Raman Spectrum

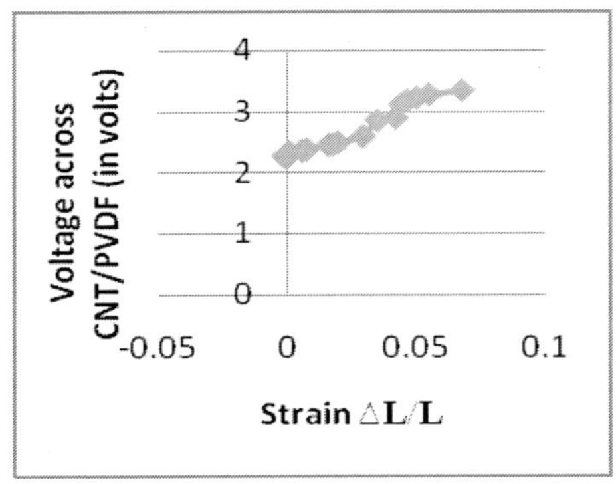

Fig. 5: Strain (VS) Voltage of CNT/PVDF Film

For Raman Analysis, the wavelength of laser light used is 488nm. In the Fig. 4 the dark (black) line indicates the spectrum of Raman scattering on CNT/PVDF nanocomposite (3 wt %) without strain. Slightly dim line indicates the spectrum of Raman scattering on the same MWCNT/PVDF film with strain. The peak position obtained for D-band is at 1368 cm^{-1}.The peak position obtained for G-band is at 1589 cm^{-1}. Intensity of D-band increases for the strained specimen when compared to the unstrained specimen. The Intensity of G-band decreases for the strained specimen. This indicates that as strain is introduced in the film the defect also starts increasing and the ratio of I_G to I_D starts decreases [9].

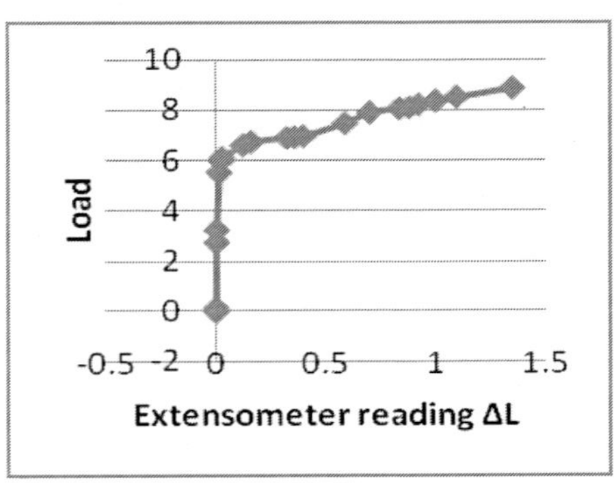

Fig. 6: Load vs. Extensometer Reading

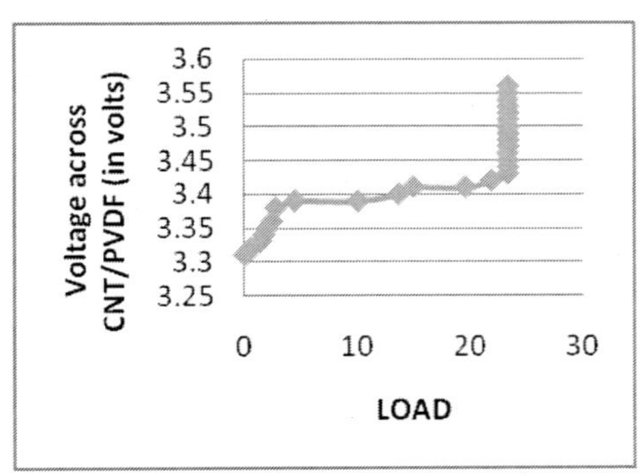

Fig. 7: Load vs. Voltage Across the Film

By applying a constant current across one probe, the change in voltage across the CNT/PVDF film due to tensile load is measured and recorded. Voltage variation is actually due to change in resistance of the film with the change in length of the film. Change in resistance is mainly due to the piezoresistive property of the film. Input current applied is 300 μA. The value of resister used is 47kΩ. Initial resistance across CNT/PVDF film is 11.3kΩ. The Change in resistance of the film used and the Change in length of the film are noted. The ratio of two gives gauge factor [10]. The value of gauge factor determines the sensitivity of the sensor. Three different Calibration curves are plotted. Figure 5 represents Change in voltage in CNT film as a function of strain in the specimen.Figure 6 represents Change in extensometer reading as a function of load. Change in voltage across the film as function of load applied is represented in Fig. 7. Similarly calibration curves are drawn for other materials.

DEFECT IDENTIFICATION

A metal plate is taken and defects with known depth and known length are introduced at various positions.MWCNTs/PVDF films are taken and are cut to equal sizes. These films are pasted on the defects induced region using a nonconductive adhesive and resistance variations in the films are tested at various location. The resistance value increases to a great extent even for a fractional increase in depth of the defect. The test shows that resistance in the zero defect region is 77 Ω. Resistance at 0.7 mm defect region is 4.5K. Resistance at 1.5 mm defect region is 8k Ω. From the test, it is easy to predict the variation of resistance with that of the depth in the defect. The variation in the resistance can be due to the electron scattering. Calibration curves are drawn and one can predict the defect size and length from the variation of resistance.

CONCLUSION

There is linear relationship between change in voltage and the strain observed in the test specimen. This is confirmed in materials like Zr alloy and carbon steel. The result indicates the change in voltage is mainly due to change in resistivity of the film. Defect identification is possible by obtaining a calibration curve as a function of defect size and resistance. So there is a possibility that this technique can be used for real time monitoring of defects and strain measurement on the engineering components at critical locations in any manufacturing plants and Industries. Future work in this field would be to use graphene instead of MWCNT which could increase the sensitivity of the sensor.

REFERENCES

[1] B. Chellappa, Vaidehi Ganesan, T. Balasubramanian and P. Chellapandi, "Structural Integrity Assessment of Scaled Down PFBR Main Vessel Which Had Undergone CDA Test - Using Different NDE Techniques", *E-Journal of Non-Destructive Testing*, pp. 180-184, 2009.

[2] P. Kumar Raja, Vaidehi Ganesan, T. Saravanan, S. Bagavathiappan and Krishnan Balasubramaniam, “Detection of Micro-Cracks in the HP Turbine Blades Using Phased Array and Radiography Techniques and Enhancement of Radiographic Images by Image Processing,”, *E-Journal of Non-Destructive Testing*, pp. 298-303, 2009.

[3] X. Li, C. Levy and L. Elaadil, “Multiwalled carbon nanotube film for strain sensing.,” *Nanotechnology*, vol. 19, no. 4, p. 045501, Jan. 2008.

[4] Varrla Eswaraiah, Venkataraman Sankaranarayanan, and Sundara Ramaprabhu, “Inorganic nanotubes reinforced polyvinylidene fluoride composites as low-cost electromagnetic interference shielding materials”, *Nanoscale research letters*, vol. 6, no. 1, p. 137, Jan. 2011.

[5] G. Mago, D. M. Kalyon and F. T. Fisher, “Membranes of Polyvinylidene fluoride (PVDF) and PVDF Nanocomposites with Carbon Nanotubes via Immersion Precipitation”, *Journal of Nano Materials*, March 2008.

[6] X. Shan, P. Wu, L. Zhang and Z. Cheng, “Crystalinity Properties of Carbon Nanotube-Polyvinylidene Fluoride Composites”, *Material Research Society Symposium Proceedings*, vol. 1134, 2009.

[7] P. Dharap, Z. Li, S. Nagarajaiah and E. V. Barrera, “Flexural strain sensing using carbon nanotube film”, *Sensor Review*, vol. 24, no. 3, pp. 271-273, 2004.

[8] P. Dharap, Z. Li, S. Nagarajaiah and E. V. Barrera, “Nanotube film based on single-wall carbon nanotubes for strain sensing” *Nanotechnology*, vol. 15, no. 3, pp. 379-382, March 2004.

[9] G. Washer and F. Blum, “Raman Spectroscopy for the Nondestructive Testing of Carbon Fiber”, *Research Letters in Materials Science*, vol. 2008, pp. 1-3, 2008.

[10] Ayoola Ike Akinwande Jr, “Nanocomposite Strain Sensors: Study of Electrical and Thermal Properties” MSc Thesis, Rice University, Houston Texas, United States of America, April 1998.

DESIGN AND DEVELOPMENT OF CARBON NANOTUBE FOR PROTECTIVE APPLICATION

S. KUMARAVEL, A. VENKATACHALAM[1] AND THILAK VADICHARLA[2]

Department of Textile Technology, PSG College of Technology, Coimbatore
[1]Department of Textile Technology, Bannari Amman Institute of Technology, Sathy
[2]Department of Fashion Technology, Bannari Amman Institute of Technology, Sathy
E-mail: Kumaravel_s@sify.com; a_venkatachalam@rediffmail.com; thilak.vadicherlaa@gmail.com

ABSTRACT

The Single wall nano tube is interlaced with nylon filament and bullet proof woven fabric will be made. When compare to Kevlar, Carbon nano tube has 17 times of toughness value that used in bullet proof vest.[2] The warp way will be nylon yarn and carbon nano tube material has selected as a weft way yarn. The potential application of developed material with existing material will be analyzed. The ballistic performance of existing and developed material will be analyzed by using hand gun and long rifle.

INTRODUCTION

Requirements of Ballistic Protective Clothing

- Spreading the impact load over a large area
- Dissipation of energy by deformation and breaking the protective material
- High work of rupture
- High tensile strength
- Light weight and comfortable
- High modulus and elongation at break[1]

V. Rajendran, P. Manivasakan and K.E. Geckeler (eds.)
Advanced Nano Materials for Industrial Applications, pp. 19-22 (2012).

Physical Requirements for Military Textile

- Light weight
- Low bulk
- High durability and dimensional stability
- Cleanable
- Good handle and drape
- Low noise emission
- Antistatic [3]

Textile Materials for Ballistic Protection

Ballistic protection involves arresting the fight of projectiles in as short of distance as possible. This requires use of high modulus textile fibres that is those having very high strength and low elasticity. Woven textiles are most commonly used form.[3]

Principles of Ballistic Protection

The principle of ballistic protection by cloth armor is to dissipate the energy of fragment through stretching and breaking the yarns in the many layers of high performance woven fabrics in ballistic vests. Each layer reduces energy of projectile.[4]

Weave Structure Preferred for Ballistic Protection

Plain weave. (To maximize number of cross over points)

Bullet proof vest are consist of three quilted layers. In term of strength, one fabric layer is sufficient to stop a pistol bullet and other two layers are a safety factor, and the air-filled interstices contribute to slowing down the projectile. [5]

Preparation of Carbon Nano Tube Material

The carbon nanotubes are produced using four methods. One of the methods is catalytic chemical vapor deposition method, which involves the decomposition of hydrocarbons.[6]

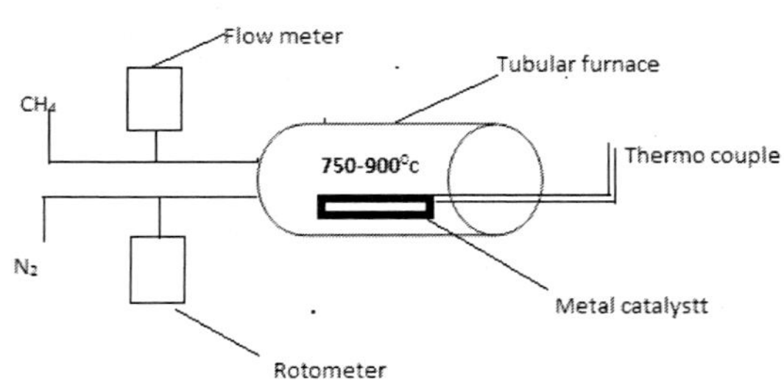

Experimental set up of CVD method for producing Carbon Nano Tube

PRACTICAL SIGNIFICANCE/USEFULNESS

Draw back of Kevlar in Bullet Resistance

Kevlar has a few disadvantages. The fibers themselves absorb moisture, so Kevlar composites are more sensitive to the environment than glass or graphite composites. Although tensile strength and modulus are high, compressive properties are relatively poor. This makes kevlar composites weak under compressive loads.

Kevlar is very difficult to cut. You will need special scissors for cutting dry fabric and special drill bits for drilling cured laminates. Cutting of cured laminates without fraying is very difficult.

Advantages of Carbon Nanotube for Bullet Proof Material

1. Less weight
2. Extra ordinary strong
3. High toughness [6]

EXPERIMENTAL PLAN

Nylon filament yarn sourcing and development of Carbon nano tube

↓

Development of Nylon/Carbon tube fabric

↓

Analysis of fabric characteristics

↓

Development of Bullet proof jacket and case study

CONCLUSIONS OR SUMMARY

1. The Nylon/Carbon nanotube fabric can able to resist bullet when compare to Kevlar fabric.
2. Because of less weight, Nylon/Carbon nanotube fabric has influence in handle property.

ACKNOWLEDGEMENTS

I wish to express my sincere gratitude to my beloved principal Dr.R.Rudramoorthy for providing an opportunity and necessary facilities for carrying out this research work. I also wish to express my sincere thanks to Dr.G.Thilagavathi, HOD,Textile Technology,PSG College of Technology for the encouragement and support that she has extended towards this research work. My sincere thanks are due to research supervisor Dr.A.Venkatachalam, Dean-Department of Textile Technology,Bannari Amman Institute of Technology,Sathy whose valuable guidance and continuous encouragement for this work.

REFERENCES

[1] Kothari V.K., "Protection against bullets", *Asian Technical Textiles*, Vol. 2 No. 2, April-June 2008.

[2] Ray Baughman, "Super tough composite carbon nanotube fibres for electronic textiles", *AAPPS Bulletin*, August 2003.

[3] Horrocks A.R. and Anand S.C.,"Handbook of Technical Textiles", Woodhead Publishing Ltd, England, 2008.

[4] Sabit Adanur, "Handbook of Industrial Textiles", Technomic publishing, Swizerland, 1995.

[5] Puspa Bajaj & Sengupta A.K., "Protective Clothing", Textile progress, Manchester, Vol. 22 No. 2/3/4,1992.

[6] Kulkarni K., "Nanotechnology: Principles and practices", Capital publishing company, NewDelhi, 2008

CHARACTERIZATION AND OPTIMIZATION OF *VITEX NEGUNDO* NANOPARTICLES FOR THE DEVELOPMENT OF ANTIMICROBIAL TEXTILES

V. Rajalakshmi, R. Rajendran and R. Radhai

PG & Research Department of Microbiology, PSG College of Arts & Science, Coimbatore
E-mail: radhai.research@gmail.com

ABSTRACT

Nanoparticles are of great scientific interest due to their wide variety of application. Nanotechnology imparts high durability for fabrics because nanoparticles have an outsized surface area, thus leading to an increase in durability of the function. In this present study an effort was made to synthesize nanoparticles loaded with plant extract which is further applied on textile fabrics, for the development of antimicrobial textiles. Herb loaded nanoparticles were synthesized using controlled gelification of alginate through different optimization parameters like concentration of chitosan and calcium chloride, centrifugation time and speed. The optimization was done using Dynamic Light Scattering. The nanoparticles thus synthesized under optimization parameters were then characterized physically using TEM. The herb loaded nanoparticles were then coated on textile fabrics and tested for their antimicrobial efficiency using standard AATCC microbiological methods. The TEM micrographs revealed the nano size and structure of the herb loaded nanoparticles which were further substantiated using DLS. The fabrics loaded with the nanoparticles exhibited excellent antibacterial efficiencies.

INTRODUCTION

Nanotechnology can be best described as activities at the level of atoms, molecules that have applications in the real world. Nanoparticle researches are presently an area of intense scientific

V. Rajendran, P. Manivasakan and K.E. Geckeler (eds.)
Advanced Nano Materials for Industrial Applications, pp. 23-28 (2012).

awareness due to a wide variety of potential applications in biomedical, textiles, optical and electronic fields. Nanoparticles often possess unexpected optical properties as they are small enough to confine their electrons and produce quantum effects. Encapsulation can be achieved by several methods, e.g. interfacial polymerisation, microemulsion polymerization, precipitation polymerisation and diffusion. According to Pinto Reis *et al.*, [1] usually nanospheres have a matrix type of structure. Nanoparticles have been used for the controlled delivery and stabilization of various antimicrobial agents. According to Jain *et al.*, [2] mostly nanoparticles reveals completely new or improved properties based on explicit characteristics such as size, distribution and morphology. The herb loaded nanoparticles using antimicrobial finishes defeat the drawbacks of chemical finishes. They will not cause damage to the fabrics, are eco-friendly, non-toxic, non allergic because naturally occurring herbs are used, the cost factor is also reasonable.

METHODOLOGY

Collection, Processing and Extraction of Medicinal Herb

The leaves of *Vitex negundo* were collected in and around Coimbatore, Tamil Nadu, India, then air-dried and powdered. About 20 grams of the herbal powder was balanced in 100 ml ethanol and incubated overnight. The supernatant was filtered twice using Whatman No.1 filterpaper and the filterate was further condensed using rotary vacuum evaporator and stored.

Preparation of Herb Loaded Nanoparticles

Ethanolic extracts of *Vitex negundo* loaded alginate nanoparticles were prepared according to the procedure of Zahoor Ahamed *et al.*,[3] The herb loaded nanoparticles were prepared in different concentration sodium alginate/chitosan (1:0.5, 1:0.3, 1:0.2). and purified by centrifugation at various speeds (6000, 7000 and 8000 rpm). Followed by the centrifugation, the pellet containing the leaf extract loaded alginate nanoparticles was purified by washed thrice and then suspended in distilled water and stored at 4°C for further analysis. Alginate nanoparticles controls were also prepared by the same procedure without loading herbal extract.

Physical Characterization of the Leaf Extracts Loaded Nanoparticles

The herb loaded nanoparticles synthesized under optimized conditions were characterized using DLS (Malvern Instruments version 2.2) and HR-TEM (JEOL-JEM 2100).

Assessment of Antibacterial Activity of *Vitex negundo* Leaf Encapsulated Nanoparticles Loaded Fabrics: Parallel Streak Method (AATCC 147-2004) [4]

A loop full of test culture (*E.coli, S.aureus, K.pneumoniae, P.vulgaris, S.marcescens, S.typi, A.baumannii,* and *B.subtilis*) was loaded and transferred to the surface of the AATCC agar plate by making five parallel inoculum streaks without refilling the loop. The sample was gently pressed across the five inoculums of streaks and incubated at 37°C for 18-24 hours. Following incubation, the average width of the zone of inhibition around the test specimen was calculated in mm using the equation (1).

$$\text{Zone of inhibition (mm)} = (T-I)/2 \qquad \text{... (1)}$$

where T-width of zone of inhibition; I-width of specimen

RESULTS

Optimization of Herb Loaded Nanoparticles

In this study, herb loaded nanoparticles were optimized using two different parameters i.e., centrifugation speed and sodium alginate/chitosan ratio. The alginate chitosan nanoparticles formed by the interaction of positively charged alginate and negatively charged chitosan molecules. The ultimate potential formulation of the nanoparticles were achieved, when the concentration of alginate/chitosan of about 1:02. The minimum particles size and maximum zeta potential (–30.5 mV) were obtained, when the chitosan concentration of the nanoparticles was 0.02 gm. This relatively high negative value was due to cationic nature of Chitosan and ratio of Alg/Chi. An increase in centrifugation speed from 6000 rpm to 8000 rpm, the size of the nanoparticles were reduced from 523.8 to 182.6 nm. This could be due to the fact that, as the centrifugal force or as the rpm increases the coarse particles present will be converted into finer particles. It was observed that decreasing the concentration of the chitosan resulted in the formation of nanoparticles without aggregation.

According to Parveen *et al.,* [5] decreasing the concentration of chitosan in addition to sodium alginate resulted in the formation of particles without aggregation.

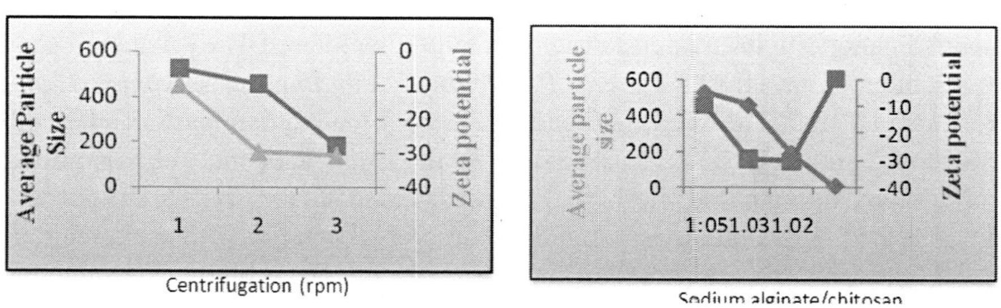

(A) Centrifugation Speed. (B) Ratio of Sodium Alginate/Chitosan Concentration

Fig 1. Optimization of Herb Loaded Nanoparticles

Average Size and Zeta Potential of Herbal Extract Loaded Nanoparticles

The average size of the herbal extract loaded nanoparticles was found to be 182.6nm. The average particle size and zeta potential of herbal extract loaded nanoparticles were shown in

Fig. 2. The zeta potential of the test sample was found to be -30.3 mV. The results were inferred to be good for the average particle size and Zeta potential.

According to Azizi *et al.*,[6] the stability of many colloidal systems was directly related to the magnitude of their zeta potential. In general, if the value of the particle zeta potential is large, the colloidal system will be stable. On the other hand, if the particle zeta potential is relatively small, the colloidal system will agglomerate. The surface charge of the particles is of substantial importance in all the production steps of these particles because the efficiency of the different steps is directly related to the establishment of electrostatic interactions.

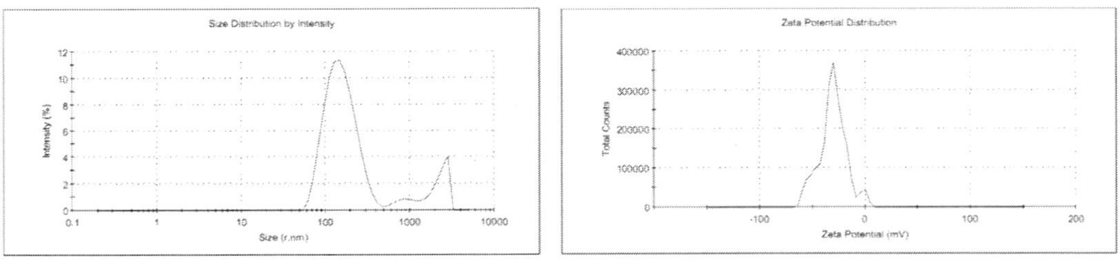

Average Particle Size – 182.6 nm Zeta Potential – 30.3 mV

Fig. 2: Figure Showed the Size Distribution and Zeta Potential of the Optimized Herb Loaded Nanoparticles

Qualitative Assessment-Parallel Streak Method (AATCC 147)

The antimicrobial efficacy of the cotton fabric was determined by parallel streak method. The cotton fabric was treated with *Vitex negundo* encapsulated nanoparticles. The area of antimicrobial activity was measured and recorded. The inhibitory level was found higher in the herb loaded nanoparticles treated fabric. Methanolic extract of *Vitex negundo* showed the maximum inhibitory zone aginst *S. aureus, P. vulgaris, B. subtilus* and *S. typhi* with an average zone size of about 9 mm. The *Vitex negundo* leaf extract loaded nanoparticles treated fabrics showed maximum antibacterial activity against all the test organisms. When compared with the control sodium alginate and the herbal extract loaded fabric, the herbal extract loaded nanoparticles treated fabric exhibited a higher antimicrobial activity. The use of chitosan as an antimicrobial agent in cotton fabrics was manifest from the outcome of Zhang *et al.*, [7].

It was relatively evident from the results of quantitative antibacterial assessments that the *Vitex negundo* nanocapsules treated fabrics possessed higher antimicrobial activity than the alginate chitosan (control) treated fabrics. [8,9] The enhancement of antimicrobial activity could be attributed to the fact that nanoparticles have a large surface area to volume ratio and high surface energy, thus presenting better affinity with the fabrics.

Table 1: Antibacterial Activity of the Treated Fabrics Parallel Streak Method (AATCC 147)

S.No	*Test Organisms*	*Antibacterial Activity Zone of Inhibition (mm)*	
		Herb loaded Nanoparticles Treated Fabric	*Alginate/Chitosan Loaded Nanoparticles Treated Fabric*
1	*S.aureus*	8 mm	3 mm
2	*P.vulgaris*	10 mm	7 mm
3	*K.pneumoniae*	5 mm	3 mm
4	*S.marcescens*	5 mm	4 mm
5	*S.typhi*	9 mm	2 mm
6	*E.coli*	4 mm	3 mm
7	*A.baumannii*	8 mm	0 mm
8	*B.subtilis*	10 mm	3 mm

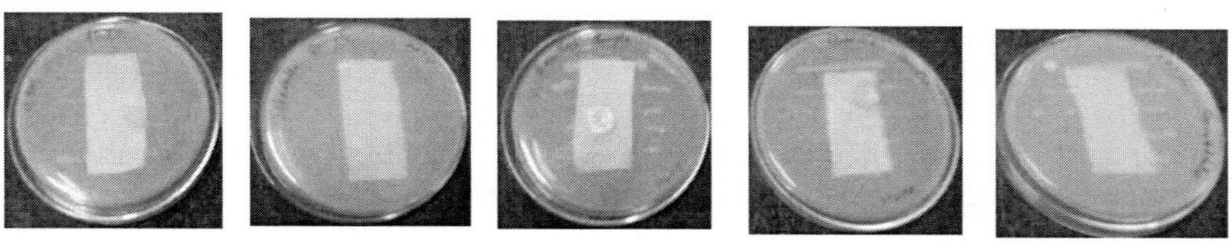

Fig. 3: Antibacterial Activity of Treated Fabrics-A-*S.aureus*, B-*S.typi*, C-*B.subtilis*, D-*K.pneumoniae*, E-*A.baumannii*

HR-TEM

The HR–TEM monographs of the Herb-loaded nanoparticles synthesized using sodium alginate/ chitosan method were represented in Fig 4. High Resolution Transmission Electron Microscopy (HR-TEM) has depicted the morphology and size details of the herb-loaded nanoparticles synthesized by using alginate gellification method. The morphology of the nanoparticles was roughly spherical in shape and the size of 20-50 nm. They were found to be monodispersed.

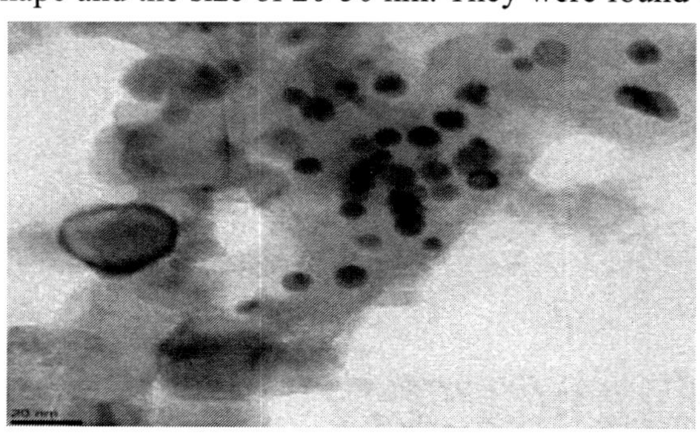

Fig. 4: Figures have Shown that the Nanoparticles Were in the Size Range of 20 – 50 nm

The nanoparticles were well alienated and no agglomeration was observed. The results of TEM correlated with the results obtained in DLS analysis hence proving the size of the nanoparticles.

DISCUSSION

In this research, herb loaded nanoparticles were used to finish fabrics which showed maximum activity against all test bacteria. Rajendran *et al.* (2012), also have reported that the higher antimicrobial activity of the nanoparticles was due to the smaller particle size and controlled release of the active ingredients from nanoparticles. Rajendran *et al.* (2013), proved that the nanoparticles loaded with herbal extracts having maximum antimicrobial activity against *P. aeruginosa, S. aureus and E. coli.*

CONCLUSION

Nanoparticles of the plant extracts were prepared by ionic gelation method using chitosan. In the present study, it was found that the ethanolic extract of *Vitex negundo* leaf showed better level of antimicrobial activity when compared to all the other solvents by agar diffusion method. Further the herb loaded nanaoparticle were prepared, synthesized and optimized. The antimicrobial activity of sample loaded nanoparticles were substantiated using Parallel streak method (AATCC 147).

REFERENCES

[1] Pinto Reis *et al.*, Nanoencapsulation I. Methods for preparation of drug-loaded polymeric nanoparticles, *Nanomedicine: Nanotechnology, Biology, and Medicine*, 2, 8–21, 2006.

[2] Jain *et al.*, synthesis of plant-mediated silver nanoparticles using Papaya fruit extract and evaluation of their anti microbial Activities, *Digest Journal of Nanomaterialsand Biostructures* 4: 3, 557–563, 2009.

[3] Z.Ahmad *et al.*, "Alginate Nanoparticles as Antituberculosis Drug Carriers: Formulation Development, Pharmacokinetics and Therapeutic Potential", *The Indian Journal of Chest Diseases & Allied Sciences., 2006, 48.*

[4] AATCC technical Manual, Vol., 76, 2001.

[5] Parveen *et al.*, Enhanced antiproliferative activity of carboplatin-loaded chitosan–alginate nanoparticles in a retinoblastoma cell line, *Acta Biomaterialia* 6: 3120–313, 2010.

[6] Azizi *et al.*, Release profile and stability evaluation of optimized chitosan/alginate nanoparticles as EGFR antisense vector, *International Journal of Nanomedicine* 5, 455–46, 2010.

[7] Zhang, Zitao, et al., Properties of cotton fabrics treated with chitosan, *Textile Research Journal,* 73, 1103-1106, 2003.

[8] Rajendran R., R. Radhai *et al*, Synthesis and characterization of Neem Chitosan Nanoparticles for development of antimicrobial cotton textiles *Journal of engineered fibers and fabrics*, 7-1,136-141, 2012.

[9] R. Rajendran., R. Radhai., *et al* Development of Antimicrobial cotton fabrics using herb loaded nanoparticles. *Carbohydrate polymers.* 91, 613-617, 2013.

SYNTHESIS AND CHARACTERIZATION OF NOVEL $MoBiGaSe_5$ THIN FILMS BY VACUUM DEPOSITION TECHNIQUE

S.P. Patil, S.S. Mali[1], R.M. Mane, R.R. Kharade, V.B. Ghanwat, V.V. Kondalkar and P.N. Bhosale

Materials Research Laboratory, Department of Chemistry, Shivaji University, Kolhapur, Maharashtra
[1]Thin Film Materials Laboratory, Department of Physics, Shivaji University, Kolhapur Maharashtra
E-mail: p_n_bhosale@rediffmail.com, svpctb@gmail.com

ABSTRACT

In the present investigation, nanocrystalline Mixed Metal Chalcogenide (MMC) $MoBiGaSe_5$ thin films with primary semispherical balls and secondary diffused nanograins have been prepared onto glass substrates by vacuum deposition technique. Ammonium molybdate $(NH_4)_6Mo_7O_{24}\cdot 4H_2O$, Bismuth nitrate $Bi(NO_3)_3\cdot 5H_2O$, Gallium nitrate $Ga(NO_3)_3\cdot xH_2O$, and Selenium Se metal powder were used as a precursors of Mo^{+4}, Bi^{+3}, Ga^{+3} and Se^{-2} ions respectively. The evaporation time was kept fixed at 7 min for each precursor. The as deposited and vacuum annealed films were characterized for their optostructural, morphological and compositional properties using different characterization techniques. Optical absorption studies reveal that as deposited and annealed $MoBiGaSe_5$ films show visible light absorption in the wavelength range of 400 to 800 nm. Scanning electron microscopic (SEM) images of as deposited films showed uniform, well defined granular morphology, besides annealed samples have diffused granular growth of uniform spheres. X-Ray diffraction (XRD) analysis of annealed thin films show, the phase transition from orthorhombic Bi_2Se_3 to stoichiometric hexagonal Bi_2Se_3 phase. Phase transition confirms the formation of hexagonal nanocrystalline $MoBiGaSe_5$ pure phase.

V. Rajendran, P. Manivasakan and K.E. Geckeler (eds.)
Advanced Nano Materials for Industrial Applications, pp. 29-34 (2012).

INTRODUCTION

Recently nanocrystalline Mixed Metal Chalcogenide (MMC) thin films have great potential in solar cell application. MMC and chalcopyrite semiconductors have been widely used as the most promising candidates for low cost solar absorbers and significant efficient thin film photovoltaic devices [1-3]. The ternary and quaternary thin films of VIB-VA-VIA, VIB-VA-IIIB-VIA and IB-IIIA-VIA groups have high conversion efficiency due to its tunable band gap in the range of 1.1 to 1.6 eV which absorbs maximum solar spectrum. Recently quaternary MMC materials are at front because of similar band gap energy as that of $Cu(In,Ga)Se_2$ (CIGS) and Cu_2ZnSnS_4 (CZTS) [4,5]. Quaternary materials are synthesized by various methods such as electrodeposition [6], spray pyrolysis [7], hydrothermal [8], chemical vapor deposition [9], hot injection method [10], sputtering [2] and arrested precipitation technique [11]. In those studies researchers paid more attention on absorption coefficient, band gap and photo response of the material. Recently S.S. Mohite et al. [12] showed the vacuum evaporation method is more promising for the synthesis of quaternary MMC thin film materials. The preparative parameters like evaporation rate, deposition time and easy of precursor selection are the key factors for achieving good quality MMC thin films.

In this investigation, we report deposition of quaternary $MoBiGaSe_5$ thin films by vacuum evaporation technique and their optostructural properties.

EXPERIMENTAL DETAILS

All the reagents are of analytical grade (AR) and were used without further purification. Ammonium molybdate $(NH_4)_6Mo_7O_{24}\cdot 4H_2O$, Bismuth nitrate $Bi(NO_3)_3\cdot 5H_2O$, Gallium nitrate $Ga(NO_3)_3\cdot x\cdot H_2O$ and Selenium metal powder were used as a precursors for Mo^{+4}, Bi^{+3}, Ga^{+3} and Se^{-2} ions respectively. The $MoBiGaSe_5$ thin films were deposited on to cleaned glass substrates by sequential evaporating Mo^{+4}, Bi^{+3}, Ga^{+3} and Se^{-2} precursors using vacuum deposition technique (HIND-HIVAC model: 15 F6-India) by resistive heating. Molybdenum boat was used as resistive heating source. Pressure was maintained at 10^{-6} mbar and time 7 minutes during deposition for all materials. The distance between the source and substrate was kept 25 cm. The thickness of the film measured using surface profilometer and was found to ~750 nm, which was in agreement with the value obtained through quartz crystal monitor mounted in the evaporation unit. Deposited $MoBiGaSe_5$ thin films were blackish in colour with reflecting surface. Optimized deposition parameters for $MoBiGaSe_5$ samples by vacuum evaporation technique are summarized in Table 1. As deposited samples were annealed at 80°C in vacuum furnace and used for further characterizations.

Table 1: Various Optimized Parameters During Vacuum Deposition of the $MoBiGaSe_5$ Films Having Substrates at Room Temperature

Element	*Precursor*	*Deposition Rate (Å/sec)*	*Pressure (mbar)*	*Current (Amp)*
Mo	$(NH_4)_6Mo_7O_{24}.4H_2O$	20	10^{-5}	105
Bi	$Bi(NO_3)_3\cdot 5H_2O$	20	10^{-5}	110
Ga	$Ga(NO_3)_3\cdot xH_2O$	20	10^{-5}	118
Se	Se metal powder	20	10^{-5}	68

CHARACTERIZATIONS

The optical absorption study was done by using a UV-Vis spectrophotometer (UV-1800 Shimadzu, Japan). The surface morphology of MoBiGaSe$_5$ thin films were studied using Scanning Electron Microscopy (SEM: JEOL-6360). The compositional analysis was carried out using Energy Dispersive X-ray analysis (EDS) attached to the SEM. The thickness of films was measured by surface profiler (AMBIOS XP-1, USA) The structural properties of as deposited and annealed thin films are analyzed by X-ray Diffraction (XRD) analysis (Bruker AXS Model D8 Advanced X-ray Diffractometer) with Cuká target having wavelength 1.542 Å, radiation (I) the 2θ ranges from 20 to 60^0.

RESULTS AND DISCUSSION

Optical Absorption Studies

The optical absorption of as deposited and vacuum annealed nanocrystalline MoBiGaSe$_5$ thin films was carried out in the wavelength range 300-1100 nm. The thickness of as deposited film is near about 750 nm while on annealing thickness of the film slightly decreases due to diffusion of the smaller grains to larger grains. The variation of optical coefficient with wavelength was further analyzed to find out the nature of electronic transition across the optical band gap. The nature of transition was determined by using the relation,

$$\alpha h\nu = A(h\nu - Eg)^{n/2} \quad \text{... (1)}$$

Where, A is a constant, hν is a photon energy, n = 1 for a direct gap material and n = 2 for an indirect gap material. The plot of $(\alpha h\nu)^2$ versus hν for the MoBiGaSe$_5$ thin films for as deposited and annealed samples are shown in fig. 1.

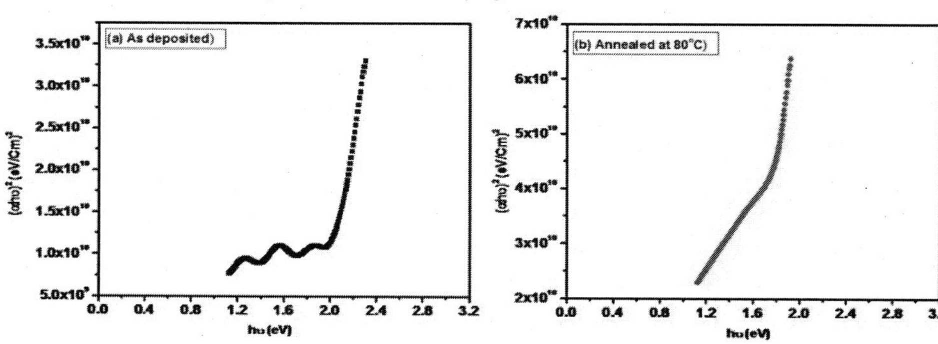

Fig. 1: Optical Absorption of the MoBiGaSe$_5$ thin Films (a) As Deposited and (b) Annealed at 80^{0}C

The nature of plot suggests direct allowed transition. The extrapolating of straight line portion to the X-axis leads to the determination of band gap energy of the material. Figure 1 clearly reveals that the band gap energy decreases for annealed sample from 1.96 to 1.60 eV.

The band gap energy decreases due to compact grain growth after annealing grain size of the film.

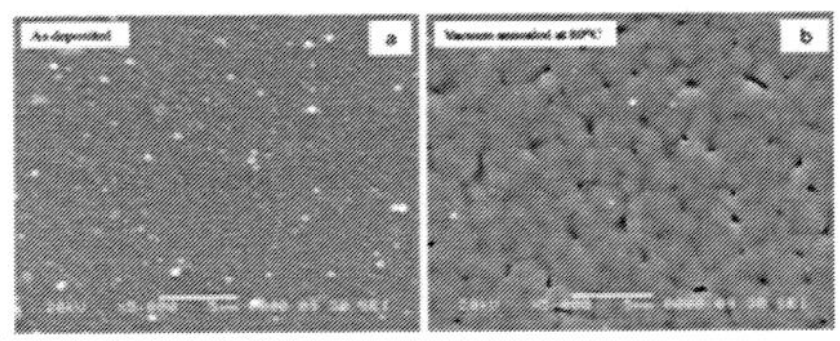

Fig. 2: SEM Micrographs of $MoBiGaSe_5$ Thin Films Deposited by Vacuum Evaporation: (a) as Deposited and (b) Vacuum Annealed at 80^0C

Scanning Electron Microscopy (SEM) Analysis

Figure 2 show SEM images of as deposited $MoBiGaSe_5$ thin film sample exhibits a uniform deposition with granular morphology. When $MoBiGaSe_5$ thin film annealed at 80^0C in a vacuum, the granular morphology converted into diffused grains. The grain size of as deposited and vacuum annealed samples is about 600 nm and 900 nm respectively.

Electron Dispersive X-ray Spectroscopy (EDS) Analysis

The spectrum reveals the presence of Mo, Bi, Ga and Se at 2.293 KeV, 2.419 KeV, 9.210 KeV and 1.379 KeV respectively which confirms the presence of Mo, Bi, Ga and Se elements in the film. The representative EDS spectrum of sample is shown in Fig. 3. The observed compositional analysis shows slightly deviation from chemical stoichiometric film. The Actual atomic percentage of Mo, Bi, Ga and Se is shown in inset table of Fig. 3.

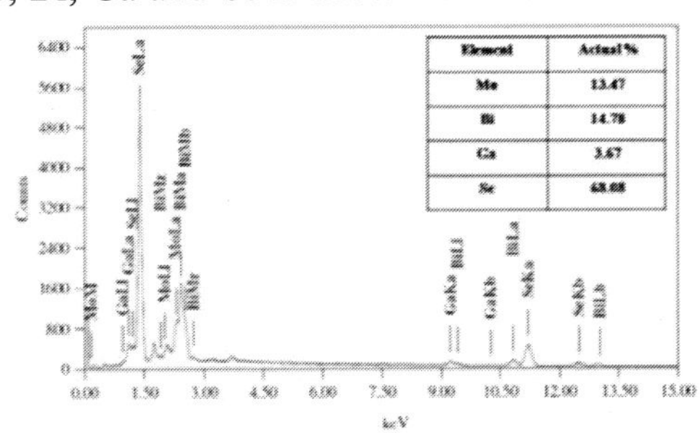

Fig. 3: EDS Spectrum of 80^0C Annealed $MoBiGaSe_5$ Thin Film

X-ray Diffraction (XRD) Analysis

For the structural analysis of as deposited and annealed $MoBiGaSe_5$ thin films, the diffraction angle 2θ is varied in the range of 20 to 60^0. Figure 4 show the XRD patterns for as deposited and vacuum annealed samples.

Different intensity peaks in X-Ray spectra are indexed and the corresponding values of interplaner spacing 'd' are calculated and compared with standard values of JCPDS data and). The X-ray diffraction pattern for as deposited thin film displayed diffraction peaks at 2θ values of approximately at 27.165^0 corresponds to (230) plane of orthorhombic structure of Bi_2Se_3 (Card No. 77-2182 and hexagonal Bi_2Se_2 (card No. 72-2182) phase, 37.863^0 corresponds to (103) plane of hexagonal structure of $MoSe_2$ (Card no. 17-0887), 22.359^0 and 32.365^0 corresponds to (004) and (103) planes of hexagonal structure of GaSe (Card No. 37-0931) phase. XRD analysis of vacuum annealed thin film shows peak at 2θ value approximately at 29.74, 43.67^0 corresponds to (104), (109) planes of hexagonal Bi_2Se_2 phase. The XRD analysis reveals that the transition from Orthorhombic Bi_2Se_3 to stoichiometric Hexagonal Bi_2Se_2 phase. Phase transition confirms the formation of Hexagonal nanocrystalline $MoBiGaSe_5$ pure phase.

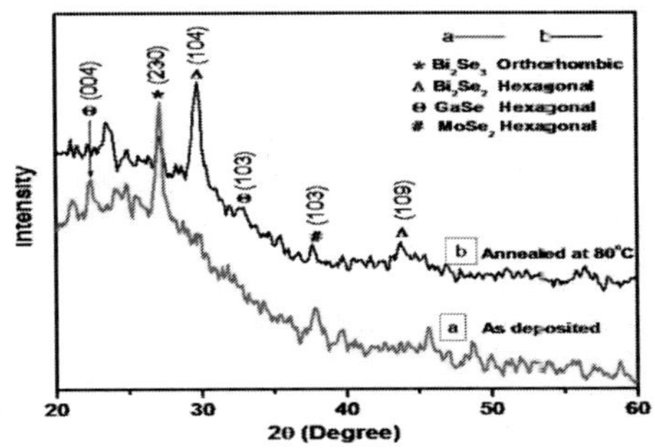

Fig.1: XRD Patterns of the $MoBiGaSe_5$ thin Films (a) As Deposited and (b) Annealed at 80°C.

The crystallite size 'D' was calculated using FWHM data and Debye Scherer formula [13]

$$D = K\lambda/\beta \cos\theta \quad \text{... (2)}$$

where, 'D' is average crystallite dimension, 'K' 0.94 is a constant 'λ' is wavelength of the x-ray used, 'β' is full width at half maximum (FWHM) and 'θ' is Bragg's angle. The crystallite size estimated for as deposited and annealed $MoBiGaSe_5$ films having (230) and (104) planes were 24 and 15 nm respectively. The small crystallite in the range of nanoscale confirms nanocrystalline nature of the films.

CONCLUSIONS

Resistive heating vacuum evaporation technique is applied to deposit stoichiometric, adherent and uniform deposition of $MoBiGaSe_5$ thin films. The samples were characterized by using different characterization techniques for optical, structural and compositional studies. As deposited films show granular morphology while on annealing grains get diffused. As deposited and vacuum annealed thin films show good absorption in the visible light. XRD analysis of annealed thin film showed phase transition from Orthorhombic Bi_2Se_3 to stoichiometric

Hexagonal Bi_2Se_2 phase. The properties of $MoBiGaSe_5$ thin films investigated in the present study showed that these thin film material can be used for fabrication of ECPV cell as a photoelectrode.

REFERENCES

[1] N.B. Pawar, S.S. Mali, M.M. Salunkhe, R.M. Mane, P.S. Patil and P.N. Bhosale, Effect of surfactant on optical and structural properties of chemically deposited $MoBi_2Se_5$ thin films, *New J. Chem.*, 36, 1807-1812, 2012.

[2] C.H. Chen, W.C. Shih, C.Y. Chien, C.H. Hsu, Y. Huie and C.H. Lai, A promising sputtering route for one-step fabrication of chalcopyrite phase Cu(In,Ga)Se_2 absorbers without extra Se supply, *Sol. Energy Mater. & Sol. Cells.*, 103, 25-29, 2012.

[3] M.M. Salunkhe, S.M. Patil, R.M. Mane, S.V. Patil and P.N. Bhosale, Photochemical cell performance of chemically deposited $MoBi_2Te_5$ thin films, *Adv. Mat. Lett.*, 3 (2), 71-76, 2012.

[4] Q. Guo, G. M. Ford, W. C. Yang, B. C. Walker, E. A. Stach, H. W. Hillhouse and R. Agrawal, Fabrication of 7.2% efficient CZTSSe solar cells using CZTS nanocrystals, *J. Am. Chem. Soc.*, 132, 17384- 17386, 2010.

[5] J.H. Shi, Z.Q. Li, D.W. Zhang, Q.Q. Liu, Z. Sun and S.M. Huang, Fabrication of Cu (In,Ga)Se_2 thin films by sputtering from a single quaternary chalcogenide target, *Prog. Photovolt: Res. Appl.*, 19, 160-164, 2011.

[6] M. Harati, J. Jia, K. Giffard, K. Pellarin, C. Hewson, D.A. Love, W.M. Lau and Z. Ding, One-pot electrodeposition, characterization and photoactivity of stoichiometric copper indium gallium diselenide (CIGS) thin films for solar cells., *Phys. Chem. Chem. Phys.*, 12, 15282-15290, 2010.

[7] K. Ramasamy, M.A. Malik and P. O'Brien, Route to copper zinc tin sulphide Cu_2ZnSnS_4 a potential material for solar cells, *Chem. Comm.*, 48, 5703-5714, 2012.

[8] H. Jiang, P. Dai, Z. Feng, W. Fan and J. Zhan, Phase selective synthesis of metastable orthorhombic Cu_2ZnSnS_4, *J. Mater. Chem.*, 22, 7502-7506, 2012.

[9] T. Washio, T. Shinji, S. Tajima, T. Fukano, T. Motohiro, K. Jimbo and H. Katagiri, 6% efficiency Cu_2ZnSnS_4- based thin film solar cells using oxide precursors by open atmosphere type CVD, *J. Mater. Chem.* 22, 4021-4024, 2012.

[10] C. Yan, C. Huang, J. Yang, F. Liu, J. Liu, Y. Lai and Y. Liu, Y. Lai, J. Li and Y. Liu, Synthesis and characterization of quaternary Cu_2FeSnS_4 nanocrystals., *Chem. Comm.*, 48, 2603-2605, 2012.

[11] R.M. Mane, R.R. Kharade, N.S. Patil and P.N. Bhosale, Optical and electrical studies on $MoBi_2Se_5$ thin films prepared by arrested precipitation technique (APT), *Arch. Apll. Sci. Res.*, 2(1), 275-283, 2010.

[12] S.S. Mohite, R.R. Kharade, S.S. Mali, D.G. Kanse, P.S. Patil and P.N. Bhosale, Low temperature synthesis of novel $MoBiCuSe_5$ nanowire thin films by vacuum deposition method and their characterization, *Res. J. Chem. Environ.*, 15 (2), 653, 2011.

[13] B.D. Cullity, The elements of X-ray diffraction first ed., Addson-Wesly Publishing company, Inc., United State of America (1956).

MAGNETIC PROPERTIES OF DOPED NiO

LAKSHMI NAIR, K. RAJENDRA BABU, DHANYA CHANDRAN AND POURNAMI VIJAYAN[1]

Department of Physics, Mahathma Gandhi College, Trivandrum
[1]Department of Physics, S.B.College, Changanacherry, Kerala
E-mail: lakshmi.nair8@gmail.com

ABSTRACT

Nanoparticles of magnetic metals and oxides have attracted great interest in recent years because of their unique physical and chemical properties. Stoichiometric Nickel Oxide exhibits insulating character and antiferromagnetic (AFM) order that remains even at rather high temperatures $T_N = 523$ K. Although pure NiO is a Mott- Hubbard insulator at room temperature, an introduction of Ni^{2+} vacancies or doping with Li^{+} will cause considerable increase in the conductivity of NiO. In doped NiO a lot of intensive investigations have been conducted for its great theoretical and technological importance due to the special electrical behaviors and related spin dynamics. This paper reports the synthesis of pure and tin doped NiO nanopowders by using a sol gel method. The sizes of the samples are calculated using XRD and the morphology are studied using TEM. The concentration of the dopants is determined from EDS. The magnetic properties of the samples are studied using VSM. The pure sample is found to exhibit ferromagnetism and tin doped sample is found to exhibit superparamagnetism.

INTRODUCTION

Diluted magnetic semiconductors are being widely investigated for their potential applications in the field of spintronics. A lot of work on preparation and investigation of ferromagnetic mechanism of these systems with transition metal ions doped have been reported.[1]DMS are formed by partial substitution of cations of the host semiconductor with small amount of magnetic transition metal (TM) ions. Nickel Oxide is a strongly co- related material which

V. Rajendran, P. Manivasakan and K.E. Geckeler (eds.)
Advanced Nano Materials for Industrial Applications, pp. 35-40 (2012).

exhibits insulating character and antiferromagnetic (AFM) order that remains even at rather high temperatures T_N = 523 K.The magnetic structure of NiO consists of ferromagnetic sheets of Ni $^{2+}$ parallel to the (111) plane with opposite spin directions in neighboring sheets.[2,3,4,5,6] Doping with transition metal ions or reduction of particle size has been shown to disturb the magnetic structure of NiO. The magnetic properties of NiO nanaoparticles (31.5 nm in diameter) have been recently studied by Kodoma et al.[7] Remarkable room temperature magnetism has been observed in Fe doped NiO.[8] In the present study we study the change the change in magnetic properties of Nickel oxide on doping with tin.

EXPERIMENTAL

Polycrystalline samples of pure (NI) and Sn doped (NISn, NISn1) Nickel oxide were prepared using a sol-gel method. The precursors used for the synthesis are Nickel nitrate hexahydrate, sodium hydroxide and Stannic chloride pentahydrate. All chemicals were of the analytical grade and were used without further purification .Nominal amounts of the starting material were weighed and dissolved separately into triple distilled water. The dopants were added in calculated amount to the prepared solutions. NaOH solution was added drop wise to nickel nitrate solution. The green gel so obtained were kept overnight and later calcined at 400°C for two hours. The crystal structure and crystalline sizes of the calcined samples were identified by X-ray powder diffractometer (XRD, Bruker AXS D8 Advance) with Cu Kα radiation (λ = 1.5406 Å). The average crystallite size was estimated according to the scherrer equation. The concentrations of the Ni and Sn ions were determined from EDS spectra. Energy dispersive X-ray (EDAX) analysis confirmed the concentrations of Sn to be close to those in the starting nominal compositions. The morphologies of the pure and 2% tin doped samples were obtained from TEM images. The magnetic properties of the powder sample were performed by Lakeshore 7410 vibrating sample magnetometer.

RESULTS AND DISCUSSION

The XRD patterns of the pure and doped samples are shown in Fig. 1. All the samples show distinct peaks at 2θ of 37, 43, 62, 75 and 79. These peaks have been identified as peaks of cubic crystalline NiO and are assigned to the [111], [200], [220], [311] and [222] diffracting planes (JCPDS card no 78-0429). Secondary phases corresponding to the dopants were not detected in the doped samples. The particle sizes were calculated using the debye – sherrer equation. The average grain size for the pure sample was found to be 18 nm. The sizes of the tin doped samples are found to be 8.8 nm and 6.1 nm for 2% (NISn) and 6% (NISn1) doping respectively. The morphology of the samples is studied from the TEM images. For TEM measurements sample is dispersed in ethanol and sonicated for 5 minutes. The dispersed solution is dropped on a copper grid and allowed to dry overnight in ambience. This is then analyzed with TEM. The TEM images of pure and 2% tin doped sample is shown in Fig. 2. It can be seen that the sizes of the particles are not homogeneous, ranging from 13 to 18 for pure NiO and from 4 nm to 10 nm for 2% Sn doped NiO. The concentration of the dopants was found out from the EDS analysis. It was examined with EDS that the samples with 4%, 10% nominal Sn concentrations have real Sn concentrations of 2% and 6%.

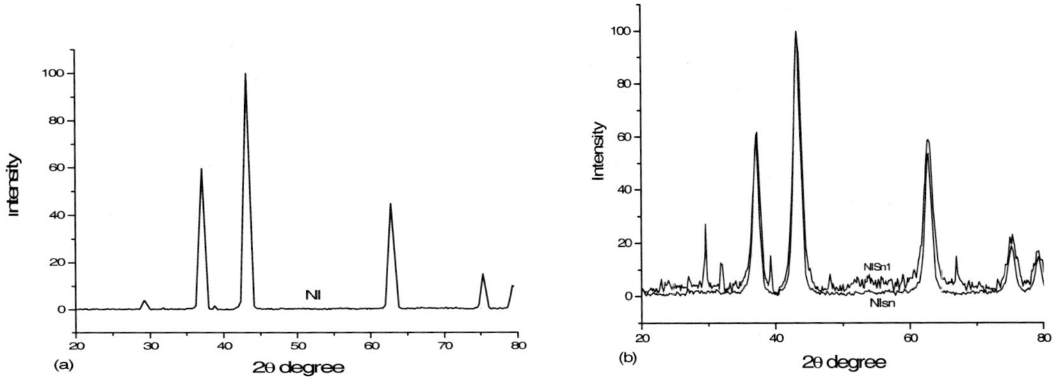

Fig 1: (a) XRD Pattern of Pure Sample (b) XRD Pattern of Sn Doped Samples

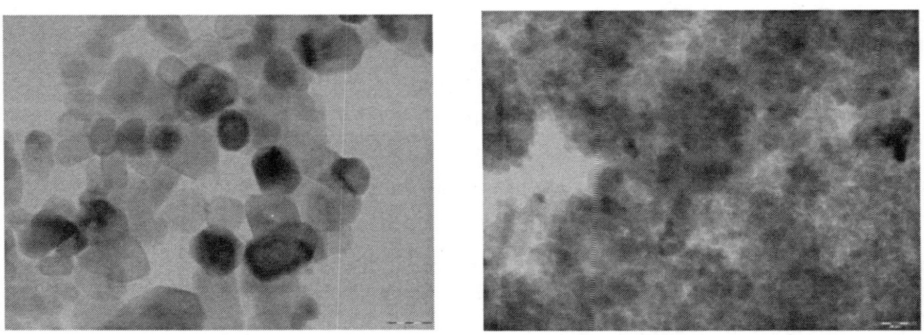

(a) (b)

Fig. 2: TEM Images of (a) Pure NiO (b) 2% Sn Doped NiO

Magnetizations as a function of field for the samples at room temperature are shown in Figs. 3 and 4. The pure samples are found to exhibit ferromagnetism. The coercivities of the pure sample is found to be 309.80 Oe .The magnetization value of the pure samples is found to be 0.39 emu/gm. Interestingly enough other authors have reported similar magnetization values for nanosized nickel oxide ≈ 2.5 emu/g for particles of sizes about 2-3 nm and 0.3 emu/gm for particles ranging between 13 and 18 nm, the particle size being in our case about 18nm.[9] The ferromagnetism of the pure NiO sample can be attributed to the finite size effect as reported by Makhlouf et al.[10] NiO nanoparticles are shown to exhibit finite size effect, where 8-, 6- or 4-sub lattice spin configuration arise due to the reduced coordination of surface spins, leading to anomalous magnetic properties like large moments and coercivities and loop shifts. In some cases finite size effects leads to ferromagnetism[11] (for particles of size D< 24 nm), superparamagnetism (D < 31.5 nm)[7] spin glass behavior (D < 10 nm) and even core-shell like structure (4 < D < 22 nm) where the core of NiO nanoparticles behaves like a ferrimagnet and the shell contains randomly oriented spins with low coordination. Bulk antiferromagnetic materials contain equal number of spins in mutually opposite directions (e.g., up and down) and thus are fully compensated. However, as the dimension is reduced the two types of spin become progressively unequal in number (especially at the surface), remain uncompensated and produce substantial magnetization at low temperatures.

For the Sn doped samples the magnetization curve is found to be composed of a ferromagnetic component and a paramagnetic component, the latter part increasing with increase in magnetic field and concentration of Sn. In the M versus H data the higher H linear behavior is due to the paramagnetic component .The hysterics curve is found to be very close to the paramagnetic content. Similar magnetization data for Mn doped ZnO and Co doped Ti O_2 has been reported in literature.[14,15,16] The magnetization values were found to decrease from 0.6825 emug$_{-}^{1}$ to 0.59 emug$_{-}^{1}$ with increase in the percentage of Sn ions .This shows that the super paramagnetic behavior may be due to the presence of tin ions. With increase in tin concentration the hysterisis was found to become more linear. A large coercivity value of 327.14 Oe has been reported previously for 5nm NiO particles.[17] In our study the coercivity value were found to be much lower for the sample with approximately the same size. This may be due to the presence of secondary phases of Sn in the sample which is not evident in the XRD pattern. For the sample with real doping concentration as 6% the nominal Sn concentration is 10%. This shows that there may be many more Sn ions that doesn't exist in forms of substitutions of Ni cations. Theses Sn clusters that may remain paramagnetic at room temperature may be the reason for the decreasing ferromagnetism with increase in doping concentration. The super paramagnetic behavior may be due to the extremely small size of the 6% Tin doped samples. It has been reported that a particle of a magnetic material below a particular diameter (8 nm for Fe2O3) (Schmidt 2003) contains only one single magnetic domain so that this particle is in a state of uniform magnetization at any field without any interaction with neighbor domains in one particle or a well dispersed suspension. In fact the critical diameter is the single domain size of the material.[18] As the particle size is reduced, the surface to volume increases and the role of surface atoms in the magnetization behavior becomes significant and energy consideration favors the formation of single domain particles. This leads to a super paramagnetic behavior.

When the size is further reduced with the increase in dopant concentration of tin from 2% to 6% the magnetic structure changes from a multidomain to a single domain configuration. Accordingly the magnetization process will goes from domain wall motion to the mechanism of magnetization rotation which would leads to increase in Hc.[19] This trend of increase of the coercivity with decreasing size of single domain nanoparticles (above a critical size) has been reported by various authors.[20, 21] Chen et al[20] explain this behavior in nanoparticles in terms of an energy barrier (ΔE) for the rotation of the moments that depends on the surface area

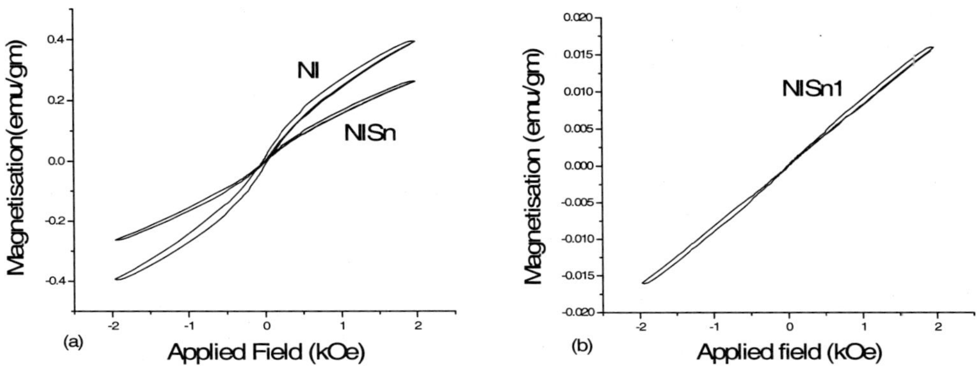

Fig.3. Magnetization of the Pure and Sn Doped NiO as a Function of Field Measured at Room Temperature

of the particle, $\Delta E = K_s S$. Here K_s is the surface anisotropy constant and S is the surface area of the particle. This has been shown to lead to a coercivity that decreases with the size of the particle as $H_c \sim 1/d$. Thus we understand that the observed increase in the coercivity of the tin doped samples with decreasing size reflects the increasing role of the surface in determining the anisotropy and hence the coercivity. The observed low values of coercivity for Sn doped samples cannot be attributed to the possible difference in the concentrations of Ni^{3+} due to doping since a detailed study by Richardson et al had established that the effect of valency of Ni on the magnetic response of nanostructured NiO sample is negligible.[22] Further studies are required to find the cause of the super paramagnetic behavior. A RTFM has been previously reported for Ca^{2+} doped CeO[23]. The ferromagnetism observed was discussed in terms of the oxygen vacancies and defects at the surface/interfaces. But in our study of doping NiO with a nonmagnetic ion this effect was not observed. The Sn doped samples were found to exhibit very weak ferromagnetism much lower than the pure samples and the magnetization decreased with increase in doping. The weak ferromagnetism can be due to the size effect than to the presence of Sn ions.

CONCLUSION

In this paper we report the synthesis of pure, Mn and Sn doped Nickel Oxide nanopowders through a sol-gel method. The pure sample was found to exhibit room temperature ferromagnetism. This may be due to the size effect. The Sn doped samples were found to exhibit superparamagnetism with decrease in ferromagnetism with increase in concentration of Sn. This can be attributed to the change in magnetic structure from a multi domain to a single domain due to the extremely small size of the nanoparticles (<10 nm) and due to the presence of secondary phases of tin.

REFERENCES

[1] W. Prellier, A. Fouchet and B. Mercey, *J. Phys. Condens. Matter* 15, R1583 (2003)

[2] P. Mallick, C. Rath, A. Rath, A. Banerjee, N.C. Mishra, *Solid State Commun.* 150 (2010), 1342-1345.

[3] Y.H. L in, J. Wang , J. Cai, M. Ying, R. Zhao, M. Li , C.W. *Nan, Phys. Rev. B* 73, (2006) 193308

[4] J. Wang, J. Cai, Y.H. Lin, C.W. Nan, *Appl. Phys. Lett.* 87(2005) 202501.

[5] J. Nowotny, M. Rekas, *Solid State Ion* 12(1984) 253.

[6] P.D. Battle, A.K.Cheetham, G.A. Gehring, *J. Appl. Phys.* 50(1979) 7578.

[7] R.H. Kodoma, S.A. Makhlouf, and A.E.Berkowitz, *Phys. Rev. Lett.* 79, 1393 (1997).

[8] Y.H. lin, J. Wang, J. Cai, M. Ying, R. Zhao, M. Li, C.W. Nan, *Phys, Rev.* B 73(2006) 193308.

[9] Khadar A.M, Biju V, Ivanove A, *Mater. Res. Bull.* 2003, 38, 1341.

[10] S.A. Makhlouf, F.T. Sparker, F.E. Spada, A.E.Berkowitz, *J. Appl. Phys.* 81(1997) 556.

[11] L. Li, L. Chen, R. Qihe, G. Li, *Appl. Phys. Lett.* 89, (2006) 134102.

[12] S.D. Tiwari, K.P. Rajeev, *Phys. Rev.* B 72 (2005), 104433.

[13] S. Thota, *J. Kumar, J. Phys. Chem. Solids* 68 (2007) 1951.

[14] C. Jing, Y. Jiang, W. Bai, J. Chu, A. Liu, *J. Magn. Magn. Mater* 16 (2010) 322.

[15] D Rubi, J. Fontcubert, *Phys. Rev.* B 75 (2007), 155322.

[16] Ji. Guang Li, R. Biichel, M. Isobe, T. Mori and T. Ishigaki, *J. Phys. Chem.* C 2009, 113, 8009-8015.

[17] H. BI, S.L. Yzhang, Y. Du, *J. Magn. Magn*, Mater. 277 (2004) 363-3670.

[18] Yunyu Shi, Superparamagnetic Nanoparticles for Magnetic Resonance Imaging (MRI) Diagnosis, Ph. D. Thesis, School of Chemical Engineering, The University of Adelaide, 2006.

[19] H. Wang, X. Kou, L. Zhang and J. Li, *Maetr. Res. Bull.* 43 (2008) 3529-3356. 20 J.F. L.

[20] J.F. Löffler, H.B. Braun, W. Wagner, G. Kostorz and A. Wiedmann, *Mat. Sci and Engg.* A304-306 (2001) 1050.

[21] Chen Chen, Osamu Kitakami, Yutaka Shimada, *J. Appl. Phys.* 84(1998) 2184. G. Herzer, *J. Magn.* Mater. 112 (1992)258)

[22] J.T. Richardson, W.O. Milligan, *Phys. Rev.* 102 (1956) 1289. 23 X. Chen, G. Li, Y. Su, X. Qiu, L. Li and Z. Zou, *Nanotechnology* 20 (2009) 115606 (8pp).

STUDIES ON MECHANICAL, THERMAL PROPERTIES AND CHARACTERIZATION OF NANOCOMPOSITES OF NYLON-6 – THERMOPLASTICS POLY UREATHANE RUBBER (TPUR) BLEND

CEASAR GUNA SINGH, S. SOUNDARARAJAN AND K. PALANIVELU

Central Institute of Plastics Engineering and Technology, Chennai
E-mail: bssrajan@yahoo.co.in

ABSTRACT

Nylon 6–TPU(15%) blended with two types of Nanoclay Cloisite Na+ and 30B(1, 3, 5%) using a Twin screw extrude As nanoclay concentration increases the tensile strength and modulus, flexural strength and modulus were increasing upto 5%. The elongation decreases. Impact strength was increases upto 1% with cloisite 30B and upto 3% with cloisite Na+ and and then decreases. Generally higher tensile Strength, flexural strength and modulus were obtained with Cloisite 30B than that of Closite Na+ since Cloisite 30B has more compatibility due to –OH groups present in the organic modifier. The thermal stability were improved for both types of nano composites as determined by TGA(thermo gravimetric analysis). The flammability tests shows that closite Na+ filled composites were having UL 94 V2, but cloisite 30B filled composites were not meeting the UL94 standard. DSC and SEM show the TPUR–Nyllon6 with nanoclays are compatible.

INTRODUCTION

Polymer nanocomposites based on Mont Morillonite(MMT) have been successfully prepared previously with improved stiffness and strength of polymers. The extent of property of

V. Rajendran, P. Manivasakan and K.E. Geckeler (eds.)
Nano Material Synthesis and Characterisation, 41-46 (2012).

improvement depends on many factors such as aspect ratio of the filler, extent of dispersion and adhesion and at the filler matrix interface (1) Mica based phyllosilicates such as mont morillonite has generated great deal of interest. Nano clays are a form of high purity aluminium silicate mineral derived from MMT which have a sheet or plate type structure and the layers are about I nm (Nano meter) thick with plate dia 100-500 nm. A surface treatment confers an affinity between the MMT and the polymer choice and also expands the gaps between the plates by as much as six times. Since Inorganic materials donot have good interaction with organic polymer to achieve good dispersion and adhesion, hence requires surface treatment of clays for good dispersion (2).

The hydroxyl groups and carboxyl on quarternary ammonium provides better binding energy of the MMT with the polymers (3). Improved performance of nanocomposites is usually related to the large interface between inorganic fillers and organic matrix even at lower filler (1-5%) loading (4) than conventional fillers (30-40%).

Automotive applications have seen the biggest take up of nano plastics so far car trims, and panels, for examples, display grater tensile strength, impact strength and scratch resistance and engine parts gain higher HDT. Polyamides, Polypropylene, Polyesters and TPU have been studied with nanoclays and are useful in automotive application due lighter weight.

Nylon-6/clay nanocomposites are usually more brittle than pure nylon-6 although dramatic improvement is observed in modulus, strength and heat resistance(5)

The nylon 6–TPU (85 : 15 wt%) blend will have optimum mechanical properties with better impact strength, lower water absorption and good electrical insulation properties. However the elastic modulus will be lowered, if nano filler is used (1%, 3% & 5%) the modulus will be improved along with good toughness. Hence, in this present study our interest is to know the effect of nanofiller montmorillonite on the mechanical and thermal properties of nylon-6 TPU blend (85:15%).

EXPERIMENTAL

Materials

The Nylon-6 used in this work AKULON® K222D (DSM), TPU was DESMOPAN® 345 (BAYER). The filler was sodium mont-morillonite (CLOISITE® $Na^{+)}$ and organically modified mont-morillonite (CLOISITE® 30B) with Bis (2-hydroxyl-ethyl) methyl tallow ammonium (Southern clay products Inc. USA).

Twin Screw Compounding and Test Specimen Preparation: Nylon-6 was dried at 80 C in an air circulated oven for 6 hrs before compounding. TPU was dried at 110°C for 6 hrs. Cloisite Na+ and Cloisite® 30B were dried for 4hrs in the oven at 80°C.

The direct melt blending was carried out in a co-rotating twin screw extruder (Berstorff, FRG) The nylon–6 + nano clay were fed in the main hopper, partially melted,and the TPU was introduced in the sixth zone and the extrudate was cut into granules. The residence time for TPU was reduced to half of that of nylon-6 and was about minute and the materials are not foaming. The nylon-6/TPU blend composition was 85/15 by weight. Clay was added in the ratio 1%, 3%, and 5% by weight. The compounding temperature was in the range of 190°C–230°C and screw rpm was 150. The extruded strands were passed through a water bath and then pelletized.

The nanoblends were further dried at 110°C before injection molding the test specimens for mechanical and thermal properties. SP 130 Clacknar windsor Injection moulding machine was used for the test specimen preparation. The process temperature range was 190-240°C and the injection pressure was 75 Mpa.

Testing and Characterization

Tensile properties evaluation (ASTM D 638) was carried out on a UTM (universal testing machine) SHIMADOZU 50KN Shimadzu Corp., Tokyo. The flexural properties was performed as per ASTM 790 with an UTM Lloyd LR 100k U.K. The Rockwell hardness was carried out as per ASTM D785 using digital hardness tester RASN-E, FIL-INDIA. The notch impact strength (ASTM D 256) was performed using ATS FAAR impact tester,FRG, at room temperature. Fractured surfaces of the notched impact test specimen were studied using SEM (JEOL 200M). X-ray diffraction studies were carried out on a PAN Anatica equipment. TGA and DSC were done using Perkin Elmer Pyris 1 TA instrument with a heating rate of 20 C/min. Flammability test was carried out as per UL 94 Vertical burning standard.

Results and discussion

The Test results are shown in the Tables 1-4. and the SEM figures are shown in Figues 1-4.

Table 1: Tensile Properties of PA6 + TPU + Cloisite 30B

Name	*Max_Force kgf*	*Max_Stress Kgf/cm^2*	*Break_strain %*	*Elastic Modulus Kgf/cm^2*
PA6	245	600	29	21695
PA6+TPU(85:15%)	158	454	185	15107
PA6+TPU+Cloisite 30B 1%	195.4	468	71.9	14294
PA6+TPU+Cloisite 30B 3%	205.1	496	39.2	15133
PA6+TPU+Cloisite 30B 5%	264.0	636	4.14	17585

Mechanical Properties

As Nano clay concentration increases the the tensile strength & tensile modulus were increasing(Table1 & 2). In the case of Cloisite Na+, the 5% formulation has higher the tensile strength (Figure 1) and modulus. In the case of Cloisite 30B, the 5% formulation has higher Tensile strength and modulus (Table 1). The elongation at break was decreasing in both cases.

Generally higher tensile modulus were obtained with Cloisite 30B than that of Closite Na+ since Cloisite 30B has more combatibilty due to the polar interaction of the -OH groups present in the organic modifier. The flexural strength and modulus were increasing as the clay concentration was increasing. The rigidity was increasing hence the above trend was obtained. The results were shown in the Table 3. Higher flexural modulus were obtained in the case of Cloisite 30B than that of closite Na+. Similarly the hardness was increasing in both type of nanocomposites.

Table 2: Tensile Properties of PA6 + TPU + Cloisite Na+

Name	*Max_Force kgf*	*Max_Stress Kgf/cm2*	*Break_Strain %*	*Elastic Modulus Kgf/cm2*
PA6+TPU+Cloisite Na+ 1%	192.0	461	79	8336
PA6+TPU+Cloisite Na+ 3%	193.8	466	76	11561
PA6+TPU+Cloisite Na+ 5%	217.1	522	52.	11617

Table 3: Flexural properties and Hardness of PA6 + TPU with CloisiteNa+ and Cloisite 30B

Name	*Flexural Strength Kg/cm^2*	*Flexural Modulus Kg/cm^2*	*Rockwell Hardness*
Cloisite Na+			
1%	420	11800	75
3%	443	12666	78
5%	557	14900	87
Cloisite 30B			
1%	500	14433	81
3%	510	14800	85
5%	608	18800	95

Table 4: Izod Impact Strength Values(Kgcm/cm) of Nylon6/TPU Blend with Nanofiller

Sample Name	*Cloisite Na+*	*Cloisite 30B*
0%	11.41	11.41
1%	12.14	15.19
3%	14.43	10.86
5%	12.39	10.82

The Nylon 6-TPU (85: 15wt %) was in good compatibility due to the polar functional groups present in these polymers. The izod impact strength of nylon was improved from 4.5 Kgcm/cm to 11.41 Kg cm/cm when TPU was added. In the case of Cloisite Na+, due to the nano filler toughening notched izod impact strength was increased for 1% and 3% and then lowered for 5% due to more rigidity (Table 4). In case of Cloisite 30B the impact was improved for 1% due to toughening due to the-OH functional groups interaction with both amide groups and isocyanate groups; but lowered for 3% & 5% since more interaction leads to higher rigidity and hence hardness and flexural modulus were higher and hence impact strength were little lowered (Table 4). Also these values are higher than that of nylon6. Hence super tough materials were obtained with optimum mechanical properties.

Fig. 1: SEM of PA6 + TPU

Fig 2: SEM PA6 + TPU with 5% Na+

Fig. 3: SEM of PA6 + TPU with Cloisite Na+

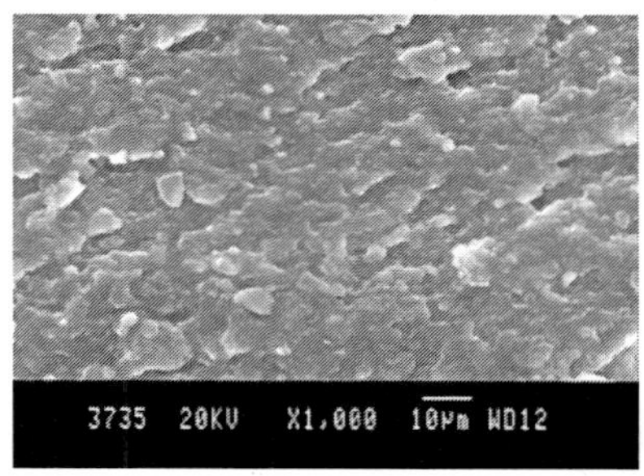

Fig. 4: SEM of PA6 + TPU with 3% 5% Cloisite 30B

Morphology (DSC, SEM AND X-RAY Differaction)

DSC shows the Nylon6-TPUR is compatible in presence of Nano clays. The SEM microphotograph shows good compatibility of the TPU with Nylon 6 Fig (1) and of both cloisite Na+ and 30B at all concentrations in Nylon6-TPU (Fig. 2-4). At 5% level the nanofiller may be saturated. The XRD were taken for all formulations. The XRD indicates the intercalation of CLOISITE 30B at 1%, 3% and 5% clay incorporation. The XRD is the clear indication of intercalation of clay platelets happening in PA6 and TPU matrix.

TGA (Thermo Gravi Metric Analysis)

From the TGA study the thermal stability was evaluated. The test results show the thermal stability was improved due to the inorganic nature of the nanoclays. Higher Thermal stability was achievied with 5% Na+ and with 1% 30B. in the case of cloisite 30B due to the organic nature of the modifiers the thermal stability is not much improved.

Flammabilty

The UL-94 vertical flammability test indicates that UL 94 V2 was obtained with Na+ Nano clay. The TPU in the Nylon 6 makes the material dripping and making the cotton to catch fire. Hence V2 results were obtained. However, the test specimen ceases the fire within 10 seconds. But, in the case of 30B clay the test specimens are burning upto the holding clamp with dripping and making the cotton to catch fire. The organic modifier may make the nanocomposites more flammable.

CONCLUSION

Optimum mechanical properties were obtained with 5% Cloisite Na+ and 5% Cloisite 30Bfilled nylon6-TPU blends i.e higher tensile strength and flexural modulus with medium impact strength than that of nylon6-TPU blend. Generally the Cloisite 30B provides higher tensile and flexural properties. 1% cloisite 30B shows highest impact strength. DSC and the SEM micro photographs shows the nylon6-TPU blend is compatible and also in presence of the Nanofillers the nylon6-TPU blends were compatable.

REFERENCES

[1] Sinha S, Okamoto, *Progress Polym* Sci, 28, 1538-1641, 2003.

[2] Fornes T.D.,Hunter D.I., Paul D.R., *Polymer 45*, 2321-31, 2004.

[3] Maurizio Fermeglia, Margo Ferrone and Sabrina Prict Fluid Phase Equillibria 212, 315-329, 2003.

[4] M. Muriel A Duabult and J L Halary, *Polym Int* 56, 214-223, 2007.

[5] K. Yang and R. Ozisik, *Polymer* 47, 2849-2855, 2006.

COMPARISON AND ANALYSIS OF PHYSICAL PROPERTIES OF CARBON NANOMATERIAL DOPED POLYMER COMPOSITES

Smrutisikha Bal and Snigdha Bhattacharjee

Department of Physics, N.I.T, Silchar
E-mail: smrutisikha_bal@yahoo.com

ABSTRACT

Carbon nanofibers (CNFs) are mainly differentiated from carbon nanotubes (CNTs) by the orientation of the graphene planes; the graphitic layers are parallel to the axis in nanotubes, while nanofibers can show a wide range of orientations of the graphitic layers with respect to the fibre axis. With the purpose to investigate the multifunctionality caused by CNTs/CNFs, mechanical properties such as flexural modulus & hardness and electrical property such as conductivity of various nanocomposites were determined. These results were supplemented and complimented by SEM and TEM. Both mechanical and microscopic observation indicated that aligned CNT composites show a significant reinforcement in the epoxy matrix along with bridging mechanism. Lower mechanical results were obtained in CNF composites that attributes to breaking of nanofibers. All nanocomposites were showing electrical percolation that occurs due to network formation of nanoparticle inside the matrix.

INTRODUCTION

Carbon nanotube (CNT)/Carbon nanofibres (CNF) based polymer composites have become popular in structural applications because of their unique atomic structure, very high aspect ratio and extraordinary mechanical and physical properties like strength and flexibility [1,2].

V. Rajendran, P. Manivasakan and K.E. Geckeler (eds.)
Advanced Nano Materials for Industrial Applications, pp. 47-52 (2012).

The high bond strength of the constituent carbon–carbon bonds of multi-walled carbon nanotubes (MWNTs) are the reason behind its outstanding mechanical properties. In the polymer field, epoxy resins are well established thermosetting matrices of advanced composites, displaying a series of interesting characteristics, which can be adjusted within broad boundaries [3-5]. Several studies related to the enhancement of mechanical and electrical properties of epoxy matrix by the introduction of CNT/CNF have been conducted [3,6–9].

In the present work, the nanocomposites involving various nanomaterials of low content (0.75 wt%) have been prepared for the first time at low temperature under refrigeration process. The purpose of this study is to examine the effect of nanomaterials on mechanical and electrical properties of prepared composites which were again supplemented by microscopic studies.

MATERIALS AND METHODS

All the nanoparticles used for this experiment are >95% purity. Multiwall carbon nanotubes (MWNTs) are of 10–50 ìm length and 20–70 nm diameter, Functionalised carbon nanotubes (FCNT) are of 10-30 ìm length and 10-20 nm diameter and Aligned carbon nanotubes (ACNT) are of 5-15 ìm length and 10-20 nm diameter. Carbon nanofibers (CNFs) used is 200–500 nm in diameter, 10–50 μm long. Polymer matrix was prepared by mixing epoxy resin (Ciba-Geigy, araldite LY-556 based on Bisphenol A) and hardener HY-951 (aliphatic primary amine).

Nanocomposites were prepared by dispersing CNTs (MWNTs, FCNTs and ACNTs) and CNFs in the epoxy resin through ultrasonication followed by low temperature curing process. Mechanical characterizations (flexural modulus & hardness) and electrical conductivity of nanocomposite samples was determined. Dispersion behaviour and fracture surface study of the above were observed through SEM and TEM.

RESULTS AND DISCUSSION

Mechanical Characterization

Flexural modulus and hardness of CNT & CNF nanocomposites are presented in Table 1. Increase in flexural modulus was more pronounced in ACNT composite samples in comparison to other samples. There is 41% enhancement in the value of flexural modulus in MWNT and more in case of FCNT (51%) & ACNT (67%) in comparison to CNF composite. Study reveals that CNFs are more prone to agglomeration that ultimately affects the modulus [10,11]. A considerable improvement of hardness of 28.2 MPa was observed in ACNT nanocomposite which was 68% more than the value obtained for CNF (16.7 MPa). In case of FCNT and MWNT, 55% and 11% enhancement in hardness was observed respectively compared to that of CNF. Reinforcement of high strength nanotube contributes to an increase in hardness while reduction in hardness is attributed to presence of microvoids during fabrication of composites [12]. Hardness results agreed well with flexural test results with the confirmation that aligned and functionalised CNT composite samples had improved mechanical properties than MWNT and CNF composite samples.

Table 1: Flexural Modulus and Hardness of CNT/CNF Composites

Sample	*Flexural Modulus (Mpa)*	*Hardness (MPa)*
MWNT	3343	18.5
FCNT	3589	25.0
ACNT	3954	28.2
CNF	2370	16.7

Electrical Characterization

The electrical conductivity of nanocomposites was in the range of 2.7×10^{-3} S/cm to 2.35×10^{-2} S/cm (Fig. 2). Conductivity values of ACNT & FCNT composite samples were higher compared to that of MWNT & CNF samples. From CNF to ACNT, conductivity seems to rise only by one order. This improvement was due to the fact that aggregated phases form a conductive 3-dimensional network throughout the whole sample. The electrical result reveals that although flexural modulus and hardness of CNT samples are much higher than CNF samples, conductivity in those samples offer little variation thus achieving percolation. The percolation limit for most polymers filled with carbon black is typically found between 5% and 20% [13], but the amount of CNFs/CNTs utilized here is only 0.75%.

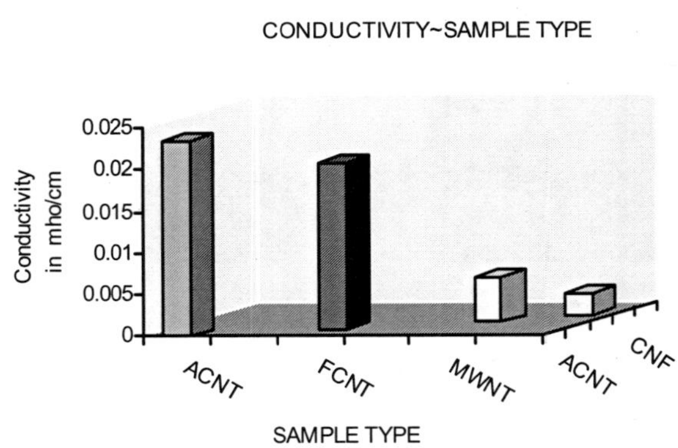

Fig. 2: Conductivity of CNT and CNF Composites

Microscopic Studies

Fig. 3a shows uniform dispersion of nanofillers inside epoxy matrix. Fracture surface (Fig. 3b) revealed the formation of network inside the matrix by the aggregate phases of nanotubes.

Bridging of a crack surface by a CNT was observed at higher magnification (Fig. 3c). The nanofibers (Fig. 5a) were found to be aligned perpendicularly and broken (shown by arrow mark); that might be the reason for lower value of flexural modulus in comparison to CNT composites [14].

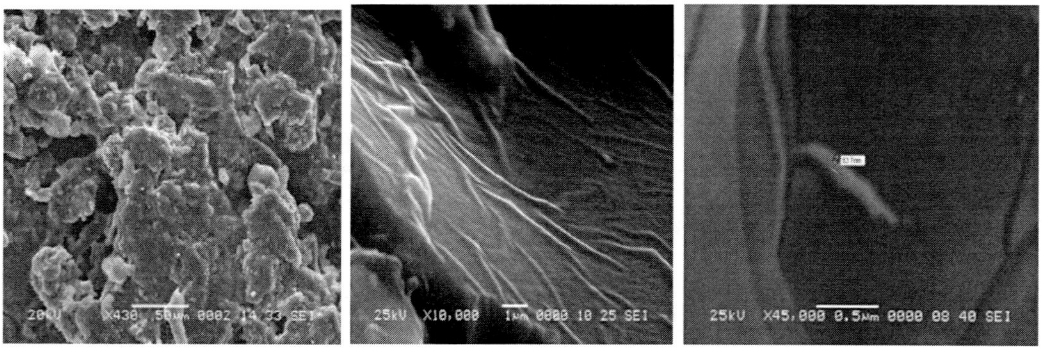

(a) (b) (c)

Fig. 3: Microstructures of CNT/Epoxy Composite Showing (a) Dispersion, (b) Network Formation & (c) Bridging

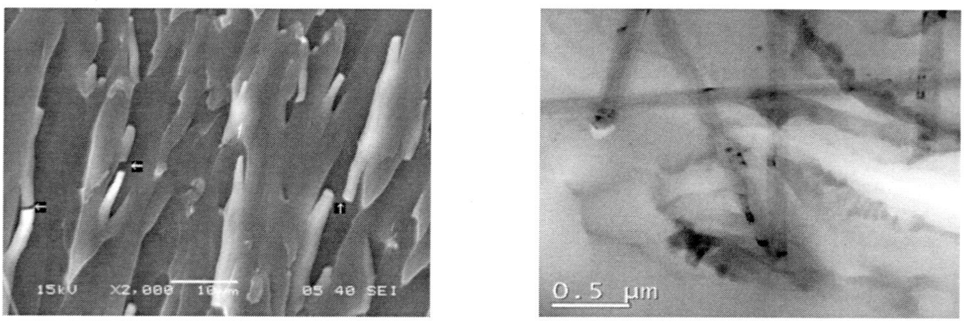

(a) (b)

Fig. 5: (a) SEM of CNF Composite (b) TEM of CNF Composite

TEM of the same sample (Fig.5b) revealed the interaction of fiber with epoxy matrix at the interface along with network formation. So the results obtained from microscopic study are well agreement with the mechanical and electrical studies.

CONCLUSIONS

Epoxy composites filled with a little wt% of aligned CNTs and functionalized CNTs yield better mechanical and electrical properties than MWNT and CNF samples due to improvement in dispersion and strong bonding of carbon nanotubes with the matrix. Relative low mechanical properties in CNF composites were due to breaking of nanofibers. Enhancement in electrical conductivity was attributed to aggregated phases forming a conductive 3D network throughout the whole sample.

REFERENCES

[1] Wagner, H.D., Lourie, O., Feldman, Y. and Tenne, R., Stress-Induced Fragmentation of Multiwall Carbon Nanotubes in a Polymer Matrix, *Appl.Phys. Lett.,* 72, 188-190, 1998.

[2] Maruyama, B., Alam, K., Carbon nanotubes and nanofibers in composite materials, *SAMPE J,* 38(3), 59–70, 2002.

[3] Schadler, L.S., Giannaris, S.C. Ajayan, P.M., Load transfer in carbon nanotube epoxy composites, *Appl. Phys.Lett.*, 73, 3842-3844, 1998.

[4] Sandler, J., Shaffer, M.S.P., Prasse, T., Bauhofer, W. Schutle, K. et al., Development of a dispersion process for carbon nanotubes in an epoxy matrix and the resulting electrical properties, *Polymer*, 40, 5967– 5971, 1999.

[5] Cooper, C.A., Young, R.J., Halsall, M., Investigation into the deformation of carbon nanotubes and their composites through the use of Raman spectroscopy, *Composites A*, 32, 401-411, 2000.

[6] Gojny, F.H., Wichmann, M.H.G., Köpke, U., Fiedler, B. Schulte, K., Carbonnanotube reinforced epoxy composites: enhanced stiffness and fracture toughness at low nanotube content, *Compos. Sci. Technol.*, 64, 2363-2371, 2004.

[7] Kotaki, M., Wang, K., Toh, M.L., Chen, L. Wong, S.Y., et al. Electrically conductive epoxy/clay/vapor grown carbon fiber hybrids, *Macromolecules*, 39(3), 908–911, 2006.

[8] Lee, H., Mall, S., He, P. Shi, D.L., Narasimhadevara, S., Characterization of carbon nanotube/nanofiber-reinforced polymer composites using an instrumented indentation technique, *Compos Part B*, 38(1), 58–65, 2007.

[9] Choi, Y.K., Sugimoto, K.I., Song, S.M., Gotoh, Y. Ohkoshi, Y., et al. Mechanical and physical properties of epoxy composites reinforced by vapor grown carbon nanofibers, *Carbon*, 43, 2199–2208, 2005.

[10] Xu, L.R. Rosakis, A.J., Impact failure characteristics in sandwich structures: Part II. effects of impact speed and interfacial strength, *Int J Solids Struct*, 39, 4237–4248, 2002.

[11] Xu, L.R., Huang, Y.Y. Rosakis, A.J., Dynamic crack deflection and penetration at interfaces in homogeneous materials: experimental studies and model predictions, *J Mech Phys Solids,* 51, 425–460, 2003.

[12] Gojny, F.H. Schulte, K., Functionalisation effect on the thermo-mechanical behaviour of multi-wall carbon nanotube/epoxy-composites, *Compos Sci Technol*, 64, 2303–2308, 2004.

[13] Kotsilkova, R., Nesheva, D., Nedkov, I., Krusteva, E. Stavrev, S., Reological, electrical and microwave properties of polymers with nanosized carbon particles. *J Appl Polym Sci*, 92, 2220–2227, 2004.

[14] Bal, S., Experimental study of mechanical and electrical properties of carbon nanofiber/ epoxy composites, *Materials & Design*, 31, 2406-2413, 2010.

STRUCTURAL AND OPTICAL PROPERTIES OF BORON CARBIDE NANOPARTICLES FOR LIGHT WEIGHT ARMOR

Navid Hussain

Centre for Converging Technologies, University of Rajasthan, Jaipur
E-mail: naveedqumar@gmail.com

ABSTRACT

In this paper we have prepared nanoparticles of B_4C using chemical method. A mixture of boron powders, boron oxide powder and carbon powder were placed in the bottom of a graphite crucible. The crucible was then inserted into a high temperature tube furnace. The furnace was vacuumed then filled with flowing argon (50~200 sccm). The furnace temperature was raised quickly to 1000~1150°C and held at this point for 3~ 4 hour under argon atmosphere. After the reaction, a grayish product was collected and examined. A vapor-liquid-solid (VLS) mechanism may be involved: The phase and orientation of the sample was characterized by X-ray diffractometry using CuK_α (λ = 1.5407 Å) radiation. The microstructure was examined by scanning electron microscope. The optical properties were investigated using spectrophotometer. Due to light weight, high strength, hardness it is being used as light weight armor.

INTRODUCTION

Boron carbide is a hard material after diamond and boron nitride and also known as black diamond. It is lightweight (theoretical density: 2.52 g/cm^3) and has a 4-point bending strength of at least 300 MPa [1]. In addition to having a Knoop hardness of 2800 kg/mm^2 [2], using a 100 g load. It has a high melting point (2450°C), high neutron absorption cross-section, and is

V. Rajendran, P. Manivasakan and K.E. Geckeler (eds.)
Advanced Nano Materials for Industrial Applications, pp. 53-56 (2012).

also resistant to chemical agents. Due to having properties like low specific weight, high hardness and impact resistance it is useful for body and vehicle armor.

Boron carbide was first prepared in 1883 but the stoichiometric composition B_4C, known as 'technical boron carbide', was not assigned until 1934 [3].The purity of boron carbide produced by early researchers was less than 75% but Ridgway claimed to have produced crystalline B_4C of 90% purity by carbothermic process in 1933 [4]. Different processes to produce nanoparticles of boron carbide are known. Boron carbide can be synthesized from elemental boron and carbon directly. But this method has become unattractive due to high cost of elements involved in the process. Other synthesis methods include metallotermic and carbothermic processes such as carbon thermal reduction of boron oxide (B_2O_3) over 1000°C, reduction of BCl_3 by CH_4 at a temperature of 1500°C with laser, thermal decomposition of a mixture of pure carbon and boron trichloride in an atmosphere of hydrogen, and gaseous reaction between BCl_3 and methane- hydrogen mixture in r. f. argon plasma [5-8]. Recently, boron carbide nanoparticles have been synthesized by co-reduction route using BBr_3 and CCl_4 as the source materials at a low temperature of 480° *C* and new selective synthesis routes to boron carbide and boron nitride based on the use of new polymer systems, poly[2-(vinyal) pentaborane] and poly [B-vinylborozine] as precursors. It was reported that $2(H_2C{=}CH)\ B_5H_8$ will undergo thermal polymerization to vinly-pentaborane oligomers and that these species can be converted to pure boron carbide with high ceramic yields under mild conditions. Isao et al., reported that boric acid and 2-hydroxy benzyl alcohol (HBA) were used as the B and C sources of B_4C. H_3BO_3 and HBA in DMF solution resulted in a phenolic resin-B_2O_3 hybrid. Among the methods, carbothermal reaction and magnosoisothermal reaction are widely used as main techniques for the production of B_4C powders. The mostly used industrial method to produce B_4C is reduction of B_2O_3 in presence of carbon using Magnesiothermic techniques.

EXPERIMENTAL

Boron powder (Loba Chemie, 200 mesh, 99.00% pure), carbon powder (Fisher Scientific), di-boron tri-oxide powder (B_2O_3, Search Chem) and Cobalt powder (Molychem, 99.50% pure) were used without further changes as purchased. The metal like cobalt (Co) was used during reaction as catalyst to facilitate the formation of boron carbide nanorods. Boron oxide, boron amorphous powder and carbon were used as the precursor agents. Before weighing, carbon powder was baked at 150-200 °C in vacuum. Different ratios of $B/B_2O_3/C$ were mixed for the reaction. Cobalt metal powder was used as the catalyst. Reactions were carried out in a high-temperature tube furnace with a maximum working temperature of 1200°C. Dry powders were mixed by roll milling and put into a covered graphite crucible. The crucible was then placed into the center reaction zone of the tube furnace. The reactor was pre-heated under vacuum at about 200°C for 2-3 hours. Ultrahigh purity argon was then flushed through the furnace several times before filling the reactor with argon at a constant flow of 200 sccm. The furnace was then heated to 400°C for 30 minutes. After this period, the temperature was rapidly raised to 1000- 1150 °C it took about 30 min and was maintained at these temperatures for 3~4 hours before it was turned off and cooled down to room temperature. After the reaction, the as-formed product inside the crucible was collected and analyzed. The as prepared sample was also characterized that confirmed the existence of crystalline boron carbide.

RESULTS AND DISCUSSION

The phase and orientation of sample were characterized by X-ray diffractometry using CuK_{α} (λ = 1.5407 Å) Figure 1. The XRD pattern confirms the formation of B_4C. The crystalline size of the sample was calculated using Scherrer's formula

$$\mathbf{T = 0.91\lambda / \beta \cos \theta}$$

where λ = wavelength of x-rays (1.5407 Å), β = Full width at half of the maximum peak (FWHM), θ = Angle of maximum intensity. The calculated crystallized size is about 25 nm.

The surface topography and microstructure were examined by scanning electron microscope (SEM) Figure 2. The average particle size is observed about 36 nm.

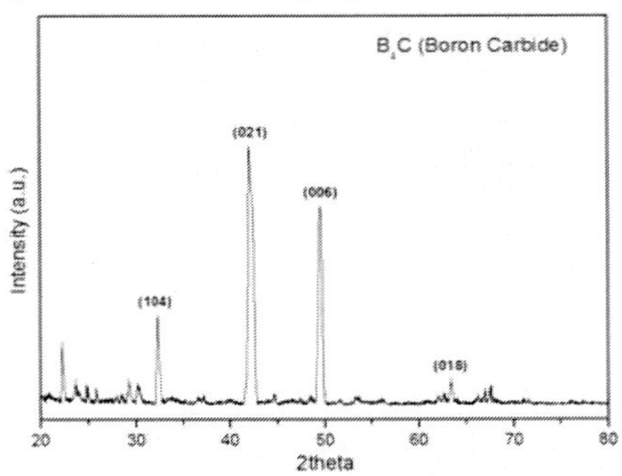

Fig. 1: XRD Data of Sample

Fig. 2: SEM Image of Sample

The optical absorption spectra of Boron Carbide was observed using spectrophotometer. The relation between the absorption coefficient, α, and the incident photon energy, hν, can be written as [9]:

$$\alpha h\nu = A\left(h\nu - E_g\right)^n$$

where α is the absorption coefficient, h is Planck's constant, ν is the photon frequency and E_g is the optical band gap, A is a constant depending on the transition probability and *n* is an index that characterizes the optical absorption process and is theoretically equal to n = 1/2 and 3/2 for direct allowed and direct forbidden transition, respectively, while n = 2 and 3 for indirect allowed and indirect forbidden transition, respectively. To calculate the direct band gap of B2O3, plot of $(\alpha h\nu)^2$ versus hí is shown in Figure-3. An extrapolation of the linear region of a plot of $(\alpha h\nu)^2$ on the y-axis versus photon energy ($\eta\nu$) on the x-axis gives the value of the optical band gap E_g. So, the optical band gap of sample (Boron Carbide nanoparticles) was found to be 2.39 ev.

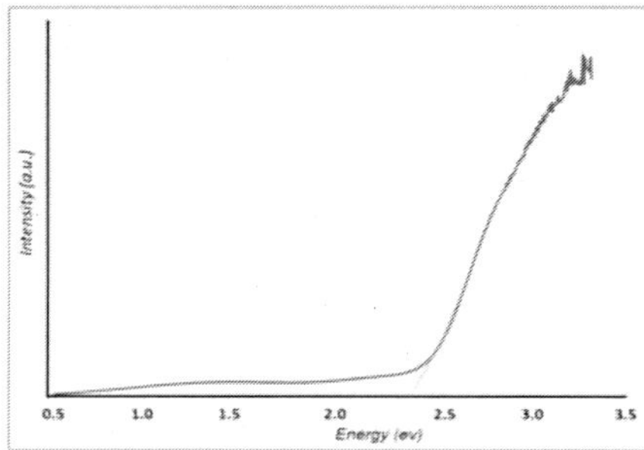

Fig. 3: Shows Energy Band Gap for Boron Carbide Nanoparticles

CONCLUSIONS

Different methods are available for fabrication of boron carbide nanomaterials like mechanical milling, carbothermal processes, sol-gel method, classical chemical VD, plasma-enhanced CVD. Carbothermal reduction of boric acid which has been the commercial method for the production of boron carbide was used. Cobalt metal powder was used as the catalyst. A smooth range of nanoparticles of boron carbide were synthesized and characterized using SEM images and XRD data. The optical properties of sample were also looked into. Due to having properties like low specific weight, high hardness it is being used for body and vehicle armor.

REFERENCES

[1] F. Thevenot, "A Review on Boron Carbide," *Key Engineering Materials*, 56-57 59-88 (1991).

[2] J. F. Shackelford and W. Alexander, CRC Materials Science and Engineering Handbook, CRC Press, Boca Raton, Fl., 1991.

[3] A. Joly, C.R. Hebd, *Seances Acad. Sci.* 97 (1883)

[4] R. R. Ridgway: 'Boron carbide and the method of making the same', US patent no. 1897214, 1933.

[5] A. M. Hadian and J. A. Bigdeloo," The Effect of Time, Temperature and Composition on Boron Carbide Synthesis by Sol-gel Method," Journal of Materials Engineering and Performance, vol. 17[1], pp. 44-49, 2008

[6] S. Mondal and A. K. Banthia, "Low-temperature synthetic route for boron carbide," *Journal of the European Ceramic Society*, vol. 25, pp. 287-291, 2004.

[7] A. Sinha, T. Mahata and B. P. Sharma, "Carbothermal route for preparation of boron carbide powder from boric acid-citric acid gel precursor," *Journal of Nuclear Materials*, vol. 301, pp. 165-164, 2002.

[8] K. Yamada, "Boron carbide particles formed from an amorphous boron/ graphite powder mixture using a shockwave technique," *J Am Ceram Soc., vol.* 79[4], pp. 1113-1116, 1996.

[9] J. Tauc, Amorphous and Liquid Semiconductors, Plenum Press, New York, 1974.

EFFECTS OF 2-METHOXY ETHANOL ON THE STRUCTURAL AND OPTICAL PROPERTIES OF ZnO NANOPARTICLES PREPARED BY SOL-GEL CO-PRECIPITATION METHOD

A.S. Haja Hameed, C. Karthikeyan, S. Sasikumar and G. Ravi[1]

PG & Research Department of Physics, Jamal Mohamed College, Tiruchirappalli, Tamil Nadu
[1]School of Physics, Alagappa University, Karaikudi, Tamil Nadu
E-mail: hajahameed2001@gmail.com

ABSTRACT

In the present study, ZnO nanoparticles were synthesized by sol-gel co-precipitation method using 2-methoxy ethanol as a solvent. From the X-ray diffraction studies, the hexagonal phases of the ZnO nanoparticles were found out. The sizes observed from the above analysis were found to be consistent with the sizes of the ZnO nanoparticles (32 nm) estimated from the XRD peak broadening of the nanoparticles. The HRSEM images clearly showed that all the synthesized ZnO nanoparticles retained the hexagonal shape. The particles possessed smooth surface and average particles grain size of 20-35 nm. From their UV-VIS transmission spectrum, the ZnO nanoparticles had good transparencies with 80% of transmission. The photoluminescence studies indicated that the samples exhibited the strong blue emissions with a major peak located at around 424 nm due to Zn vacancy. The second peak located at around week blue-green band at 485 nm was due to the transition between oxygen vacancy and oxygen interstitial defects. The third peak was observed as a green emission band at 504 nm because of a large fraction of oxygen vacancies. The fourth peak was observed to be a very weak green emission band at 530 nm corresponding to the singly ionized oxygen vacancy in ZnO.

V. Rajendran, P. Manivasakan and K.E. Geckeler (eds.)
Advanced Nano Materials for Industrial Applications, pp. 57-62 (2012).

INTRODUCTION

In recent years, research interest on nanomaterials has been greatly spurred due to their essential significance in science and potential technological applications[1]. ZnO is an II-VI group of highly conductive transparent oxide, semiconductor material, which has attracted much attention due to its important electric, piezoelectric, opto-electronic properties [2]. ZnO nanoparticles have direct wide band gap of 3.37 eV with large exciton binding energy of 60 meV [3]. ZnO has excellent chemical and thermal stability [4]. Sol-Gel chemistry has recently evolved into a powerful approach for preparing inorganic materials for such applications [5]. The various methods such as thermal evaporation, precipitation, solvo-thermal and hydrothermal methods, etc., have been applied to growth of the ZnO nanoparticles. ZnO forms in the wurtzitehexagonal structure with the lattice parameters a = 3.25 Å and c = 5.21 Å, where zinc (Zn) atom is tetrahedrally coordinated to four oxygen (O) atoms [6]. The native defects such as Zn interstitials and oxygen vacancies are believed to be a factor which produces n-type conductivity of intrinsic ZnO material [7]. The XRD studies on the ZnO nanoparticles prepared by the various methods are reported by many researchers [8]. The photoluminescence (PL) spectrum of ZnO exhibits a broad defect related visible emission. This emission is most commonly green luminescence and blue emission is also been observed. For device applications such as high efficiency UV light emitting devices, it is important to suppress the visible emission. On thegreen luminescence, its origin is still controversial and a number of suggestions have been proposed. The green luminescence has been attributed to various types of defects such as oxygen vacancies [9], zinc vacancies [10], as well as donor–acceptor pairs [11]. The situation is complicated since it has been reported that the spectral position and the intensity of the visible emission also depend on the fabrication process [12]. A plausible explanation is that the species of impurities as well as the concentration of intrinsic and extrinsic defects are related to the growth procedure.

In the present work, the sol-gel co-precipitation method was used to synthesis ZnO nanoparticles by using 2-methoxy ethanol as a solvent. The ZnO nanoparticles were characterized by X-ray diffraction, UV-VIS spectroscopy, High resolution scanning electron microscopy and Photoluminescence (PL) spectroscopy.

EXPERIMENTAL

The ZnO nanoparticles were prepared by known sol-gel co-precipitation method. For a typical synthesis, a solution was prepared by using zinc acetate dihydrate and 2-methoxy ethanol in the presence of monoethanolamine as a stabilizer. NaOH was used as a precipitation agent and 5.6 ml of NaOH was added into the solution. Then, a white precipitate was obtained. The solution with the white precipitate was stirred at a particular temperature for 20 minutes. This solution was used as a 'sol'. In order to prepare a required 'gel', 5 g of polyethylene glycol was fully dissolved in double distilled water. The gel solution was slowly added in to sol mixture. The mixtures were stirred at 60°C for 5 hours where any evaporation of solvents was avoided. After the formation of the sol-gel, it was washed several times and was filtered by using a filtration unit. The precipitate was dried at 90°C. Thus, the ZnO nanopowders were obtained. The ZnO nanopowders were annealed at 500 °C for 1 hour because the energy from the heat will enhance the vibration and diffusion of lattice atoms for atomic rearrangement. Furthermore, the annealing helped to remove the residual impurities. The annealed ZnO nanoparticles were finally cooled into room temperature and they were used for further studies.

The nanoparticles were characterized by X-ray diffractometer (XRD, model; XPERT PRO). The diffraction patterns were recorded in the range of 20°-80° where the monochromatic light of wavelength 1.54 Å was used. The UV-VIS spectrum was recorded in the range of 200-800 nm using a Ultra Violet-Visible-Near Infrared spectrometer (UV-VIS-NIR, model:5E). The morphology and size of the particles were observed by a High Resolution Scanning Electron Microscope (HRSEM, model : FEI- QUANDA 200F). The Photoluminescence (PL) measurement was performed by a Hitachi F-4500 spectrometer and the corresponding spectrum was recorded in the range of 400-700 nm for the ZnO nanoparticles.

RESULTS AND DISCUSSION

Figure 1 shows the typical XRD pattern of the as-obtained ZnO nanoparticles using 2-methoxyethanol. All the diffraction peaks are well indexed to the hexagonal phase of ZnO. The lattice parameters a and c values are found to be 0.3252 nm and 0.5209 nm respectively. Its lattice parameters are found to be consistent with the reported values available in JCPDS card (No: 36-1451, a = 0.3249nm and c = 0.5206 nm). No impurities are observed in these patterns. The obtained ZnO nano particles are in wurtzite hexagonal phase. The sharp peaks reveal that the products are well crystallized. The grain sizes of the particles are estimated by using the Scherer's formula relation

$$D = K\lambda/\beta cos\theta$$

where K is the shape factor with a value of 0.9. λ is the wavelength of X-ray related to CuKα (1.54Å). θ is the diffraction angle. The FWHM (β) values are estimated as 0.2647 and 0.2408 radian for the hkl values of (100) and (002) respectively. The grain sizes of the particles are found as 30.28 nm and 34.56 nm for the above respective hkl values.

Figure 2 shows UV-VIS transmission spectrum of the ZnO nanoparticles recorded in the range of 200 to 800 nm. In the visible spectral region, the ZnO nanoparticles have good transparencies with 80% of transmission at 800 nm. The transmission increases sharply near the visible spectral region.

The HRSEM images shown in Fig. 3 reveal the morphology and size distribution of the ZnO nanoparticles. All the particles are found to be uniform and hexagonal in shape. The particles possess the smooth surfaces and particles grain size of 20-35 nm. The sizes of the nanoparticles arefound to be a good agreement with the estimated average ZnO nanoparticles grain size of 32 nm from the XRD peak broadening of the nanoparticles.

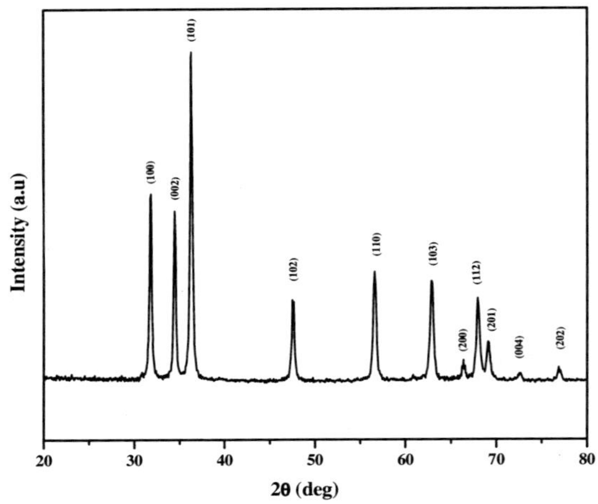

Fig. 1: X-ray Powder Diffraction Pattern of the ZnO Nanoparticles Prepared using 2-Methoxy Ethanol

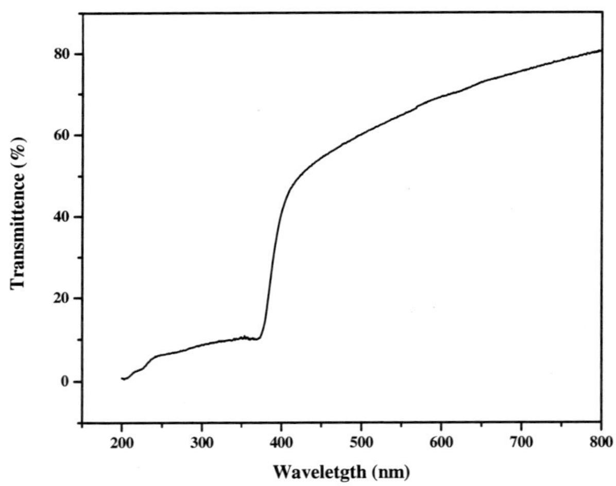

Fig. 2: UV-vis Transmittance Spectrum of the ZnO Nanoparticles Prepared using 2-Methoxyethanol

The photoluminescence spectrum of as-synthesized ZnO sample shown in Fig. 4 was recorded with the excited wavelength of 370 nm. In the photoluminescence (PL) spectrum of ZnO, there are some emission bands in the UV and visible regions. The UV peak is usually considered as the characteristic emission of ZnO and attributed to the band edge emission or the exciton transition. Generally, the UV emission in ZnO disappears in two cases. First, if the excitation energy is considerably lower thanits band gap energy [13] and second, if the intensity of visible emission is much higher due to increased defect density [14]. In the present study,

Fig. 3: HRSEM Images of the ZnO Nanoparticlesprepared using 2-Methoxy Ethanol

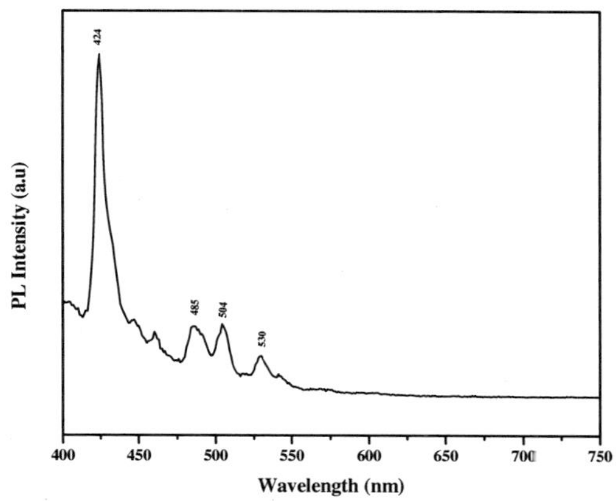

Fig. 4: Photoluminescence (PL) Spectrum of the ZnO Nanoparticles Prepared using 2-Methoxy Ethanol

the UV emission does not appear at the excitation wavelength of 370 nm. It is reported that the PL emission of the ZnO nanoparticles in the blue region (420 nm) is due to Zn vacancy [15]. The four emission peaks are observed in the spectrum. The samples exhibit strong blue emissions with a major peak located at around 424 nm due to Zn vacancy. The second peak is observed as weak blue–green band at 485 nm. The weak blue-green emissions are possibly due to surface defects in the ZnO nanopowders corresponding to the transition between oxygen vacancy and oxygen interstitial defects.The third peak is observed as green emission band at 504 nm. The strong intensity of the green emission band indicates a large fraction of oxygen vacancies existing in ZnO nanostructures. The fourth peak is observed as very weak green emission band at 530 nm which corresponds to the singly ionized oxygen vacancy in ZnO, and this emission results from the recombination of a photo generated hole with the singly ionized charge state of the oxygen vacancy. These defects indicate non-stoichiometric characteristic of the ZnO nanoparticles structures and it is believed that they are reasonable for n-type conductivity.

CONCLUSIONS

The ZnO nanoparticles were synthesized by sol-gel co-precipitation method using 2-methoxyethanol. From the X-ray diffraction studies, the hexagonal phases of the ZnO nanoparticles were found. The sizes observed from the above analysis were found to be consistent with the sizes of the ZnO nanoparticles (32 nm) estimated from the XRD peak broadening of the nanoparticles.In the visible spectral region, the ZnO nanoparticles have good transparencies with 80% of transmissionat 800 nm. The HRSEM images clearly showed that all the ZnO nanoparticles were in hexagonal shape. The particles possessed smooth surface and average particles grain size of 20-35 nm. The photoluminescence studies indicated that the samples exhibited the strong blue emissions.

ACKNOWLEDGEMENTS

One of the authors (A.S.H) is grateful to University Grant Commission (UGC), New Delhi, India for sanctioning the financial assistance (UGC MRP F. No. 38-97/2009(SR)).

REFERENCES

[1] X. Duan, Y. Huang, R. Agrawal, C.M. Lieber, *Nature* 421 (2003), 241.

[2] Z. Sofiani, B. Derkowsk, P. Dalansinski, M. Wojdula, S. Dabosseignon, M. Alaouilamrani, L. Dghoughi, W. Bala, M. Addou and B. Sahraoui, *Optics Communications* 267 (2006)1303.

[3] S. Hingorani, Y. Pillai, P. Kumar, M.S. Multani., D.O. Shah, *Mater. Res. Bull.*, 28 (1993), 1303.

[4] P. Tonto, O. Mekasuwandumrong, S. Phantanasri, V. Pavarajan and P. Praserthdam, *Ceram. Int.*, 34 (2008), 57.

[5] L.H. Larry, K.W. Jon, *Chem, Rev.*, 90 (1990) 33.

[6] D.R. Lide (Ed), CRC Handbook of Chemistry and Physics, CRC Press, NewYork, 1992.

[7] H.S. Kang, G.H. Kim, S.H. Lim, H.W. Chang, J.H. Kim, S.Y. Lee, *Thin solid Films*, 516 (2008), 3147.

[8] C. Wu, Q. Huang, *Journal of Lumina.*, 130 (2010), 2136.

[9] H.S. Kang, J.S. Kang, J.W. Kim, S.Y. Lee, *J. Appl. Phys.*, 95 (2004), 1246.

[10] F. Tuomisto, K. Saarinen, D.C. Look, G.C. *Farlow, Phys. Rev. B Condens. Mater.Phys.*, 72 (2005) 085206.

[11] D.C. Reynolds, D.C. Look and B. Jogai, *J. Appl. Phys.*, 89 (2001) 6189.

[12] D. Li, Y.H. Leung, A.B. Djurisic, Z.T. Liu, M.H. Xie, S.L. Shi, S.J. Xu, W.K. Chan, Appl.Phys. Lett., 85(2004)1601.

[13] K. Vanheusden, W. L. Warren, C. H. Seager, D. R. Tallant, J. A. Voigt,and B. E. Gnade, J. *Appl. Phys.*, 79 (1996) 7990.

[14] W.C. Zhang, X.L. Wu, H.T. Chen, J. Zhu, and G.S. Huang, J. Appl. Phys., 103(2008) 093722.

[15] K. Sharda, S. Jayanthi, Chawla. *Appl. Sur. Sci.*, 256 (2010), 2635.

EFFECT OF SUBSTRATE TEMPERATURE ON THE MAGNETO-OPTIC PROPERTIES OF $Bi_{1.8}Lu_{1.2}Fe_{3.6}Al_{1.4}O_{12}$ GARNET FILMS PREPARED BY RF SPUTTERING

M. NUR-E-ALAM, MIKHAIL VASILIEV AND KAMAL ALAMEH[1]

Electron Science Research Institute, Edith Cowan University, Joondalup, WA, 6027, Australia
[1]Department of Nanobio Materials and Electronics, Gwangju Institute of Science and Technology, World-Class University (WCU), Gwangju Institute of Science & Technology (GIST), Gwangju 500-712, South Korea
E-mail: m.nur-e-alam@ecu.edu.au

ABSTRACT

In this paper we present a study conducted on RF sputtered garnet films of composition type $Bi_{1.8}Lu_{1.2}Fe_{3.6}Al_{1.4}O_{12}$ prepared by using low and high substrate temperatures during the deposition process inside the vacuum chamber of sputtering system. Comparatively low coercive force is achieved in garnet films prepared at high substrate temperature of 680°C simultaneously with high MO quality and almost in-plane magnetization direction, which are the properties desired in various MO sensing, switching and imaging applications.

INTRODUCTION

Bismuth-substituted iron garnets are very well known to be the best magneto-optical (MO) materials for various applications in integrated optics and nano-photonics due to their strong MO behaviour and fast magnetization switching response [1]. A number of important application

V. Rajendran, P. Manivasakan and K.E. Geckeler (eds.)
Advanced Nano Materials for Industrial Applications, pp. 63-68 (2012).

areas require MO thin films with very high MO quality and also low coercivity with either the in-plane-oriented easy axis of magnetization or having a strong in-plane magnetization component. Multiple variations in the material properties of garnets are required for different applications in integrated optics and in photonics, which can be engineered by adjusting the material composition, however every composition requires its own optimized deposition and annealing regimes. We were the first group that synthesized a garnet material of composition type $Bi_{1.8}Lu_{1.2}Fe_{3.6}Al_{1.4}O_{12}$ using the sputter-deposition technique and oven-annealing, and achieved films that possessed very high MO performance simultaneously with strong in-plane magnetization component and magnetically-soft switching behavior [2, 3].

In this paper, we investigate the effects of substrate temperature (used during the deposition of garnet layers using RF magnetron sputtering) on the materials properties and compare the results achieved using either "cold (250°C)" or "hot (680°C)" substrates.

FILM GROWTH AND CHARACTERIZATION

We fabricated highly Bi-substituted lutetium iron-aluminium garnet thin films (several batches from 650 nm to 1000 nm in thickness) using RF magnetron sputtering technology. The oxide-mix-based sputtering target of nominal stoichiometry $Bi_{1.8}Lu_{1.2}Fe_{3.6}Al_{1.4}O_{12}$ (prepared using 99.9% pure oxides) was used to deposit amorphous garnet-type layers onto different substrates (GGG, Corning Eagle XG and silicon) which crystallized into ferrimagnetic phase after annealing. As-deposited garnet layers were sputtered using two widely different substrate temperatures (250°C and 680°C as measured by a thermocouple sensor placed next to SiC heater) at high vacuum (1-2E^{-6} Torr), but the pure-argon (Ar) plasma was always kept at 1 mTorr. All the initially-amorphous films were crystallized using optimized annealing process within a conventional oven annealing system. The film thicknesses were monitored during the deposition processes using in-situ laser reflectometry and also re-measured after the deposition using the iterative fitting of their transmission spectra obtained with a UV/Visible spectrophotometer (Beckman Coulter DU 640 B) to the modeled transmission spectra. The measurements of Faraday rotation and hysteresis loop of Faraday rotation were performed with a well calibrated measurement setup using Thorlabs PAX polarimeter and a custom-made electromagnet. A transmission mode polarization microscope was used to observe the magnetic domain patterns in the annealed thin garnet films.

RESULTS AND DISCUSSION

Crystallized garnet films annealed using temperatures above 620°C (prepared on GGG substrates) showed the best optical and MO properties across the visible spectral region (532 nm and 635 nm polarized laser sources were used to characterize the MO performance of these garnet samples). The derived absorption spectra of samples (as-deposited and annealed) prepared onto GGG substrates using 250°C and 680°C deposition are plotted in Fig. 1. It is important to note that actual substrate surface temperatures were estimated to be about 30%

lower than these measured by thermocouple sensor, which is the likely reason that even samples grown on "hot" substrates were not crystallized in-situ.

Derivations of these absorption spectra were performed using the transmission spectra of samples and fitting software written by the second author, Dr. Mikhail Vasiliev, which correctly accounts for the effects of refractive index dispersion as well as reflection spectrum. All films (sputtered at either 250°C or 680°C) had rather low optical absorption and significantly high Faraday rotation, which led to obtaining high MO figures of merit ($2\Theta_F/\alpha$). The maximum measured values of Faraday rotation per unit film thickness of the garnet films prepared at 250 °C were very close to these obtained in films prepared at 680°C substrate temperature, but the lower absorption losses achieved in films prepared at higher temperature compared to other samples helped obtain a better MO figure of merit (42.8° at 635 nm) than that of the sample sputtered at 250°C. The best obtained values of MO quality factors of $Bi_{1.8}Lu_{1.2}Fe_{3.6}Al_{1.4}O_{12}$ garnet layers deposited onto GGG (111) substrates using 250°C and 680°C substrates temperature are shown in the inset of Fig. 1.

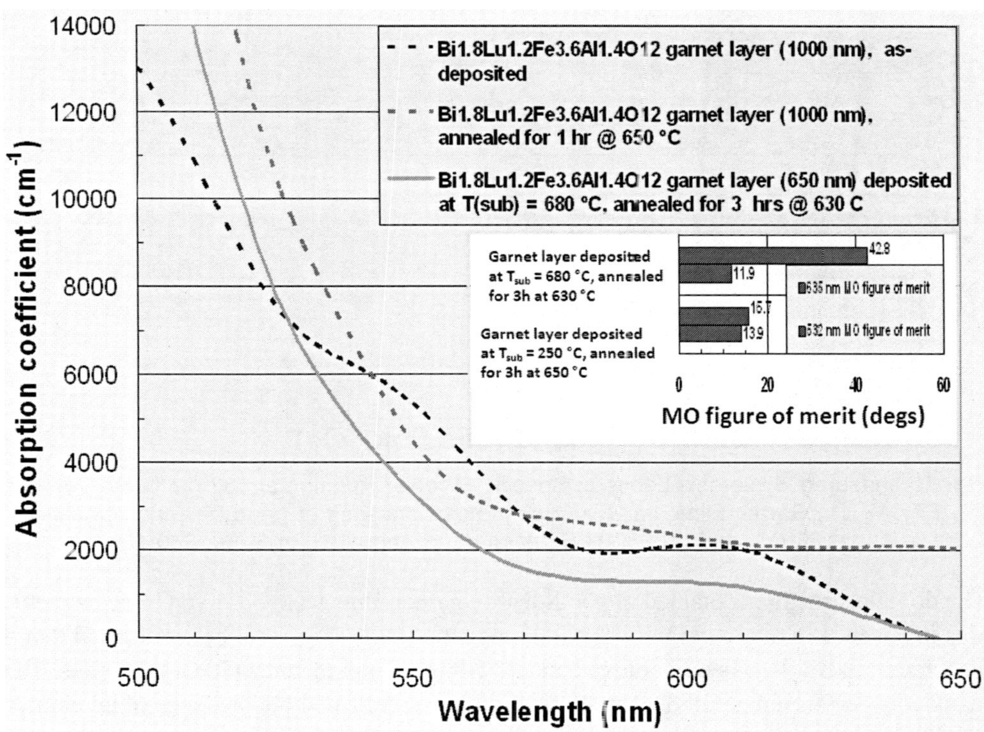

Fig. 1: Derived Absorption Spectra in Films Sputtered onto GGG Substrates using 250 °C and 680 °C Substrate Temperatures. The Values of MO Figure of Merit at 532 and 635 nm Obtained in Both Film Batches After Annealing are Shown in Inset

The magnetic films possessed very low coercive force values, as was observed during the measurements of hysteresis loop of Faraday rotation (made using a 532 nm polarized light source). Figure 2 shows the obtained hysteresis loops of specific Faraday rotation in films prepared using 250°C (green color) and also 680°C (red color) substrate temperatures. The measured coercive force values were about 45 Oe for the garnet films sputtered at 250°C and 10 (±2) Oe for the films sputtered at 680°C. Both film batches showed very strong in-plane magnetization component (weak uniaxial magnetic anisotropy), which was confirmed by the almost-linear character of their magnetization curves below the saturation observed in the measured hysteresis loops of Faraday rotation.

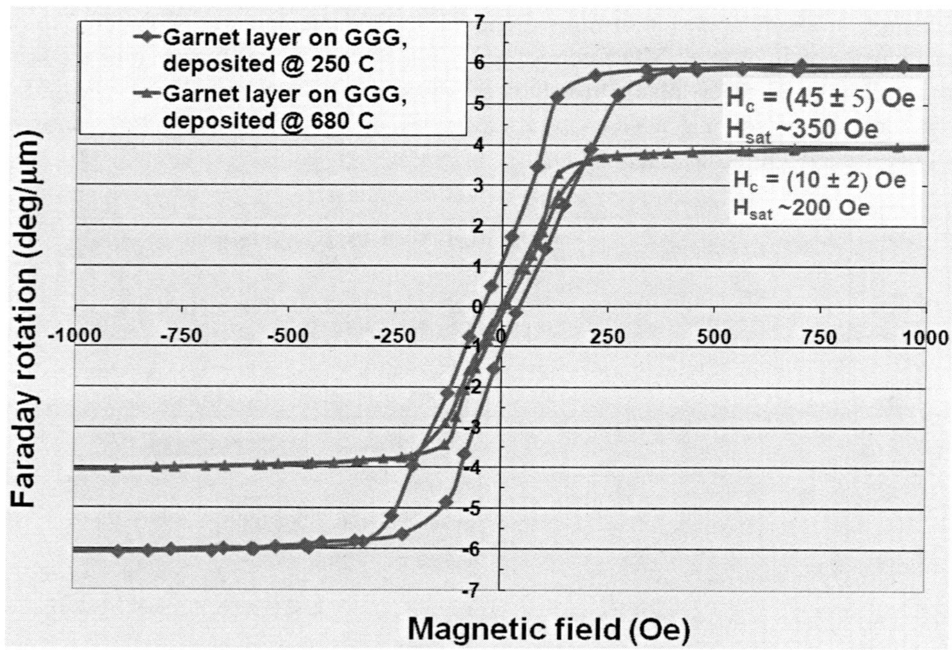

Fig. 2: Measured Hysteresis Loops of Specific Faraday Rotation at 532 nm in Sputtered $Bi_{1.8}Lu_{1.2}Fe_{3.6}Al_{1.4}O_{12}$ Garnet Films on GGG Substrates Deposited at (a) 250°C and Annealed for 1 h at 650°C, and (b) 680°C Substrate and Annealed for 3 h at 630°C.

The domain structures obtained in good-quality garnet films were observed in demagnetized samples in absence of any externally applied magnetic fields (Fig. 3 (a, b)) confirmed that the good surface quality has been achieved in the films sputtered onto GGG substrates. EDX microanalysis performed at GIST characterization labs confirmed the presence of all elements expected to be present within films (measured element concentrations in atomic % are shown Fig. 3(c)). Based on these measurement data, we derived the averaged composition of thin film to be $Bi_{1.68}Lu_{0.656}Fe_{4.294}Al_{1.184}O_{12}$ which might be expected considering the sputtering target's nominal stoichiometry and possible Bi content loss occurring during layer growth.

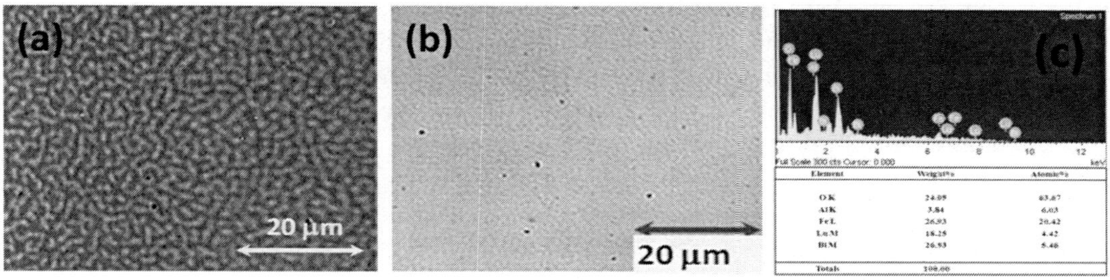

Fig. 3: Magnetic Domain Patterns Observed in Films Sputtered Onto GGG (111) at 250°C, Annealed for 1 h at 650°C (a); Deposited at 680°C and Annealed for 3 h at 630°C (b); and EDX Composition Measurement Data Obtained from Films Grown at 680°C Substrate Temperature

CONCLUSION

Magneto-optic garnet thin films of composition type $Bi_{1.8}Lu_{1.2}Fe_{3.6}Al_{1.4}O_{12}$ have been prepared using two different substrate temperatures during the deposition process and the suitable annealing process parameters for films of this type were found, leading to obtaining high quality magneto-soft thin films with excellent MO properties. Significantly low coercive forces, almost in-plane magnetization direction, and high MO performance have been observed in these garnet thin films which are attractive for MO imaging and sensing devices as well as for the future development of garnet waveguides and non-reciprocal integrated-optics components.

ACKNOWLEDGEMENTS

We acknowledge the support of Edith Cowan University and the World-Class University Program funded by the Ministry of Education, Science, and Technology through the National Research Foundation of Korea (R31-10026).

REFERENCES

[1] Zvezdin, A.K., and Kotov, V.A., Modern Magnetooptics and Magnetooptical Materials, Institute of Physics Publishing. Bristol and Philadelphia (IOP Publishing Ltd.) 1997.

[2] Nur-E-Alam, M., Vasiliev, M., Kotov, V.A., and Alameh, K., Highly bismuth-substituted, record-performance magneto-optic garnet materials with low coercivity for applications in

integrated optics, photonic crystals, imaging and sensing, *Opt. Mat. Express* 1, 2, 413-427, 2011.

[3] M. Nur-E-Alam, M. Vasiliev, V.A. Kotov and K. Alameh, "High-quality RF-sputtered magneto-optic garnet films of $Bi_{1.8}Lu_{1.2}Fe_{3.6}Al_{1.4}O_{12}$ with low coercivity for applications in integrated optics, imaging and sensing devices", *7th International symposium on High-capacity Optical Networks and Enabling Technologies Conference, Honet* 2010, pp. 62-66, Dec. 2010.

DIELECTRIC STUDIES OF NANOCRYSTALLINE MANGANESE TUNGSTATE

SUNNY JOSEPH, PRIYANKA K.P., SMITHA THANKACHAN[1], JASEENTHA O.P., MOHAMMED E.M.[1], GEORGE K.C.[2] AND THOMAS VARGHESE

Department of Physics, Nirmala College, Muvattupuzha, Kerala
[1]Department of Physics, Maharaja's College, Ernakulam, Kerala
[2]Department of Physics, S.B. College, Changanacherry, Kerala
E-mail:nanoncm@gmail.com

ABSTRACT

The dielectric properties of monoclinic manganese tungstate have been studied as a function of frequency and temperature. It is found that the dielectric constant and dielectric loss for all temperatures has high values at low frequencies which decreases rapidly as frequency increases and attains a constant value at higher frequencies. The ac conductivity increases as frequency is increased conforming small polaron hopping. As temperature is increased the values are shifted to higher values. Also the values are shifted to higher values when the particle size is decreased. These properties make the nanosized $MnWO_4$ as a promising material for fabricating humidity sensors.

INTRODUCTION

Nanostructured tungstate materials have aroused much interest because of their luminescence behaviour, structural properties and potential applications. Manganese tungstate ($MnWO_4$) is a promising material for fabricating humidity sensors due to its appropriate bulk electrical conductivity. Humidity sensors are important in many areas like meteorology, medicine, food production, agriculture, industrial and the domestic environment [1-3]. It is an antiferromagnetic material with a magnetic moment of 5.44 μ_B and Neel temperature 14.4 K [4]. The various

V. Rajendran, P. Manivasakan and K.E. Geckeler (eds.)
Advanced Nano Materials for Industrial Applications, pp. 69-74 (2012).

methods employed for the synthesis of the material include sol-gel method [2], chemical reaction in a molten salt solution, grinding in a vibrating mill, thermal treatment of the product of a metathesis reaction [5] and solvothermal method [6]. In the present work, we report a successful room temperature synthesis of manganese tungstate nanoparticles, adopting careful control of the reaction kinetics of aqueous precipitation.

EXPERIMENTAL

Manganese chloride and sodium tungstate (99.9% purity) were purchased from Sigma Aldrich Chemicals and was used without further purification. Nanosized powders of manganese tungstate were prepared by reacting aqueous solutions of manganese chloride and sodium tungstate (0.001 M each) at room temperature keeping the pH =7. The precipitate formed was centrifuged, filtered and washed with distilled water a number of times, and dried in an oven to get fine powders of manganese tungstate. The precursor is annealed in air at 450°C, 600°C and 750°C for 3 hours. The structural characteristics of the synthesized manganese tungstate nanoparticles have been studied by X-ray powder diffraction (XRD) using Bruker D8 Advance X-ray diffractometer (λ = 1.5406 Å). The powder was consolidated in the form of cylindrical pellets of diameter 11 mm and thickness d =1.2 mm at a pressure of ~ 5 GPa using a hydraulic press. Both the faces of the pellets were coated with air drying silver paste. Dielectric measurements as a function of frequency in the range of of 100Hz ~ 15 MHz are measured at various selected temperatures from 300 K – 425 K were carried out using a LCR meter (Wayne Kerr H-6500 model) in conjunction with a portable furnace and temperature controller (±1K) using four probe method. The dielectric constant and ac conductivity (σ_{ac}) were calculated by using Eq. (1) and (2) respectively.

$$\varepsilon' = Cd/\varepsilon_0 A \quad \text{... (1)}$$

$$\sigma_{ac} = \varepsilon'\varepsilon_0\ \omega \tan\delta \quad \text{... (2)}$$

where A is the face area, C is the measured capacitance of the pellet ε_0 is the permittivity of vacuum, ω is the angular frequency and tan δ, the loss tangent.

RESULTS AND DISCUSSION

Figure 1 displays the powder X-ray diffraction pattern of the samples annealed at 450 °C, 600°C and 750 °C for 3 hours. From XRD patterns using Scherrer equation, the sizes of the particles corresponding to different annealing temperatures are estimated to be 25 nm, 30 nm, and 40 nm for samples M1, M2 and M3 respectively. As annealing temperature increases XRD patterns show much sharper peaks and are due to the increase in size of the particles due to coarsening of the particles at higher temperatures. The principal'd' values taken from the JCPDS file No. 80-0135 for $MnWO_4$ are in close agreement with the observed 'd' values.

Figure 2(A) shows the variation of dielectric constant with frequency for temperatures from 303 K to 423K of sample M3. It can be seen that the real part of dielectric constant ε'

for all temperatures has high values at low frequencies which decreases rapidly as frequency increases and attains a constant value at higher frequencies. For 303K the value of ε′ is 4.07 at 100 Hz which decreased to 0.057 at 10MHz. At 363K the values are 22.2 and 0.16. The corresponding values for 423K are 57.1 at 100Hz and 0.186 at 10MHz respectively. The inset figure 2(A) shows a similar variation for all the samples with different particle sizes, at 303K but the values are shifted upward when the grain size is decreased. The value of ε′ changed from 9057 at 100Hz to 52.7 at 10 MHz for sample M1 while it is from 2357.6 to 17.02 for sample M2 and from 63.16 to 0.19 for sample M3 respectively.

In dielectric nanostructured materials, interfaces with large volume fractions contain a large number of defects, such as dangling bonds, vacancies, vacancy clusters, and microporosities, which can cause a change of positive and negative space charge distribution in interfaces. When subjected to an electric field these space charges move. When they are trapped by defects, a lot of dipole moments are formed. At low frequency region these dipole moments are easy to follow the variation of electric field [7-8]. So the dielectric loss and hence the dielectric constant shows large values at low frequency. As temperature increased more and more dipoles are oriented, resulting in an increase in the value of dielectric constant for a given value of frequency [9].

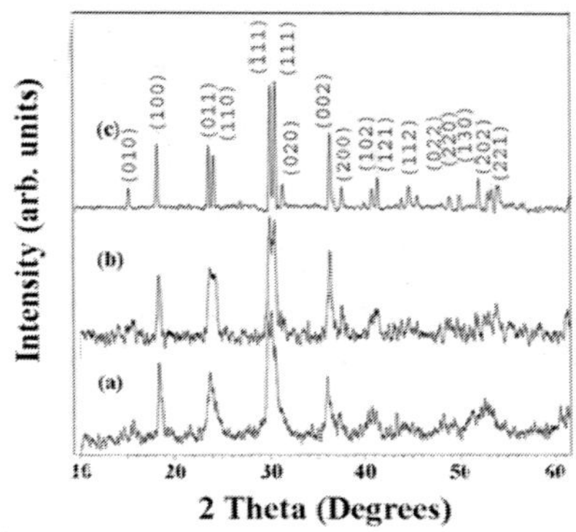

Fig. 1: Shows the XRD Pattern of the samples Annealed at (a) 450°C (b) 600°C and (c) 750°C

At very high frequencies (MHz) region the charge carriers would have started to move before the field reversal occurs and ε′ falls to a small value at higher frequencies. The nature of variation is similar for the other samples also, but the values are shifted upwards when the grain size is decreased. The amorphous nature of the surface, high surface energy, micromechanical stress, surface domain depolarization, domain wall effects are considered as the reasons for this variation. Space charge polarization and reversal of the direction of polarization contributes much to the ε′ [10-11]. With the increase in volume of the particle the volume of the interfaces decreases. When particle volume increases, the contribution to ε′ by electronic relaxation polarization appearing mainly inside the particles increases. Due to decrease in the interface volume the contribution to ε′ by space charge polarisation and rotation of the direction of polarisation occurring mainly in interfaces diminishes. As a result ε′ decreases as size of the particle increases.

The variation of tan δ with frequency of sample M3 is shown in Fig. 2(B). The variation of tan δ is similar to that of ε′. For 303K, tan δ has a value of 4.72 at 100Hz which decreases slowly at higher frequencies to 0.736 at 10 MHz. At 363K the corresponding variation is from 8.63 to 0.73. For 423K, tan δ has a value of 43.3 at 100Hz which decreases slowly at higher frequencies to 3.38 at 10 MHz. The variation of tan δ with frequency at 303K for different grain sizes is shown in Fig. 2(B). At 100 Hz the value is 71.4 for M1 which decreases to 2.20. For M2 the values are 43.3 and 3.38. For M3 the values are 4.72 and 0.736 respectively. In nanophase materials the inhomogeneities present in the interface layers produce an absorption current resulting in dielectric loss. This absorption current decreases with increase in frequency of the applied field. The hopping probability per unit time increases with increase in temperature. Correspondingly the loss tangent also increases with increase of temperature [12-13]. From X-ray results it is found that the particle size and crystallanity increases as the sintering temperature is increased. The volume percentage of interface boundaries, defects and imperfections are greatly reduced as the sintering temperature is increased. This explains the increase of loss tangent as the particle size is decreased. The loss in $MnWO_4$ can be explained by the electronic hopping model, which considers the frequency dependence of the localised charge carriers hopping in a random array of centre. This model is applicable for materials in which the polarisation responds sufficiently rapidly to the appearance of an electron on any one site so that the transaction may be said to occur effectively into the final state [14]. In the high frequency region tan δ becomes almost constant because the electron exchange interaction (hopping) between the Mn^{+2} and Mn^{+3} cannot follow the alternatives of the applied ac electric field beyond a critical frequency.

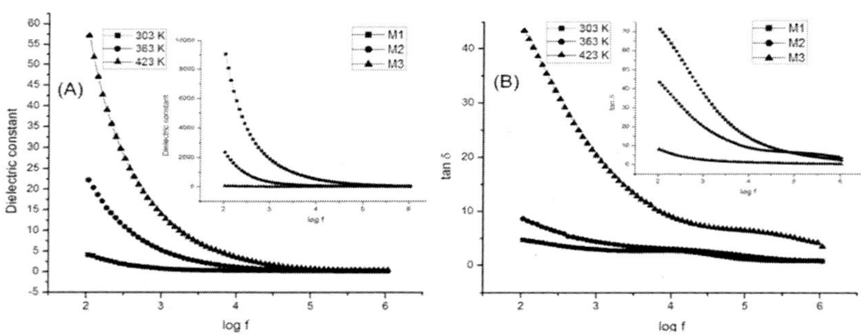

Fig. 2: (A) The Variation of Dielectric Constant with Frequency and (B) the Variation of tan δ with Frequency for Temperatures from 303 K to 423K of Sample M3

The variation of ac electrical conductivity as a function of frequency for sample M3 is shown in Fig. 3. At low frequencies σ_{ac} has a small value which increases at higher frequencies. The nature is similar for other temperatures but the values are shifted upwards as the temperature is raised. For 303 K, σ_{ac} has a value of 0.0373×10^{-3} S/m at 100Hz which increases slowly at higher frequencies to 0.434×10^{-3} S/m at 10 MHz. At 363 K the corresponding variation is from 0.123×10^{-3} S/m to 0.71×10^{-3} S/m. For 423K the values are 1.62×10^{-3} S/m at 100 Hz and 3.78×10^{-3} S/m. The variation of σ_{ac} with frequency at 303K for different grain sizes is shown in Fig. 3. At 100 Hz the value is 3.83×10^{-3} S/m for M1 which increases to 6.88×10^{-3} S/m at 10 MHz. For M2 the values are 0.0826×10^{-3} S/m and 1.15×10^{-3} S/m, for M3 0.0373×10^{-3} S/m and 0.434×10^{-3} S/m respectively.

It is clear from the figures that the conductivity increases as frequency is increased conforming small polaron hopping. Small polaron formation takes place in these materials whose conduction band belongs to incomplete 'd' or 'f' orbital. In $MnWO_4$, Mn^{+2} and W^{+6} have incomplete 'd' orbitals which are responsible for the formation of small polarons. Also there is a possibility of conduction due to impurities at low temperature. It is reported that the thermoelectric power of $MnWO_4$ increases with temperature [4]. So the ac conductivity increases with increase of temperature for lower values. Dissanayeke et al., observed a similar variation but at higher temperatures σ_{ac} is independant of frequency [15]. In $MnWO_4$ the conduction is due to the exchange interaction (hopping of electrons) between the Mn^{+2} and Mn^{+3}. According to Elliot's barrier hopping model, the ac conductivity is given as Eq. (3),

$$\sigma_{ac} = n\pi^2 NN_p \varepsilon' \omega R_\omega^6/24 \quad \dots (3)$$

where n is the number polarons involved in hopping, NN_p proportional to the square of concentration of states, ε' is the dielectric constant and R_ω is the hopping distance [16]. When the grain size of the sample is reduced the hopping distance increases which increases the conductivity. The conduction in manganese tungstate is due to the hopping and ionic conduction of Mn^+ ions. The high values of σ_{ac} for particles with small grain size are a direct confirmation of the theory.

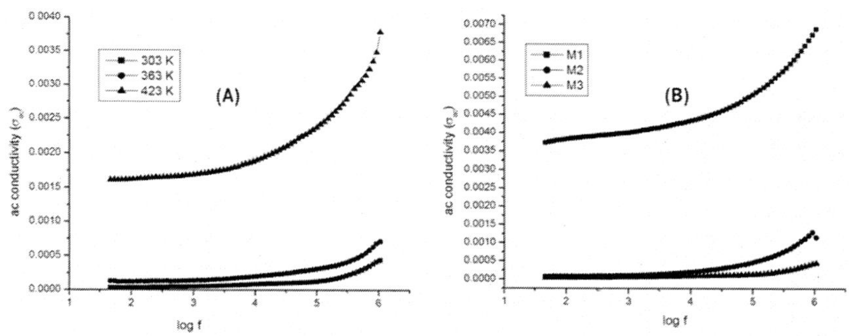

Fig. 3: The Variation of ac Electrical Conductivity as a Function of Frequency

CONCLUSION

$MnWO_4$ nanoparticles with different average sizes were synthesised. The dielectric properties of $MnWO_4$ were determined as a function of frequency from 100 Hz to 10 MHz for temperatures ranging from 303K to 423K. At lower frequency ε' and tan δ have higher values while at higher frequencies the values reach steady lower values. Similar variation is observed when the temperature is raised but the values are shifted upwards. Conductivity increases as frequency is increased conforming small polaron hopping. The values are shifted upwards when the temperature is raised. The values of ε', tan δ and σ_{ac} shows considerable increase as the particle size is reduced.

ACKNOWLEDGEMENT

The authors are grateful to KSCSTE, Thiruvananthapuram, Kerala for financial support.

Nomenclature

Gpa: Giga pascal, (XRD): X-ray powder Difraction, ε': Dielectric constant, ε_0: Permittivity of vacuum, σ_{ac}: Ac conductivity, tan δ: Loss tangent, ω: Angular frequency, R_ω: hopping distance

REFERENCES

[1] Traversa, E., Ceramic sensors for humidity detection: the state of the art and future developments, *Sens. Actuators* B, 23, 135-156, 1995.

[2] Qu,W., Wlodarski, W., and Meyer, J.U., Comparative study on micromorphology and humidity sensitive properties of thin-film and thick film humidity sensors based on semiconducting $MnWO_4$, *Sens. Actuators* B, 64, 76-82, 2000.

[3] Suresh Raj, A.M.E., Mallika, C., Sreedharan, O.M., and Nagaraja, K.S., MnO -manganese tungstate composite humidity sensors, *Mater. Lett.* 53, 316-320, 2002.

[4] Bharati, R., Singh, R.A., and Wanklyn, B.M., Electrical conduction in Manganese Tungstate, *J. Phys. Chem. Solids*, 43, 641-644, 1982.

[5] Jang, M., Weakley, T.J.R., and Doxsee, K.M., Aqueous crystallization of Manganese (II) Group 6 Metal Oxides, *Chem. Mater.*, 13, 519-525, 2001.

[6] Chen, S.J., Chen, X.T., Xue, Z., Zhou, J.H., Li, J., et al., Morphology control of $MnWO_4$ nanocrystals by a solvothermal route, *J. Mater. Chem.*, 13, 1132-1135, 2003.

[7] Parvatheeswara Rao, B., and Rao, K.H., Electrical properties of nanocrystalline $AlPO_4$, *J. Mater. Sci.,* 32, 6049-6054, 1997.

[8] Mo, CM., Zhang, L., and Wang, G., Characteristics of dielectric behavior in nanostructured materials, *Nanostruct. Mater.* 6, 823-826, 1995.

[9] Kar, T., Choudhary, R.N., Sharma, S., and Singh, K.S., Structural and electrical properties of Ba2Na3RNb10O30 Ceramics, *Indian J. Phys*, 73A (4), 453-459, 1999.

[10] Jiang, B., Peng, J.L., Busil, L.A.,and Zhong, W.L., Size effects on ferroelectricity of ultrafine particles of $PbTiO_3$, *J. Appl. Phys.,* 87 (7), 3462-3467, 2000.

[11] Ravinder, D., and Vijayakumar, K., Dielectric behaviour of erbium substituted Mn-Zn ferrites, *Bull. Mater. Sci.,* 24(5) 505-509, 2001.

[12] Potty, S.N., Khader, M.A., Dielectric Properties of nanophase Ag_2HgI_4 and Ag_2HgI_4-Al_2O_3 nanocomposites, *Bull. Mater. Sci.*, 23(5), 361-367, 2000.

[13] Mathew, J., Kurien, S., Sebastian, S., and George, K.C., Dielectric studies of nanocrystalline copper orthophosphate, *Ind. J. Phys.*, 78(9), 947-950, 2004.

[14] Murthy, V., and Sobhanadri, Dielectric properties of some nickel-zinc ferrites at radio frequency, *Phys Status Solid* (a), 36, K133, 1976.

[15] Dissanayake, M.A.K.L., Ileperuma, O.A., and Darmasena, P.A.G.D., AC conductivity of $MnWO_4$, *J. Phys. Chem. Solids*, 50, 359-361, 1989.

[16] Shukla, R., Electrical properties of $AgInTe_2$, Ind. *J. Pure Appl. Phys.* 31, 894-898, 1993.

STRUCTURAL COMPOSITES WITH NANOMATERIALS: MANUFACTURING, MECHANICAL PROPERTIES AND ELECTRICAL CONDUCTIVITY

ANOOP ANAND, RAHUL HARSHE AND MAKARAND JOSHI

Composites Research Center, Research and Development Establishment (Engineers), DRDO - Ministry of Defense, Pune
E-mail: anoop.rde@gmail.com

ABSTRACT

In the present investigation, we have adopted Resin Film Infusion (RFI) towards developing advanced structural composites with near-zero void fractions. Composites with continuous unidirectional E-glass fabric and epoxy have been generated with varying loading fractions of fillers such as carbon nanotubes (CNTs), carbon nanofibers (CNFs), Nanoclay, Nanosilica and Graphene. While improving the matrix dominated mechanical properties of composites to a substantial extent, some of the nanofillers have induced surface electrical conductivity of up to 10^{-3} S/cm. In order to achieve a better dispersion on the surface layer of composites, a low gsm fabric is used at the surface of laminates that are used for conductivity measurements. Resin film infusion also offers possibility of varying the concentrations of nanofillers through the thickness of the composite to have gradient electrical conductivity in thick structural composites.

INTRODUCTION

Composite materials have provided effective and enhanced defense application systems, the field of military services strive to design their equipment to such a standard where each machine

---

V. Rajendran, P. Manivasakan and K.E. Geckeler (eds.)
Advanced Nano Materials for Industrial Applications, pp. 75-80 (2012).

or paraphernalia possess the primary properties required during times of inevitable warfare and various other national emergencies. Earlier when the domain of materials was dominated by metals and alloys, the major setbacks faced by the defense were in accuracy, low strength to weight ratios and inadequate thermal stability.

All these problems were rectified with the entry of composites into the arena of manufacturing, finding their way through various defense applications. These materials are consumed in almost every area of modern warfare systems ranging from deployable bridges to complex structures of helicopters and fighter crafts.

Large heavy duty structural applications, armored body panels, light weight bridges for rapid transportation through unfriendly situations, tubular components, radar domes, radar transparent composites, sonar domes, etc., are being manufactured with composite materials. They are also used in the rotor blades of helicopters for the air force and due to their light weight; they are noted as the best possible candidates for the design and manufacture of numerous components of a fighter aircraft's body.

Common manufacturing techniques of structural composites for load bearing applications include the following: Hand lay-up, Filament winding, Compression molding, Pultrusion, Resin transfer molding, and Vacuum Assisted Resin Transfer Moulding. In this article, we report on Resin Film Infusion (RFI) process for the realization of load bearing composites that are hybridized with nanofillers.

EXPERIMENTAL

RFI is one of the most promising methods for composites in aerospace, automotive and military applications. It has been developed as a cost-effective technique for the fabrication of complex shaped parts resolving several critical concerns of conventional liquid composite molding methods. RFI ensures near zero void fractions because of better compaction and local flow of the polymer resin. The methodology involved is as follows (Fig. 2). Epoxy resin formulations are developed that can be calendared to films. These films are cut to the desired dimension and transferred to a fabric layer (placed over a metallic mould plate) ensuring that the film just sticks to the fabric. The process is repeated to yield sandwiches of resin films with two fabric layers on both sides. Such sandwiches are placed on one another to build desired thickness. The sandwich stack over the mould is then cured by vacuum bagging technique inside an oven. A thermocouple is placed on top of the job to continuously monitor the temperature. Vacuum is applied to the job and measured via digital sensor. The job is heated at an optimized ramp to a predefined temperature, held at that temperature for a fixed duration while the resin films melt and infuse amidst the fabric plies. Curing process takes place with subsequent heating to and holding at a predetermined temperature. After completing the heating cycle, oven is switched off and the job is allowed to cool slowly to room temperature before demolding.

Hybrid composites of unidirectional E-glass/epoxy containing various nanofillers were made through an ultrasound assisted method reported elsewhere. Their properties have been evaluated in comparison to a control composite specimen prepared with pristine epoxy resin formulation containing no nanofillers.

The matrix system used in composites of the present study consists of a proprietary epoxy resin formulation developed in-house that can form films (Figure 1). This resin system consists of DGEBA based solid and liquid epoxy components and has been optimized so as to result in sufficient tackiness in order for the resin films to stick to the dry fabric during the composite manufacturing, in addition to appropriate rheological (Figure 1), thermal and mechanical properties. This formulation has a shelf life of 2 years when stored at subzero temperatures. Our cross linking agent, offers excellent mechanical properties and a straightforward processing of the finished resin formulation. A cure accelerator is also used. Unidirectional E-glass fabric of 500 gsm areal density with > 99% fibers oriented in warp direction was used as reinforcement. Nanomaterials were incorporated in resin matrix.

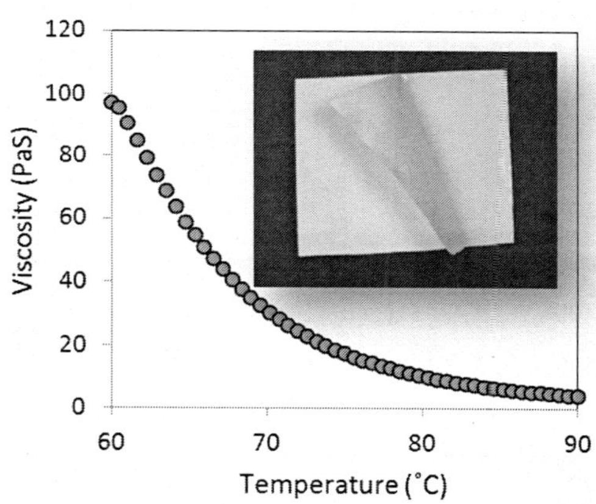

Fig. 1: Rheology of Epoxy Resin Formulation Used for RFI, Inset: Resin Film

RESULTS AND DISCUSSION

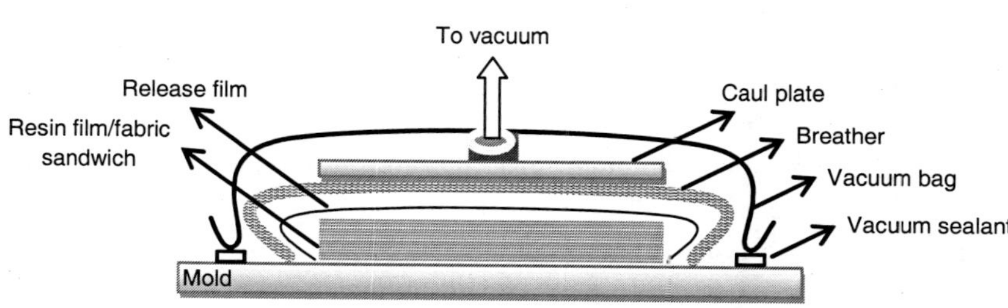

Fig. 2: Schematic of the RFI Process

The basic process of fabricating a composite by impregnating a fiber assembly over a mold with a wide range of variants in the matrix component offers a good deal of scope for focused studies. However, a very limited proportion of the effort to date has been spent on deriving hybrid composites with nanomaterials. Indeed, modifying the matrix component of composites by dispersing nanofillers leads to a plurality of constraints in traditional liquid composite molding techniques such as vacuum assisted resin transfer molding (VARTM). Increased resin viscosity, self-filtration, etc., are some of the prevailing issues that limit the use of such modified liquid resins in these processes. Nevertheless, our resin film infusion process ensured a proper dispersion of the nanofillers throughout the composite which has been substantiated by rheology of the filled polymer resins and the consistent and concentration dependant increase in matrix dominated mechanical properties of reinforced composites (results published in a separate article). Here, we report on the mechanical properties of composites containing multiwalled carbon nanotubes at a fixed concentration that are dispersed with various surface active aids.

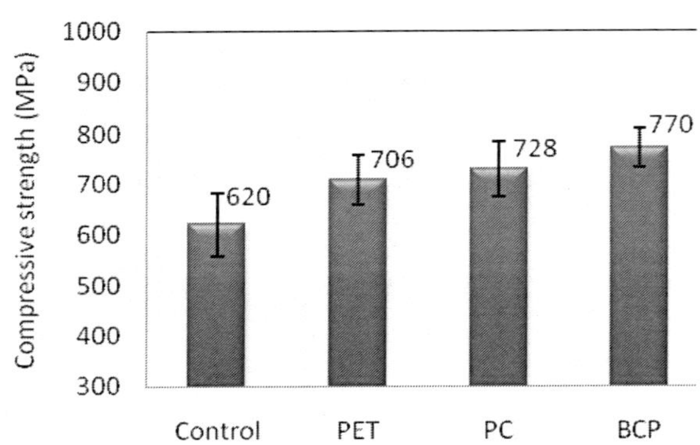

Fig. 3: Compressive Strength of Hybrid MWCNT-Composites with Various Surfactants

When subjected to tensile and compressive loads the composites showed a more pronounced enhancement in compression properties, as would be expected considering the fiber dominated nature of the tensile properties. Data in Fig. 3 depict that the compressive strength of hybrid composites with carbon nanotubes improved considerably with respect to that of the control glass/epoxy composites. At a loading fraction of 0.2 wt% CNTs, (weight fractions of CNTs in composites have been calculated from the weight fraction of epoxy which is experimentally determined as per ASTM D3171) compressive strength of hybrid composite increased by 14% or 17% with the use of PET or PC as a dispersing aid and upto 25% with block copolymer.

It has been proposed that only the outermost layer of multi walled CNTs is loaded during tensile loading, since the relatively weak (van der Waals) bonding between the concentric layers causes slippage of the inner layers, reducing its load bearing capacity. This slippage, however, seems to provide an extraordinary elastic deformation capability to multi walled CNTs when loaded in compression.

Surface electrical conductivity of unidirectional E-glass/epoxy composite laminates prepared through resin film infusion is compared with that of hybrid composites containing carbon nanotubes in Table 1. To ensure a proper distribution at the surface of the laminate, a low gsm fabric is also used as the topmost layer of the test specimen.

Table 1: Electrical Conductivity Results of Hybrid Composites

Specimen	*Surface Electrical Conductivity (S/cm)*
Pristine composite	Above the measurement limits
Composite with MWCNTs (surfactant: PET)	4.34×10^{-6}
Composite with MWCNTs (surfactant: PC)	5.96×10^{-4}
Composite with MWCNTs (surfactant: BCP)	1.21×10^{-3}

CONCLUSIONS

Resin Film Infusion is one of the best viable methods for realizing structural composites hybridized with nanomaterials. The process ensures near zero void fractions unlike conventional liquid composite molding techniques. A solvent assisted ultrasonication route has been used to disperse nanofillers in the polymer resin. Local flow of the filled polymer resin amidst the sandwiched fabric plies ensured a uniform distribution of nanofillers. Low weight fractions of multiwalled carbon nanotubes improved the compressive strength of unidirectional E-glass/epoxy composites by upto 25% and induced surface electrical conductivity in otherwise highly insulating composites.

REFERENCES

[1] R. Shvartzman-Cohen, E. Nativ-Roth, E. Baskaran, Y. Levi-Kalisman, I. Szleifer, R. Yerushalmi-Rozen. Selective dispersion of single-walled carbon nanotubes in the presence of polymers: the role of molecular and colloidal length scales, *J. Am. Chem. Soc*, 2004;126(45);14850

[2] L. Zhao, L. Gao. Stability of multi-walled carbon nanotubes dispersion with copolymer in ethanol, *Colloids Surf. Physicochem Eng Aspects*, 2003; 224; 127

[3] Q.Q. Li, M. Zaiser, V. Koutsos. Carbon nanotube/epoxy resin composites using a block copolymer as a dispersing agent, *Phys. Status Solidi* A, 2004; 201; R89-R91

[4] Marino Quaresimin, Russell J. Varley. Understanding the effect of nanomodifier addition upon the properties of fiber reinforced laminates. *Composite Science and Technology*, 2008; 68 (3-4); 718.

[5] Schadler LS, Giannaris SC and Ajayan PM: Load transfer in carbon nanotube epoxy composites, *Applied Physics Letters*, Vol. 73, 1998, pp. 3842-3844.

[6] Wagner HD et. al.: Stress-induced fragmentation of multiwall carbon nanotubes in a polymer matrix, *Applied Physics Letters*, Vol. 72, 1998, pp. 188-190.

[7] Bower C et. al. : Deformation of carbon nanotubes in nanotube-polymer composites, *Applied Physics Letters*, Vol. 74, 1999, pp. 3317-3319.

MICROFLUIDIC CHANNEL FABRICATION IN LITHIUM NIOBATE USING FOCUSED ION BEAM

M. Sridhar, D.K. Maurya[1], J.R. Friend[2], L.Y. Yeo[2] and K. Alameh[1]

Melbourne Centre for Nanofabrication, 151 Wellington Road, Clayton, VIC 3168, Australia
[1]Electron Science Research Institute, Edith Cowan University, Joondalup, WA 6027, Australia
[2]School of Electrical and Computer Eng., RMIT University, Melbourne, Victoria 3004, Australia
E-mail: d.maurya@ecu.edu.au

ABSTRACT

Microfluidic channels are fabricated in lithium niobate (LN) by Focused Ion Beam (FIB) milling. 127.68° axis rotated Y-cut (SAW grade) and Z-cut LN wafers are used. Material removal rate is 0.34±0.02 μm³/nC for Y-cut sample and 0.30 ± 0.02 μm³/nC for Z-cut sample. Experimental results show that the material removal rate decreases at high ion doses and high aspect ratios due to the increased significance of material deposition at these conditions. We also demonstrate the suitability of FIB-milled nanochannels for micro/nanofluidic applications.

INTRODUCTION

Lithium niobate (LN) represents the most common piezoelectric material used in radio frequency telecommunications including mobile phones, television, and wireless transmitters. Machining lithium niobate, an interesting and commonly used piezoelectric material, is difficult, and although Focused Ion Beam (FIB) milling [1-3] has been used to machine LN in the past, all previous works have focused on the development of photonic devices [4-5]. In recent years, piezoelectrically generated acoustic energy using LN [6] has been found to be extremely useful for microfluidics.

In this paper, we investigate the use of FIB milling for fabricating a wide range of structures and show that FIB milling of nanochannels on lithium niobate can move microfluidics towards

V. Rajendran, P. Manivasakan and K.E. Geckeler (eds.)
Advanced Nano Materials for Industrial Applications, pp. 81-86 (2012).

the next level of nanofluidics, where fluid transport in structures of feature sizes in the order of 100 nm could be easier than previously assumed.

EXPERIMENTAL

127.68° axis rotated *Y*-cut (SAW grade) and *Z*-cut LN wafers were obtained from Roditi International Corporation and diced into approximately 10 mm × 10 mm square samples. The LN samples were then coated with a thin layer (~ 25 nm) of Au by thermal evaporation to avoid charging effects and facilitate ion milling and scanning electron microscopy (SEM).

FIB milling experiments were conducted on the LN samples using a FEI Helios NanoLab 600 Dual-Beam FIB-SEM, which is an excellent machine for FIB milling. Ga^+ ions were emitted with an accelerating voltage of 30kV at normal incidence to the sample surface. The ion beam overlap was fixed to the default value of 50% for all experiments, i.e. the beam was moved through the mill area in steps equal to half the beam diameter at a particular current, to minimize the effect of the Gaussian profile of the ion beam on the profile of the milled channels. All channels were first milled, in triplicate, sequentially using the FIB, cross-sections were then cut using the FIB with a lower ion beam current than the current used to mill the channel, and finally, the milled channels were imaged and the dimensions were measured using the SEM in situ.

RESULTS AND DISCUSSION

Channels with six different volumes varying from 50 μm^3 to 250 μm^3 were milled in triplicate using the FIB in both Y- and Z-cuts LN samples. Each milled channel was cross-sectioned using the FIB, and the dimensions of the milled channels were measured using the SEM. Figure 1 shows a SEM image of the cross-section of a typical FIB-milled 2 μm wide and 600 nm deep channel. Due to the Gaussian nature of the ion beam generated by the FIB system, some of the Au conducting layer surrounding the channel also appeared to have been sputtered away during the milling process, as is evidenced by the thin gray halo region around the milled area. The step structures that are visible in Figure 1 are standard features created during the process of cross-sectioning the channel by the FIB system. The volume of each of the milled channels was then calculated by using the measured dimensions and plotted against the total Ga^+ charge incident on each channel for both the *Y*- and *Z*-cuts samples, as shown in Figure 2. As expected, the volume of LN that is sputtered away varied linearly with the number of Ga^+ ions incident on the surface of the sample for both *Y*- and *Z*-cut samples.

The gradient of a linear fit through each set of data points gives us the value for the material removal rate for each cut of LN. Using this method, we obtained an experimental material removal rate of 0.34 ± 0.02 μm^3/nC for *Y*-cut samples and 0.30 ± 0.02 μm^3/nC for *Z*-cut samples. Thus, we observed that there is no significant dependence of the material removal rate using the FIB on the surface orientation of LN. For a given channel width, we observed that the milled volume varied linearly with total incident dose at ion doses less than 100 nC and in good agreement with the material removal rate obtained from SRIM simulations. The material removal rate then tapered off at ion doses greater than 100 nC due to the effect of material redeposition at high ion doses.

Fig. 1: SEM Image of the Cross-section of a Typical FIB-milled Channel in LN. Milled Channel was 2 μm Wide and 600 nm Deep

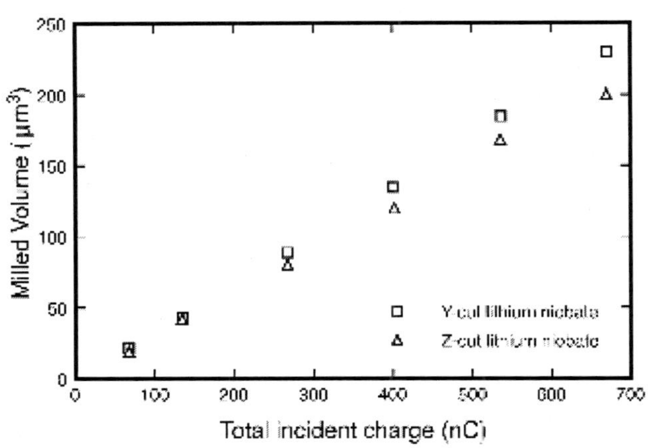

Fig. 2: Plot of Average Milled Volume of Microchannels as a Function of Total Incident Ga^+ Charge for Y- and Z-cut LN Substrates

Furthermore, we investigated the effect of material redeposition on the material removal rate of LN by milling a series of channels with varying aspect ratio, defined as the ratio of depth to width of the milled structures, while keeping the total volume of milled material constant. Figure 3 shows the plot of material removal rate observed in the FIB as a function of the aspect ratio of milled channels.

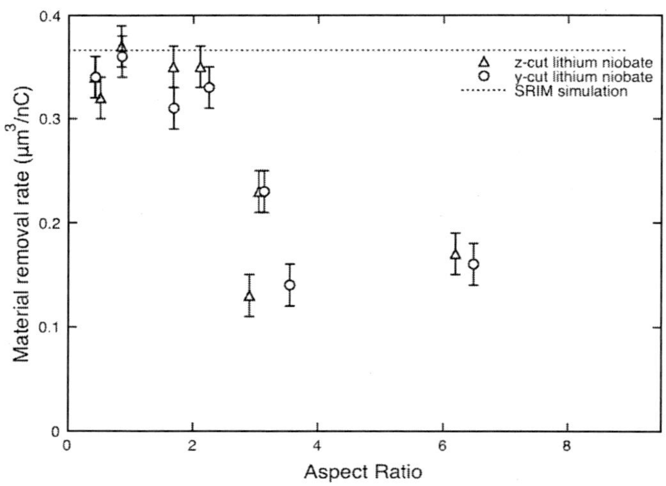

Fig. 3: Material Removal Rate as a Function of Aspect Ratio of Milled Channels for *Y*-cut and *Z*-cut LN

From Figure 3, we observe that the material removal rate in the FIB is in good agreement with the theoretical value obtained from our SRIM simulation for low aspect ratio structures. This is because material redeposition is not a significant factor in this regime. We also observe that the material removal rate decreases as a function of aspect ratio of the milled structures, decreasing to approximately 50% of the initial material removal rate at aspect ratios greater than 4. The reason for this decrease in material removal rate at high aspect ratios can be attributed to material redeposition. It becomes more and more difficult to remove material from a deep yet narrow (i.e. high aspect ratio) structure because the material has to be expelled a long way to escape the top surface of the substrate, and there is a high probability that the material will be redeposited along the sidewalls within the structure itself.

Lastly, we milled a nanochannel (W 100 nm, D 100 nm, L 100 μm) directly onto a LN sample and carried out preliminary experiments of imaging fluid inside such a FIB-milled nanochannel. Figure 4 shows a confocal microscope image of 22 nm fluorescent nanoparticles suspended in deionised water inside a FIB-milled nanochannel. The fluid appeared to fill slightly more than half the length of the nanochannel through capillary action, and illustrates the suitability of FIB-milled channels for micro/nanofluidic applications.

Fig. 4: Confocal Microscope Image of 22 nm Fluorescent Nanoparticles Suspended in Deionised Water Inside a FIB-milled Nanochannel.

CONCLUSIONS

Microfluidic channels in LN have been fabricated by FIB milling. Material removal rate is 0.34 ± 0.02 μm³/nC and 0.30 ± 0.02 μm³/nC for Y-cut and Z-cut LN, respectively. Experimental results have shown no significant dependence of the material removal rate on the surface orientation of LN. The suitability of FIB-milled nanochannels for micro/nanofluidic applications has been demonstrated experimentally.

ACKNOWLEDGMENTS

This work was performed in part at the Melbourne Centre for Nanofabrication, and supported by Edith Cowan University and the Department of Industry, Innovation, Science, Research and Tertiary Education, Australia.

REFERENCES

[1] Lacour, F., Courjal, N., Bernal, M.-P., Sabac, A., Bainier, C., Spajer, M., Nanostructuring lithium niobate substrates by focused ion beam milling, *Optical Materials*, *Vol. 27, Issue 8,* pp. 1421–1425, 2005.

[2] Nellen, P.M., Callegari, V., Brönnimann, R., FIB-milling of photonic structures and sputtering simulation, *Microelectronic Engineering*, Vol. 83, Issues 4–9, pp. 1805–1808, 2006.

[3] Sulser, F., Poberaj, G., Koechlin, M. and Günter, P., Photonic crystal structures in ion-sliced lithium niobate thin films, *Optics Express*, Vol. 17, Issue 22, pp. 20291-300, 2009.

[4] Campbell, C., Surface Acoustic Wave Devices for Mobile and Wireless Communications (Academic, NY, 1998).

[5] Hashimoto, K., Surface Acoustic Wave Devices in Telecommunications: Modelling and Simulation (Springer, New York, 2000).

[6] Friend, J., and Yeo, L., Piezoelectric Materials for Microfluidics (Encyclopaedia of Micro- and Nanofluidics), edited by D. Li (Springer, New York, 2008), Vol.1, pp. 1654–1662.

NOISE SUPPRESSION IN UNSHIELDED MAGNETOCARDIOGRAPHY: LEAST-MEAN SQUARED ALGORITHM VERSUS GENETIC ALGORITHM

VALENTINA TIPORLINI, HOANG N. NGUYEN AND KAMAL ALAMEH

Electron Science Research Institute, Edith Cowan University, Joondalup, Western Australia
E-mail: v.tiporlini@ecu.edu.au

ABSTRACT

This paper discusses adaptive noise cancellation in magnetocardiographic systems within unshielded environment using two algorithms, namely, the Least-Mean Squared (LMS) algorithm and the Genetic Algorithm (GA). Simulation results show that the GA algorithm outperforms the LMS algorithm in extracting a weak heart signal from a much-stronger magnetic noise, with a signal-to-noise ratio (SNR) of -35.8 dB. The GA algorithm displays an improvement in SNR of 37.4 dB and completely suppressing the noise sources at 60Hz and at low frequencies; while the LMS algorithm exhibits an improvement in SNR of 33 dB and noisier spectrum at low frequencies. The GA algorithm is shown to be able to recover a heart signal with the QRS and T features being easily extracted. On the other hand, the LMS algorithm can also recover the input signal, however, with a lower SNR improvement and noisy QRS complex and T wave.

INTRODUCTION

The human heart is made of conductive tissues that produce both an electric field and a magnetic field, depending on cardiac activity. Measuring the electric and/or magnetic fields enables various heart parameters as well as diseases to be diagnosed, such as heart beat rate and

V. Rajendran, P. Manivasakan and K.E. Geckeler (eds.)
Advanced Nano Materials for Industrial Applications, pp. 87-92 (2012).

arrhythmia. Electrocardiography (ECG) enables the detection of heart-generated electric fields through electrodes placed on the surface of the human body. However, magnetocardiography (MCG) has been shown to be more accurate than electrocardiography for the (i) diagnosis of atrial and ventricular hypertrophy, (ii) non-invasive location of the heart's conduction pathways, (iii) the identification of spatial current dispersion patterns, and (iv) the detection of circular vortex currents which give no ECG signal [1]. Cardiac magnetic fields surround the human body and are typically very low in magnitude (about 100 pT [2]), necessitating the use of a high-sensitivity magnetometer to measure them. Furthermore, the environmental electromagnetic noise is typically much higher (in the order of 1 nT) than the heart-generated magnetic field, resulting in an extremely low signal-to-noise ratio, if patients are examined outside a magnetic shielded room.

In this paper, we adopt the concept of adaptive noise cancellation shown in Fig. 1, and use two potential adaptive signal processing algorithms, namely the Least-Mean Squared (LMS) algorithm and the Genetic Algorithm (GA) and compare their capabilities in minimizing the mean-squared of the error signal $e(k)$ and improving the SNR performance. With respect to the LMS algorithm, the GA provides additional benefits, including (i) ability to perform parallel search for population points rather than for a single point, thus avoiding to fall into local minima; (ii) no prior information on the gradient of the signal is needed; (iii) the use of probabilistic rules instead of deterministic ones, thus ensuring the convergence to an optimum solution.

ADAPTIVE NOISE CANCELLER

Adaptive noise suppression techniques are typically based on adaptive filtering. To suppress the noise, a reference input signal is required, which is typically derived from one or more magnetic sensors placed at positions where the noise level is higher than the signal amplitude. Figure 1 shows a block diagram of an adaptive noise canceller. The primary input to the canceller, denoted $d(k)$, is the sum of the signal of interest $s(k)$ and the noise $n(k)$, which is typically uncorrelated with $s(k)$. The reference input signal of the system, $x(k) = n1(k)$, is a noise signal that is correlated in some unknown way with $n(k)$, but uncorrelated with the signal of interest $s(k)$. As shown in Fig. 1, $n1(k)$ is adaptively filtered to produce a replica of the noise $n(k)$ that can be subtracted from the primary input to eventually produce an output signal $e(k)$ equals to $s(k)$.

The objective of the noise canceller is to minimize the mean-squared error between the primary input signal, $d(k)$, and the output of the filter, $y(k)$. It is noticed from [3] that the mean-squared error (MSE) is minimum when $n(k) = y(k)$, and hence, the output signal $e(k)$ is equal to the desired signal $s(k)$.

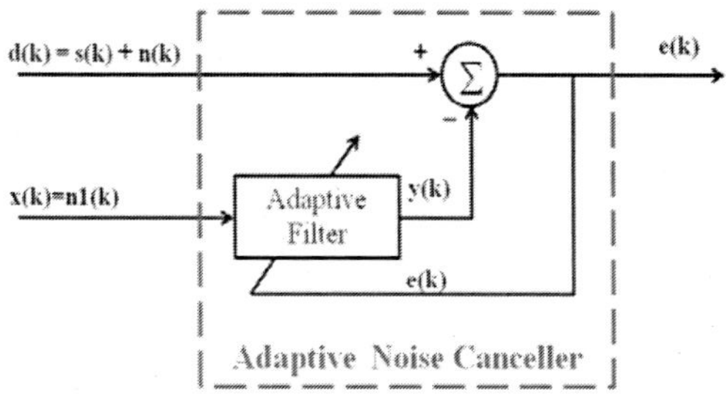

Fig. 1: Typical Block Diagram of an Adaptive Noise Canceller

The LMS algorithm aims to minimize the mean-squared error by calculating the gradient of the squared-error with respect to the coefficients of the filter. Assuming that the adaptive filter is a FIR filter of order *M,* the updating procedure is applied on coefficients b_i following the following rule [4]:

$$b_i^{(k+1)} = b_i^{(k)} + 2\mu e(k)x(k-i) \quad \text{... (1)}$$

where $i = 0,1, ..., M-1$, k is the iteration index and *m* is the step size that indicates the adaption rate of the algorithm and is usually included in the range (0,1].

Genetic Algorithms (GA) are part of the Evolutionary Algorithms, which are stochastic, population-based techniques inspired by the natural evolution process [5]. Using GA, the optimal solution is found through the minimization of a defined function, called the fitness function. For the problem of magnetic noise cancellation, the objective of the optimization process is to minimize the MSE.

SIMULATIONS RESULTS AND DISCUSSION

A measured cardiac signal taken from the MIT-BIH Arrhythmia Database (220.dat file) [6] was used to verify the ability of both LMS and GA algorithms to extract heart signals from noisy measured cardiac signals. This signal was captured by electrodes placed on the surface of a patient chest. The magnetic field and the electric field generated by human heart have similar waveforms[7]; therefore, it is possible to assume that the measured MCG signal has similar shape as the measured ECG signal but with amplitude of 100pT. Figure 2 (a) shows the heart signal with the typical cardiac features, namely, P wave, QRS complex and T wave, which correspond to atrial depolarization, ventricular depolarization and ventricular repolarization, respectively. Figure 2 (b) shows the cardiac spectrum that is mainly spread over low frequencies. The environmental magnetic noise was measured in our laboratory. The measured magnetic noise was more than 10 times higher than the heart signal, resulting in a signal-to-noise ratio (SNR) of -35.8 dB. The environmental magnetic noise was added to the heart signal to produce the input signal of the noise canceller. This signal is shown in Fig. 2 (c) and (d). It is noticed that the heart spectrum was completely encircled by the noise;

particularly strong noise peaks were exhibited at dc and 50 Hz whereas the other dominant peaks at the 60 Hz, 100 Hz and 150 Hz had lower levels. The environmental magnetic noise was linearly filtered to produce a correlated noise which was used as the reference signal input to the noise canceller, as illustrated in Fig. 1.

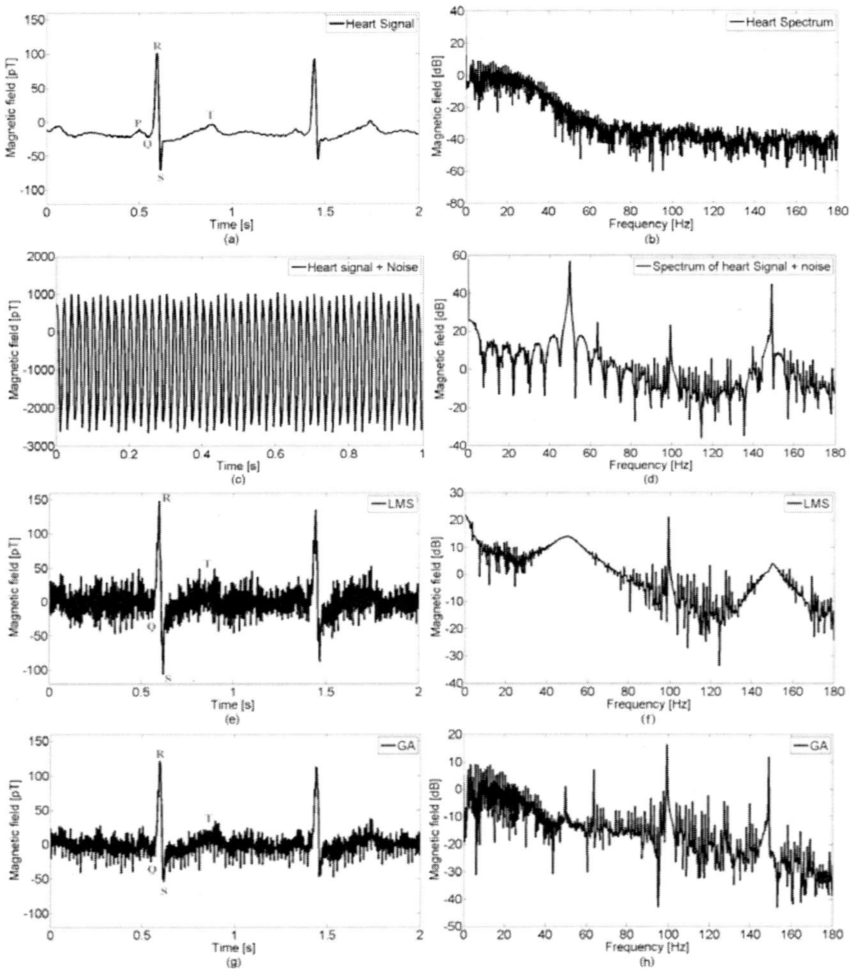

Fig. 2: (a) Measured heart signal showing the typical P wave, QRS complex and T wave, which correspond to atrial depolarization, ventricular depolarization and ventricular repolarization, respectively; (b) Spectrum of the heart signal; (c) Input signal of the noise canceller obtained by adding the heart signal to the environmental magnetic noise measured inside the laboratory; (d) Spectrum of the input signal of the noise canceller; (e) Heart signal recovered by LMS algorithm, calculated SNR improvement factor was 33 dB; (f) Spectrum of the heart signal recovered by the LMS algorithm; (g) Heart signal recovered by GA algorithm, calculated SNR improvement factor was 37.4 dB; (h) Spectrum of the heart signal recovered by the GA algorithm.

Figure 2 (e) and (g) show the heart magnetic signal recovered using the LMS and GA algorithms, respectively. It is obvious that for both recovered signals the QRS and T features are noticeable, whereas the heart magnetic signal recovered by the LMS algorithm is noisier, making the QRS complex and the T wave hardly detectable.

Figure 2 (f) and (h) show the spectra of the heart magnetic signals recovered by LMS and GA algorithms, respectively. It is important to notice that both algorithms are unable to completely cancel the noise at high frequencies; however, this is not crucial as most of the signal information lies in the low-frequency range.

CONCLUSION

The use of LMS and GA algorithms has been investigated for adaptive noise suppression and the recovering of heart signals in magnetically-unshielded environment. Measured heart signals and magnetic noise have been used to compare the performances of both LMS and GA algorithms in terms of SNR improvement and heart peaks reconstruction. Simulation results have shown that the GA algorithm attains better SNR improvement than the LMS algorithm. A measured heart signal has been recovered by the GA algorithm with a SNR improvement of 37.4 dB and the QRS complex and T wave have successfully been detected. The LMS algorithm has also recovered the input signal, however, with a lower SNR improvement of 33 dB and noisy QRS complex and T wave. The noise cancellation results shown in this paper are useful for signal processing applications where the signal to noise ratio is much below unity.

REFERENCES

[1] F. E. Smith, *et al.*, "Comparison of magnetocardiography and electrocardiography: A study of automatic measurement of dispersion of ventricular repolarization," *The European Society of Cardiology,* vol. 8, pp. 887-893, 2006.

[2] G. Bison, R. Wynards, and A. Weis, "Dynamical mapping of the human cardiomagnetic field with a room-temperature, laser-optical sensor," *Optics Express,* vol. 11, pp. 904-909, 2003.

[3] B. Widrow, *et al.*, "Adaptive Noise Cancelling: Principles and Applications," *Proceedings of the IEEE,* vol. 63, pp. 1692-1716, 1975.

[4] P.S.R. Diniz, Adaptive Filtering Algorithms and Practical Implementation: Springer, 2008.

[5] D. E. Goldberg, Genetic Algorithms in Search, Optimization, and Machine Learning: Addison-Wesley pubblishing company, inc., 1989.

[6] MIT-BIH Arrhythmia Database. Available: http://physionet.org/physiobank/database/mitdb/

[7] J. Malmivuo and R. Plonsey, Bioelectromagnetism. Principles and Applications of Bioelectric and Biomagnetic Fields. New York Osford: Oxford University Press, 1995.

THE INFLUENCES OF THE FRICTION STIR PROCESSING ON THE MICROSTRUCTURE AND HARDNESS OF AA6061 ALUMINIUM SHEET METAL

T. Prakash and P. Sasikumar[1]

Department of Mechanical Engineering, KPR Institute of Engineering and Technology, Coimbatore
E-mail: ttprakashmec@gmail.com

ABSTRACT

Friction stir processing (FSP) is an emerging surface engineering technology that can eliminate casting defects locally by refining microstructures, thereby improving the mechanical properties of the material. In this study, the influence of the Friction Stir Processing (FSP) on the microstructure and mechanical properties in terms of hardness for commercially available AA6061 Al sheet metals was studied and investigated. Samples were subjected to FSP by varying the number of passes using cylindrical geomentry type high speed steel tool fixed in the Vertical milling machine. Micro structural observations were carried out by employing optical microscopy on the modified surfaces. From the microstructural evaluation, it was observed that the grain size of the processed area was around 70% decreased as compared to unprocessed parent metal. The hardness results showed that by increasing the number of passes the hardness of the produced composite surfaces increases steadily. It was noted here that the strength of stirred surface area was around 1.75 times higher than the unstirred surface area.

INTRODUCTION

Recently friction stir processing (FSP) was developed by Mishra et al. [1,2] as a generic tool for microstructural modification based on the basic principles of FSW (Fig. 1). In this work, a

V. Rajendran, P. Manivasakan and K.E. Geckeler (eds.)
Advanced Nano Materials for Industrial Applications, pp. 93-98 (2012).

rotating pinned tool is fixed in the vertical milling machine (Fig. 2). The tool is inserted in a monolithic workpiece for localized microstructural modification for specific property enhancement. For example, high-strain rate superplasticity was obtained in commercial 7075Al alloy by FSP [1–3]. Furthermore, FSP technique has been used to produce surface composite on aluminum substrate [4], homogenization of powder metallurgy aluminum alloy [5], microstructural modification of metal matrix composites [6] and property enhancement in cast aluminum alloys [7].

Much attention has been paid to friction stir processing (FSP) as a surface modification technique based on friction stir welding (FSW) [11]. FSP is also one of a solid-state processing technique to refine the microstructure. This is carried out using the same approach as friction stir welding (FSW), in which a non-consumable rotating tool with a specially designed pin and shoulder is plunged into the plate to be processed and traversed along the line of the plate. Localized heating is produced by the friction between the rotating tool and the workpiece which increases the local temperature of the material to the range where it can be plastically deformed. The main objective of the present study is to investigate the effect of FSP on the microstructural and strength improvement of AA 6061 Al sheet metal.

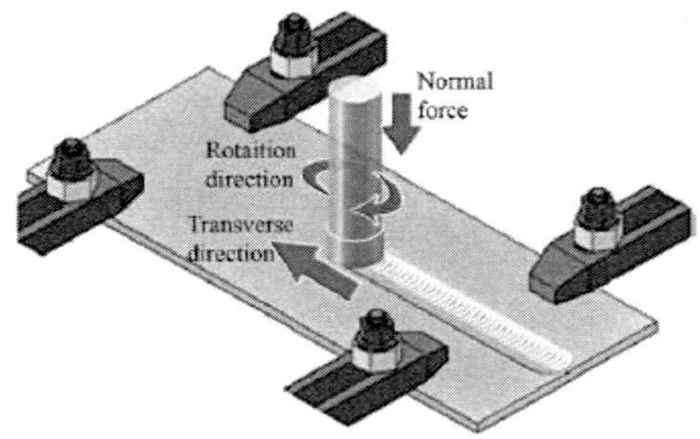

Fig. 1. Schematic Drawing of Friction Stir Processing

EXPERIMENTAL PROCEDURE

Friction Stir Processing (FSP)

The starting materials were monolithic cold-rolled plates of AA 6061aluminium alloy with a nominal composition as shown in the Table 1. The surface of plates was cleaned with grinding paper before processing. The dimensions of the workpiece were 50 mm × 100 mm × 3 mm. The plate was subjected to a FSP as shown in the macrograph in Fig. 2; the tool used to carry out the FSP is also shown as a inset (Fig.2) which is composed of a plain cylindrical shoulder with pin. The tool was made from high speed steel with the dimensions of 12 mm shoulder diameter and 4 mm pin diameter with length of 2 mm. The designed FSP tool was mounted in vertical milling machine and the workpiece was rigidly fixed on the machine table. First the pin is plunged (0.3 mm depth) into the workpiece surface to be processed and then the tool was allowed to rotate in forward direction with rotation speed (ω) of 550 rpm. The

Fig. 2: Macrograph Showing the FSP Tracks on the AA6061 Plate

traverse speed of 60mm/min was used during FSP. These parameters were optimised by carrying out several trial and error experiments. All FSP studies were carried out at room temperature.

Table 1: Chemical Composition (wt %) of Aluminium Alloy AA6061

Chemical Composistion	*Si*	*Fe*	*Cu*	*Mn*	*Mg*	*Zn*	*Ti*	*Al*
6061	0.4	0.7	0.15	0.15	0.8	0.25	0.15	Bal.

CHARACTERIZATION

The FSP samples for microstructural characterization were cut from the unstirred, shoulder stirred and pin-stirred zone of the plate. The samples were polished with emery paper up to 1200 grit followed by polishing with alumina suspension and diamond paste on a velvet cloth. Further, samples were etched with Keller reagent to distinguish the grain boundaries, identify precipitates and difference in composition. Finally, the microstructure investigation was carried out using optical microscopy.

Mechanical Property Evaluation

The mechanical property in terms of hardness tests were carried out to assess the effect of FSP on the AA6061 Al alloy sheet metal. Vickers micro hardness tests were carried out using a Wolpert Wilson Micro hardness tester. A load of 50 g and a dwell time of 20 s were used. The average micro hardness measurements were made on the FSP treated surfaces.

RESULTS AND DISCUSSION

The effect of friction stir processing on the microstructure and strength in terms of hardness for AA 6061 aluminium alloy is discussed here below:

Microstructural Evaluation

The microstructure of AA 6061 Al alloy in as-received condition (parent metal) is shown in Fig. 3. From the Fig. 3, it can be clearly seen that the average grain sizes of AA 6061 is around 85 μm for unprocessed samples (i.e. parent metal). After friction stir processing, the microstructures of AA 6061 for two pass and four pass is shown in Fig. 4. It can be noted clearly here that the material grain sizes are completely refined after FSP due to sever plastic deformation engendered by the FSP tool in the parent metal during processing. The grain size of AA 6061 after FSP is around 35 μm which was decreased by 2.4 times as compared to unprocessed parent metal for four passes. Further, the grain size of the alloy was decreased markedly with respect to number of passes. This is to the fact that the rotation speed of FSP tool and its travelling speed were expected to cause more heat input which might affects the thickness of the surface layer and as a result that refines the grain size of the material.

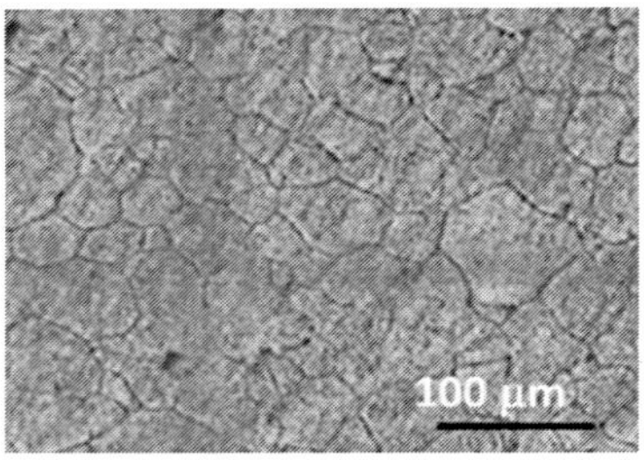

Fig. 3: Microstructure of Parent Metal AA 6061

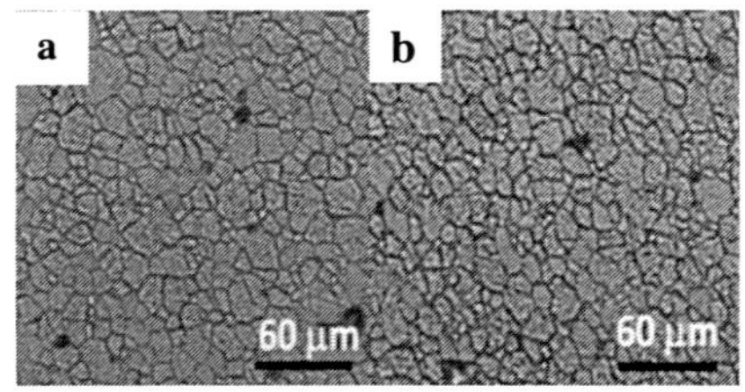

Fig. 4: Microstructure of AA 6061 after FSP: (*a*) Two pass (*b*) Four pass

Micro Hardness Evaluation

The variation of Vicker's hardness with different areas of AA6061 Al alloy processed by FSP is shown in Table 2. From the result it can be observed that in the case of AA6061 Al alloy the hardness of stirred zone is around 35%, 37%, 45% & 49% greater than unstirred zone with respect to number of passes for single, double, three & four passes respectively. This is expected due to grain refinement in the structure obtained by severe plastic deformation. Further it can be noted in the sample that as the number of passes increases the hardness of stirred zone increases rapidly first (up to three passes) and then increases slightly. This is due to the fact that the strengthening of the material reaches saturated condition up to three passes and beyond which there is no significant improvement in the strength. Hence in this work the samples are processed via FSP up to four passes. The hardness variation with number of passes for AA 6061 Al alloy is also shown in Fig. 5. For instance, the variation of hardness with distance measured from parent metal (left)-to-centre of stirred zone-to-parent metal (right) of AA 6061 al alloy for double passes is also shown in Fig. 6.

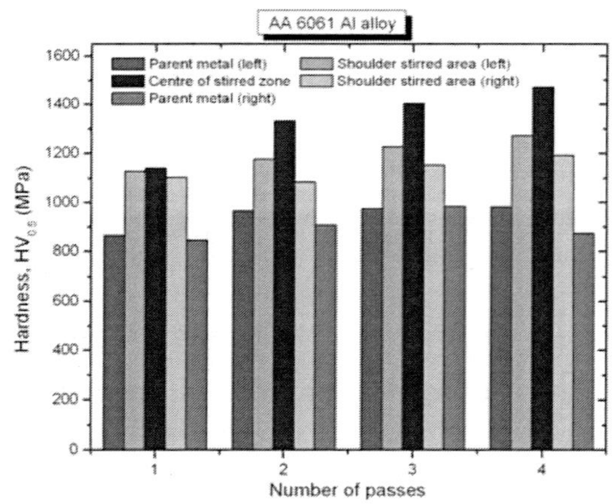

Fig. 5: Variation of Hardness with Number of Passes Processed Via FSP AA 6061 Al Alloy

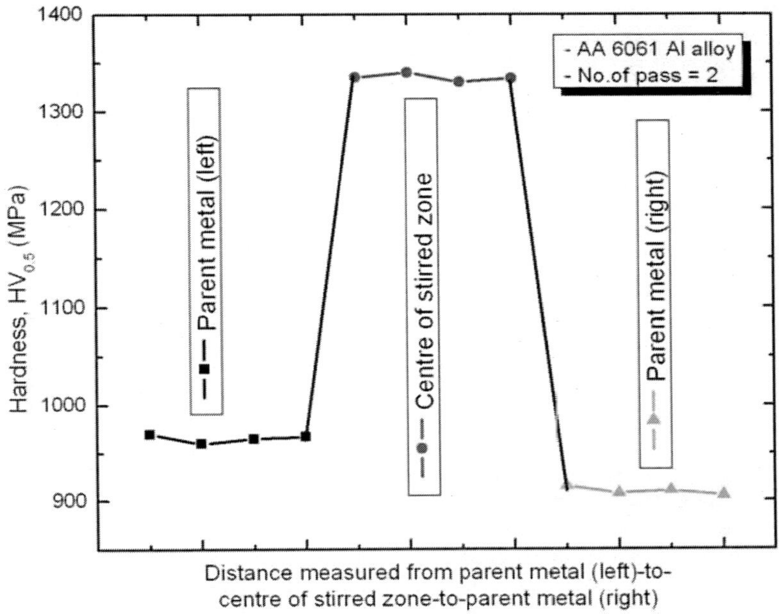

Fig. 6: Variation of Hardness with Distance Measured from Parent Metal (left)-to-centre of Zone-to-Parent Metal (right) for AA 6061 Al Alloy Processed via FSP (No. of pass = 2)

Table 2: Variation of Hardness with Different Areas of AA6061 Processed by FSP

Samples	*No. of Passes*	*Hardness, $HV_{0.5}$ (MPa)*				
		Left end (parent metal)	*Shoulder Stirred Area (left)*	*Center of Stirred Zone*	*Shoulder Stirred Area (right)*	*Right end (parent metal)*
AA6061	1 pass	865	1129	1141	1104	845
	2 pass	967	1177	1333	1085	907
	3 pass	975	1227	1405	1154	984
	4 pass	982	1273	1472	1193	875

CONCLUSIONS

In the present study, the plain AA 6061 Al-alloy surface was successfully modified by FSP. The microstructure and micro hardness were evaluated and investigated. The following conclusions were made based on present investigation:

1. It was demonstrated that FSP was an appropriate method to modify the microstructure and mechanical properties of AA 6061 Al-alloy. This is to fact that FSP decreased the grain size and increased the hardness of processed material.

2. The rotation speed of FSP tool and its travelling speed were expected to cause more heat input which affects the thickness of the surface layer which refines the grain size of the material.
3. The micro hardness of the stirred surface of AA 6061 Al–alloy was increased significantly compared with the unprocessed surface. Further, it was noted that the hardness was increased from around 160 to 185% higher than the unprocessed surface which was due to grain refinement.
4. With further research efforts and increased understanding, FSP could be conducted for improving the mechanical behaviour, like fatigue and creep response and new tool design for light weight based materials and metal matrix composites.

REFERENCES

[1] R.S. Mishra, M.W. Mahoney, S.X. McFadden, N.A. Mara, A.K. Mukherjee, High Strain Rate Superplasticity In A Friction Star Processed 7075 Al Alloy, 42 (2000).

[2] R.S. Mishra, M.W. Mahoney, Friction Stir Processing: A New Grain Refinement Technique to Achieve High Strain Rate Superplasticity in Commercial Alloys, Mater. Sci. Forum 357–359 (2001).

[3] Z.Y. Ma, R.S. Mishra, M.W. Mahoney, Superplastic deformation behaviour of friction stir processed 7075Al alloy, Acta Mater. 50 (2002).

[4] R.S. Mishra, Z.Y. Ma, I. Charit, Friction Stir Processing: A Novel Technique for Fabrication of Surface Composite, Mater. Sci. Eng. A 341 (2002) 307.

[5] P.B. Berbon, W.H. Bingel, R.S. Mishra, C.C. Bampton, M.W. Mahoney, Friction stir processing: a tool to homogenize nanocomposite aluminum alloys, 44 (2001) 61.

[6] J.E. Spowart, Z.Y. Ma, R.S. Mishra, in: K.V. Jata, M.W. Mahoney, R.S. Mishra, S.L. Semiatin, T. Lienert (Eds.), Friction Stir Welding and Processing II, TMS, 2003, pp. 243–252.

[7] Z.Y. Ma, S.R. Sharma, R.S. Mishra, M.W. Manohey, Mater. Sci. Forum 426–432 (2003). [8] R.S. Mishra, Z.Y. Ma, Friction Stir Welding and Processing, Materials Science and Engineering R 50 (2005) 1–78.

[9] C.J. Lee, J.C. Huang, P.J. Hsieh, Mg based nano-composites fabricated by friction stir processingScripta Materialia 54 (2006) 1415–1420.

[10] W. Wang, Q. Shi, P. Liu, H. Li, T. Li, A novel way to produce bulk SiCp reinforced Al-MMC by FSPJournal of Materials Processing Technology 209 (2009) 2099–2103.

[11] F. Nascimento, T. Santos, P. Vilaca, R.M.Miranda, L. Quintino, "Microstructural modification and ductility enhancement of surfaces modified by FSP in aluminium alloys", Materials Science and Engineering A 506 (2009) 16–22.

[12] C.J. Hsu, C.Y. Chang, P.W. Kao, N.J. Ho, C.P. Chang, "Al-Al3Ti nanocomposites produced in situ by friction stir processing"Acta Materialia 54 (2006) 5241–5249.

TWOFOLD LIGHT AND MAGNETIC RESPONSIVE BEHAVIOR IN NANOPARTICLE-LYOTROPIC LIQUID CRYSTAL SYSTEMS

JIJO J. VALLOORAN, STEPHAN HANDSCHIN, SREENATH BOLISETTY AND RAFFAELE MEZZENGA

ETH Zurich, Food and Soft Materials Science, Department of Health Science and Technology, CH-8092 Zürich, Switzerland
E-mail: jijo.vallooran@hest.ethz.ch

ABSTRACT

We demonstrate the dual magnetic and light responsive nature of hybrid mesophases constituted by Fe_3O_4 nanoparticles dispersed in lipid-based lyotropic liquid crystals (LC). When subjected to an external magnetic field in the mesophase isotropic state, the nanoparticles aggregate and orient along the magnetic field direction, and upon cooling the system through the disorder-order transition, the aggregates drive the orientation of the mesophase via heterogeneous nucleation[1]; furthermore order-disorder transitions in the lipidic mesophase can be triggered by Fe_3O_4–induced photothermal effect under visible light exposure[2]. Both the orientational order and the photothermal effect of the hybrid mesophase can be tuned by the nanoparticle content, offering a general route for controlled assembly of complex fluids with combined magnetic and light responsiveness.

INTRODUCTION

When swollen by water, lamellar phases result in one common symmetry of lipid-based lyotropic LC mesophases, and find diverse applications in food, cosmetics, pharmaceuticals for controlled release and delivery systems.[1] Ferromagnetic nanoparticles, on the other hand, are also getting

V. Rajendran, P. Manivasakan and K.E. Geckeler (eds.)
Advanced Nano Materials for Industrial Applications, pp. 99-102 (2012).

significant interest in a number of applications including in-vitro and in-vivo targeted drug delivery, magnetic resonance imaging (MRI) contrast agents, cell signaling, thermal imaging and in hyperthermic treatment of tumors, where researchers make use of inductive heating of magnetic nanoparticle for therapeutic purposes. The combined magnetic and light response of complex fluids such as anisotropic mesophases would be a highly desirable feature, since this could allow simultaneous switchable control over both the macroscopic alignment and the nanostructure (e.g. space group) of the complex fluid. The experimental realization of such a responsive complex fluid could open, among other things, also new possibilities in self-assembled configurable bulk metamaterial applications.

EXPERIMENTAL SECTION

The mesophase system consists of an industrial grade of monolinolein (Dimodan U/J; Danisco, Denmark), blended with water. The magnetic nanoparticles were synthesized using co-precipitation method. Magnetic and light responsive behavior was studied using Small Angle X-ray Scattering (SAXS), Small Angle Neutron Scattering (SANS), Polarized Light Microscopy (POM) and UV-vis spectroscopy.

RESULT AND DISCUSSION

In this work, we demonstrate for the first time the combined light and magnetic response of a hybrid lyotropic liquid crystal system constituted by a lipid-based mesophase within which are dispersed Fe_3O_4 magnetic nanoparticles. We chose to work with a lamellar phase to access low temperature order-disorder transitions in the body temperature range. When a magnetic field is applied upon cooling the mesophase from the isotropic phase to the lamellar phase, nanoparticles aggregates into highly elongated and oriented clusters from which aligned lamellar grains nucleate[2] (Figure 1).

The good biocompatibility of lipid based lyotropic liquid crystals and the Fe_3O_4 magnetic nanoparticles with combined magnetic and light response, make this hybrid system a promising candidate for applications in the bio-nanotechnology field, in photothermal treatment of tumors and targeted drug releases.

Moreover, by exposing the hybrid lamellar phase to visible light leads to an order-disorder transition of lamellar phase into an isotropic fluid due to the photothermal effect induced by the magnetic nanoparticles[3] (Figure 2). Increasing the nanoparticle content in the hybrid mesophase allows modulating both the macroscopic orientational order of the hybrid mesophase and the photothermal order-disorder transition.

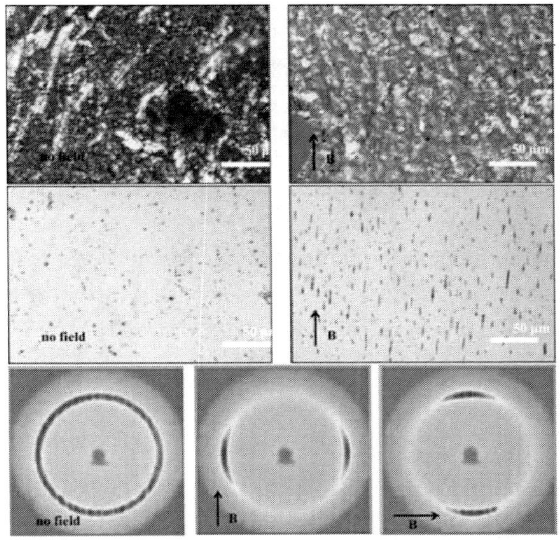

Fig. 1: Illustration of Magnetic Response of Lamellar Phase: Optical Microscopy Images Shows that the Nanoparticles are Aligned to Needle Like Clusters, in the Presence of 1.1T Magnetic Field (top and middle images). SANS Images Shows that the Macroscopic Orientational Order of Liquid Crystal Can be Stored, Erased and Re-written Completely

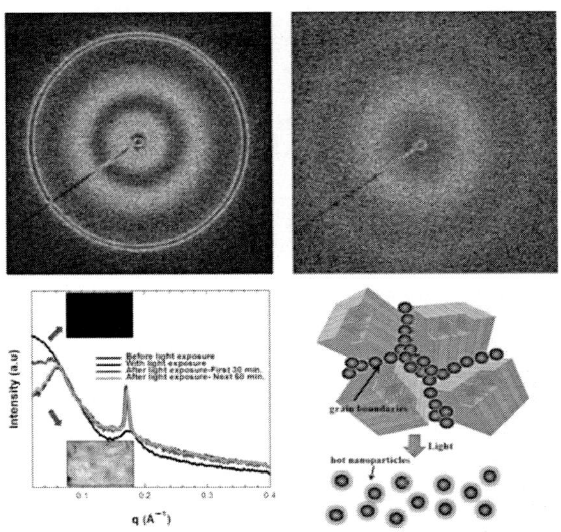

Fig. 2: Illustration of Light Response of Hybrid Mesophase: Small Angle X-ray Scattering (top images) and Polarized Optical Microscopy (inset of bottom left image) Studies Shows that the Lamellar Phase Undergoes a Order-Disorder Transition in the Presence of Visible Light with an Illuminance of 6.05 lm/cm^2

CONCLUSION

We have demonstrated combined magnetic and light responsiveness in lipid-based lamellar mesophases loaded with iron oxide nanoparticles, where both responses can be tuned by the concentration of nanoparticles in the hybrid system.

REFERENCES

[1] Mezzenga, R.; Schurtenberger, P.; Burbidge, A.; Michel, M., *Nat. Mater.* 2005, 4, 729-740.

[2] Vallooran, J.J.; Bolisetty, S.; Mezzenga, R., *Adv. Mater.* 2011, 23, 3932-3937.

[3] Vallooran, J.J.; Handschin, S.; Bolisetty, S.; Mezzenga, R., *Langmuir* 2012, 28, 5589.

IODINE DOPING STUDIES ON PERYLENE 3, 4, 9, 10-TETRA CARBOXYLIC DIANHYDRIDE (PTCDA)/ COBALT PHTHALOCYANINE (CoPc) BULK HETEROJUNCTION SOLAR CELLS

Priya A. Hoskeri, Gayathri A.G, N.H. Ayachit[1] and C.M. Joseph[2]

Department of Physics, Dayananda Sagar College of Engineering, Shavige Malleswara Hills, Kumaraswamy Layout, Bangalore, Karnataka
[2]Department of Physics, BVB College of Engineering and Technology, Vidya Nagar, Hubli, Karnataka
[3]Department of Physics, Dayananda Sagar College of Engineering, Kumaraswamy Layout, Bangalore
E-mail: cmjoseph@rediffmail.com

ABSTRACT

PTCDA (Perylene 3, 4, 9, 10-Tetra carboxylic dianhydride) films find lot of optoelectronic applications. In this work, thin films of PTCDA were deposited using vacuum evaporation technique at the rate of 10 Å/s onto clean glass substrates. Variation of conductivity, optical band gap and percentage transmission due to iodine doping for different time intervals is discussed. Organic solar cells using cobalt phthalocyanine (CoPc) /Perylene 3, 4, 9, 10-tetra carboxylic dianhydride (PTCDA) as active layers on ITO coated glass substrates are fabricated and characterized to evaluate the solar cell parameters. It is found that the doping increases the power conversion efficiency of the solar cells.

INTRODUCTION

Organic semiconductors are expected to play a vital role in both industry and research in the coming years. Perylene derivatives are commercially available as red pigments with electrical

V. Rajendran, P. Manivasakan and K.E. Geckeler (eds.)
Advanced Nano Materials for Industrial Applications, pp. 103-108 (2012).

and optical properties that allow them to be compared to conventional inorganic wide band gap semiconductors [1-2]. Recently a lot of research has been done in the study of the archetype molecular compound 3, 4, 9, 10-perylene tetra carboxylic dianhydride (PTCDA). The primary motivation for focusing on PTCDA is its promising optical and electronic properties [3-4]. PTCDA films find optoelectronic applications and serves as a model substance for the investigation of thin organic film and possible future devices [5-7]. The electrical properties of vacuum deposited films are a subject of current interest, as they may lead to useful and unusual devices.

In this work, thin films of PTCDA were deposited using vacuum evaporation technique and we have evaluated the doping effects of iodine on the electrical conductivity and optical band gap of PTCDA films. The variation of percentage transmission due to doping is also reported.

Transition metal-based phthalocyanines like cobalt phthalocyanine (CoPc) have attracted attention in the areas of fuel cells, gas sensors, biosensors and electronic devices like driver transistors and solar cells [8-11]. For doping effects on devices, CoPc active layers were used for the heterojunction.

EXPERIMENTAL

Materials used were commercially available from Sigma (Aldrich) without further purification. Thin films of PTCDA were deposited by vacuum evaporation technique using a Hind Hivac vacuum coating unit in a molybdenum boat. The glass substrates were first cleaned in an ultrasonic cleaner and then in acetone. Films were deposited at the rate of 10 Å/s in a vacuum of around 2×10^{-6} Torr. Thicknesses of the films were measured using digital thickness monitor. Doping of the thin films can be done by co-evaporation or subsequent evaporation of pure atoms on the top of the organic layer [12-13]. In this work, vacuum evaporated PTCDA thin films were doped with iodine by two methods: evaporation process (process 1) and a dip process (process 2) [14]. In the evaporation process, the iodine crystals are allowed to evaporate in a vacuum desiccator and the daily variation of electrical conductivity is measured using a Keithely 2400 model source meter. In the dip process, PTCDA thin films were doped by immersing into a solution of iodine diluted in acetonitrile (5 mg/300 ml) for different durations (< 1 hour). Then the room temperature dark conductivity was measured and the optical properties were studied. CoPc thin films were deposited using vacuum evaporation technique and were doped with iodine and also thin films of PTCDA were deposited onto these CoPc thin films using vacuum evaporation technique and were doped with iodine and we have studied the doping effects of iodine on the electrical and device properties of CoPc/PTCDA solar cells. The structure of the device fabricated was Al/CoPc:I_2/PTCDA:I_2/PEDOT: PSS (polyethylene dioxythiophene - poly styrene sulfonic acid)/ Indium tin oxide (ITO)/Glass.

RESULTS AND DISCUSSION

Conductivities of the doped and undoped PTCDA thin films are found out. It is seen that the conductivity increases from 7.6×10^{-5} $ohm^{-1}m^{-1}$ to a maximum conductivity of 9.2×10^{-4} $ohm^{-1}m^{-1}$ due to the iodine doping. The increase in the conductivity may be due to the structural variation in the PTCDA thin films while doping [15-16]. Here in this work, as expected doping effect is more and fast for the dip process.

The optical spectra were taken with %T versus wavelength α for the undoped and doped PTCDA samples. The transmission spectrum for undoped and doped (process 1 for 1 day) PTCDA thin films is shown in figure 1.

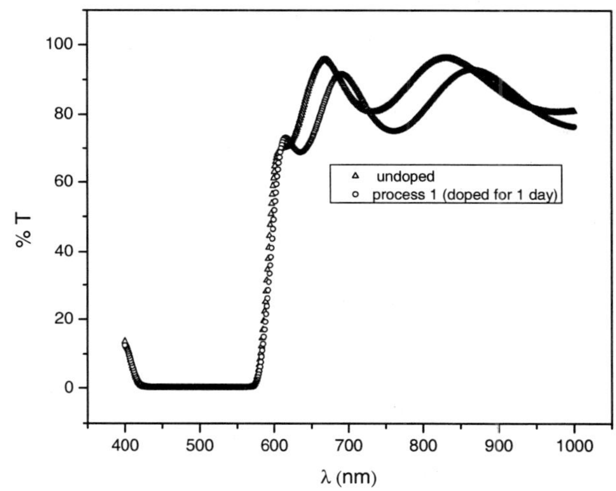

Fig. 1: %T Versus λ for the Undoped and Doped (Process 1 for 1 day) PTCDA Thin Films

Spectra show interference pattern with moderate sharp fall at the band edge which is an indication of semiconductor behavior with good crystallinity. For the undoped PTCDA thin films, a complete absorption is there in the visible range. For the doped PTCDA films the %T slightly increases in the visible range which shows that the doping of PTCDA films with iodine have an observable effect on the transmission properties as well as electrical properties of the films. This increase in % T for the doped PTCDA films is attributed to the scattering of light at the dopant atoms which were absent in the undoped films. The spectrum also reveals that doped PTCDA films have a strong absorption in the NIR region.

The absorption coefficient α is related to the photon energy $h\nu$ by,

$$\alpha = \alpha_0 (h\nu - Eg)^n \text{ where Eg is the optical band gap}$$

For the allowed transitions, $(\alpha h\nu)^2$ is plotted against $h\nu$, from which the extrapolation of the linear portion to $\alpha = 0$ near the absorption edge gives the band gap energy. For the undoped PTCDA film the optical band gap obtained from the graph is around 2.96 eV as shown in figure 2 (i). Narrowing of the band gap is observed for the doped samples.

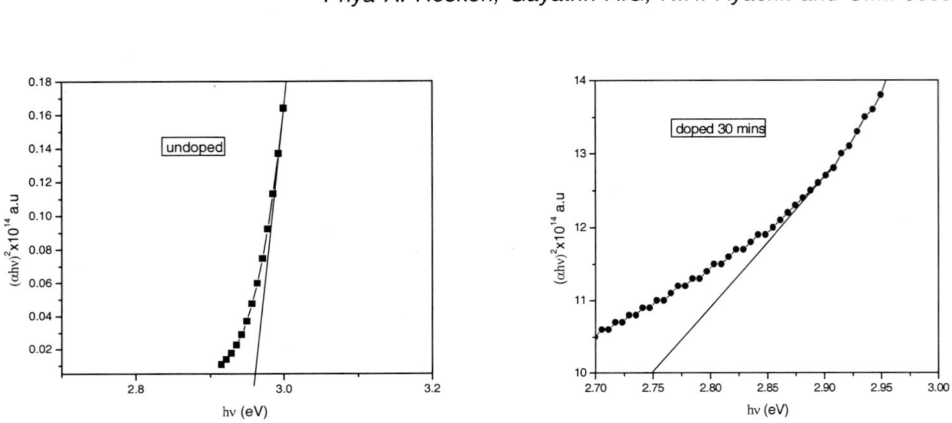

(*i*) (*ii*)

Fig. 2: $(\alpha h\nu)^2$ Versus hν Plot for (i) Undoped PTCDA Thin Film, (ii) PTCDA: I_2 Thin Film (dip process for 30 mins)

Similarly for the dip process doped sample, the band gap decreases to 2.75 eV for the first 30 minutes as shown in Figure 2(ii) and it decreases further to 2.73 eV for 45 minutes. Red shift is also observed for doped samples. The dopant establishes a highly doped surface region on top of the PTCDA layer and the excess dopant diffuses into the film as per the XPS and FTIR studies by earlier report [13].

A detailed study on the structural properties of the doping effects of iodine on these films is necessary for tuning the electrical and optical properties for device applications. Next, to evaluate the doping effects of I_2 on PTCDA devices, cobalt phthalocyanine CoPc was selected as the active layer for PTCDA solar cells. CoPc films were annealed in air and the doping with iodine was done for 20 hrs. After CoPc layer doping, PTCDA layer (undoped) was deposited. For the next set of samples, undoped CoPc layers were used as active layer and then PTCDA layer was doped. Doping of PTCDA layer with iodine was done for different timings as 4 hrs, 20 hrs, 24 hrs and 48 hrs. Finally, Al/ Ca metal electrode was coated and the devices were characterized. It is found that solar cell efficiency increases due to doping. Light characteristics of a CoPc:I_2/PTCDA organic solar cell is shown in Figure 3.

Series resistance also showed a hike with iodine doping for the devices. Maximum efficiency of 0.01% is obtained for the devices with CoPc (doped for 20 hrs)/ PTCDA and CoPc/PTCDA (doped for 20 hrs) active layers as shown in Table 1.

Table 1: Device Parameters of CoPc: I_2/ PTCDA: I_2 Solar Cells.

Device	*Voc(V)*	*Isc(μA)*	*FF*	*η (%)*	*Rs(MΩ)*
Undoped CoPc/ PTCDA	0.38	1.17	0.4	0.003	0.105
CoPc: I_2 (20 hrs)/ PTCDA	0.49	2.8	0.4	0.01	0.32
CoPc/ PTCDA: I_2 (20 hrs)	0.4	3.7	0.39	0.01	0.72

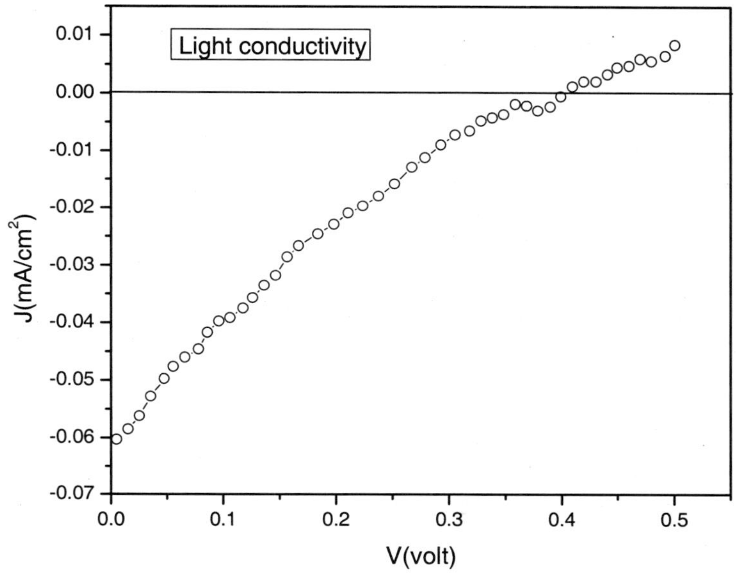

Fig. 3: Light Characteristics of a CoPc: I_2/PTCDA Organic Solar Cell

On increasing the doping time above 20 hours for the PTCDA layer, efficiency is found to decrease to 0.003%. So it can be concluded that doping of the CoPc or PTCDA can modify the properties of the organic devices [17-18].

CONCLUSION

Thin films of PTCDA were deposited using thermal evaporation in high vacuum onto cleaned glass substrates. These films were then doped with iodine by evaporation and dip process. The conductivity increases after iodine doping. Maximum conductivity is obtained for the dip processed sample (45 minutes) in a short time interval. Although, with increase in the doping time the percentage transmission is found to be decreasing in the NIR region, no much variation is observed for the optical band gap of the doped samples. It is also found that the doping effects on the properties are more evident and fast for the dip process for both thin films and the devices.

ACKNOWLEDGEMENTS

The financial support by Visvesvarya Technological University (VTU), Belgaum, Karnataka, India through a grant is gratefully acknowledged. Support by Vision Group on Science and Technology (VGST), Department of Information Technology, Biotechnology & Science and Technology, Government of Karnataka, India is also acknowledged.

REFERENCES

[1] Forrest S. R., Ultrathin organic fFilms grown by organic molecular beam deposition and related techniques, *Chem. Rev.*, 97 1793-1896, 1997.

[2] Friedrich M. et al., Optical properties and molecular orientation in organic thin films, *J. Phys. Cond. Matter.*, 15 S2699-S2718, 2003.

[3] Bulovic V. et al., Study of localized and extended excitons in 3,4,9,10-perylenetetracarboxylic dianhydride (PTCDA) I. Spectroscopic properties of thin films and solutions, *Chem. Phys.*, 210, 1-12, 1996.

[4] Lovinger A. J. et al., Structural and morphological investigation of the development of electrical conductivity in ion irradiated thin films of an organic material, *Appl. Phys.*, 55, 2, 1984.

[5] Kufazvinei C. et al., A blue light emitting perylene derivative with improved solubility and aggregation control: Synthesis, characterization and optical limiting properties, *Org. Electron.*, 10, 674–680, 2009.

[6] Leonhardt M. et al., Two-component optical spectra in thin PTCDA films due to the coexistence of α- and α-phase, *Chem. Phys. Lett.*, 313, 24-30, 1999.

[7] Hirose Y. et al., Structural and electronic properties of an organic/inorganic semiconductor interface: PTCDA/GaAs (100), *Vac. Sci. Technol.* B, 12, 4, 1994.

[8] Somashekarappa M. P. et al., *J. Pure Appl. Chem.*, 74, 1609, 2002.

[9] Leznoff C.C. and Lever A.B.P., Phthalocyanines: Properties and Applications, VCH publishing, New York.

[10] McKewn N.B., *Phthalocyanine Materials*, Cambridge University Press, UK.

[11] Joseph C.M. and Menon C.S., *Mater. Lett.*, 52, 220, 2002.

[12] Martin P. et al., Highly Efficient Organic Light Emitting Diodes by Doped Transport Layers Int. *Symp. Super-functionality Organic Devices, IPAP Conf.series* 6 pp. 108-109.

[13] Wusten J et al., Alkali metals in perylene-3, 4, 9, 10-tetracarboxylicdianhydride thin films *Journal of Applied Physics* 102, 023708, 2007.

[14] Puigdollars J. et al., Pentacene thin-films obtained by thermal evaporation in high vacuum, *Thin Solid Films* 427, 367-370, 2003.

[15] Wusten J et al., Interaction of alkali metals with perylene-3,4,9,10-tetracarboxylicdianhydride thin films, *Journal of Applied Physics* 98, 013705, 2005.

[16] Tanaka K et al., Doping effect on organic semiconductive thin film by plasma polymerization of 3,4,9,10-perylenetetracarboxylic dianhydride, *Synthetic Metals* 65, 81-84, 1994.

[17] Zhuo Z et al., Efficiency improvement of polymer solar cells by iodine doping S S. Electron., 63, 83, 2011.

[18] Mizuta S et al., Effect of iodine doping *of* phthalocyanine on the photocurrent generation in a phthalocyanine/C_{60} heterojunction, *Thin Solid Films* 520, 5761, 2012.

INTERPOLYMER COMPLEXES BASED ON NATURAL POLYSACCHARIDE GELLAN AND POLYCARBOXYLIC ACIDS

ZHUMALY A.A., CHOPABAYEVA N.N.[1] AND KUDAIBERGENOV S.E.[1,2]

Institute of Magistral and Doctoral Programs PhD of Abai Kazakh National Pedagogical University, 050010, Almaty, Zhambul Str., 25, Republic of Kazakhstan
[1]Institute of Polymeric Materials and Technology, Almaty, Republic of Kazakhstan, 050013, Almaty, Satpaev Str., 22, Republic of Kazakhstan
[2]Laboratory of Engineering Profile K.I. Satpaev Kazakh National Technical University 050013, Almaty, Satpaev Str., 22, Republic of Kazakhstan
E-mail: nazch@mail.ru

ABSTRACT

The composition of interpolymer complexes formed between natural polycaharyde gellan and polycarboxylic acids has been determined by the methods of potentiometric, conductometric and viscosimetric titration. The influence of temperature, pH and dilation on stability of interpolymer complexes was studied.

INTRODUCTION

At present special interest is paid to interpolymer reactions with participation of natural and synthetic polymers. Complexes of polysaccharides with poly(carboxylic) acids are promising materials in biotechnology and medicine for preparation of different macromolecular therapeutic systems, the prolonged medicinal forms, biocompatible coverings and etc [1,2].

Comprehensive study of the characteristics of complexation reactions involving these macromolecules, determination of the composition and properties of the resulting polycomplexes, disclosure of the mechanisms of their formation has a great theoretical and practical interest.

V. Rajendran, P. Manivasakan and K.E. Geckeler (eds.)
Advanced Nano Materials for Industrial Applications, pp. 109-112 (2012).

The aim of this work is to investigate complexation between polysaccharide – gellan and polycarboxylic acids, as well as to determine the composition and stability of formed interpolymer complexes (IPC) in dependence of various external factors.

EXPERIMENTAL PART

In this work used food gellan ("Zheijiang DMS Zhongken Biotechnology Ltd", China) with a molecular weight of 500 kDa. Polysaccharide composed of tetrasaccharide units which include interconnected angidropiranoznye units of glucose, glucuronic acid, glucose and rhamnose [3].

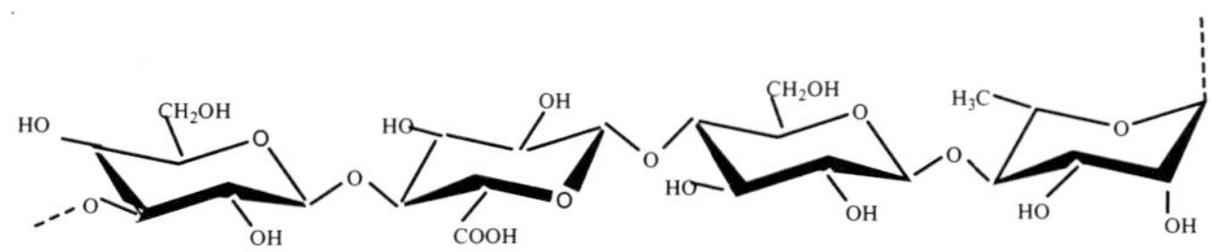

Fig. 1: Gellan Structure

Polyacrylic and polymethacrylic acids (PAA, PMAA) with a molecular weight of $450 \cdot 10^3$ were used without additional purification.

Complexation was carried out by mixing of gellan and polycarboxylic acids aqueous solutions at room temperature. Composition of the complexes were determined by potentiometric, conductometric and viscometric titration of polysaccharide solution (1·10 mol/L) with PAA or PMAA aqueous solution (1 mol/L). Value of pH and conductivity of aqueous solutions was measured by pH-meter "MPC 227" equipped with a versatile combination electrode InLab 413 "three in one", which combines the measuring electrode, reference electrode and temperature sensor in one unbreakable polymer body.

Effect of dilution, temperature and pH value on the stability of the complexes studied by viscometric method by measuring viscosity of filtered solution gellan-PAA/PMAA complexes at different temperature, pH and concentrations of IPC. Viscosity of complexes solutions were measured using a capillary viscometer of Ubellode in a thermostated chamber at 25°C.

Turbidity of the IPC solutions was determined on a spectrophotometer "Jenway-6300" (Germany).

RESULTS AND DISCUSSION

The IPC was formed by the formation of hydrogen bonds between proton-acceptor groups of gellan and proton-donor groups of polycarboxylic acids. The mechanism of IPC formation presented in Fig. 2.

Figure 3 shows the results of studying the complexation reaction between gellan and polycarboxylic acids by viscometric, conductometric and potentiometric methods. Bends in the curves allow to determine the composition of formed polycomplexes. The compositions of IPC formed between gellan and polyacrylic (PAA), polymethacrylic (PMAA) acids determined by potentiometric, conductometric and viscosimetric titration methods are equal to gellan:PAA = 2:1 mol/mol and gellan:PMAA = 3:1 mol/mol respectively.

Formation of complexes of non-stoichiometric composition is caused to the conformational state of reacting macromolecules. Presence of gellan chains in solution in the form of associated coils and spirals leads to enrich of IPC by polysaccharide links. Solubility of both complexes is obviously connected with high polymerization degree of gellan, entered in excess, compared with the polymerization degree of polycarboxylic acids.

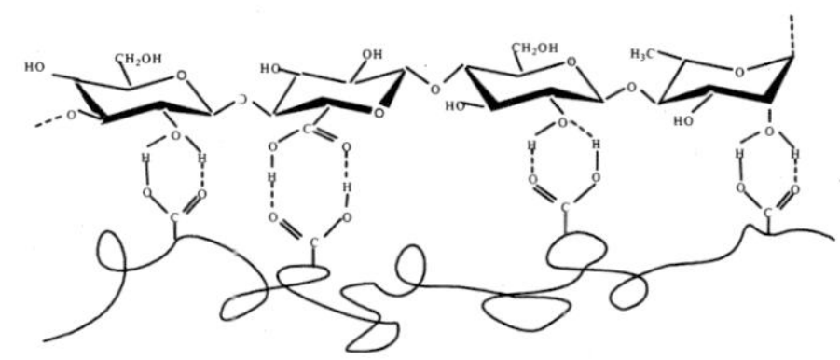

Fig. 2: Scheme of Hydrogen Bonds Formation Between Gellan and Polycarboxylic Acids

In this case, free links of gellan, which are not entered into interaction, carry out function of hydrophilic fragments contributing to the retention of IPC particles in solution.

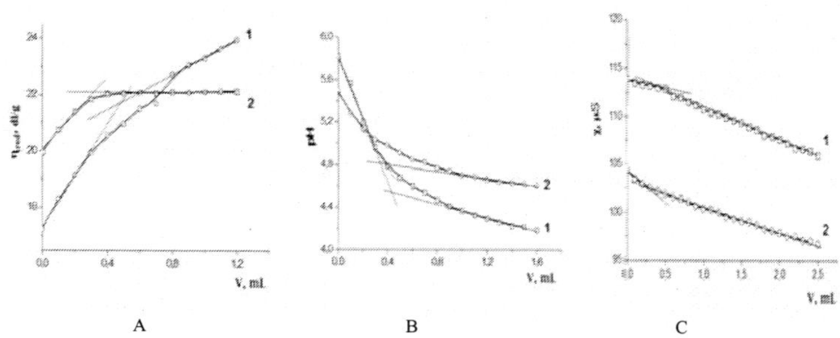

Fig. 3: Determination of Composition of Complexes Gellan-PAA (1) and Gellan-PMAA (2) by Viscosimetric (A), Potentiometric (B) and Conductometric (C) Titration

Stability of IPC on dilution, pH medium, temperature and nature of organic solvent was studied. It is well known [2] that formation of complex, which stabilized by hydrogen bonds, need to certain amount of undissociated COOH-groups. Increase of the reduced viscosity with increasing of pH is caused by destabilization of polycomplexes as a result of partial ionization of carboxylic groups (Fig.4). Destruction of IPC particles is observed at pH > 8 due to deprotonization of COOH-groups of polycarboxylic acids. In acid medium complex-forming process is accompanied by compactization of complex particles that leads to decrease of the viscosity of solutions. The sharp increase of turbidity at pH < 3 and

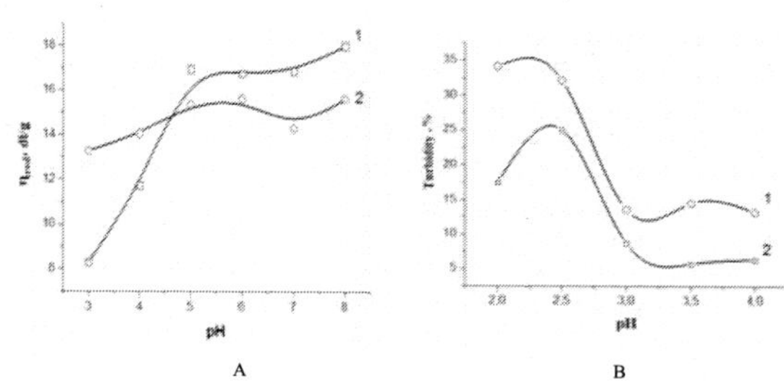

Fig. 4: Influence of pH on Reduced Viscosity (A) and Turbidity (B) of IPC Solutions of Gellan-PAA (1) and Gellan-PMAA (2)

precipitation of IPC in a narrow interval of pH (pH $\leq$ 2.5) testifies the co-operative character of complex-forming process.

Monotonous decrease of the reduced viscosity of IPC gellan-PAA with dilation is connected with absence of polyelectrolyte anomaly for gellan-PAA while both gellan and PAA behave as polyelectrolytes due to unfolding of macromolecular chains as a result of electrostatic repulsion of fixed charges in macromolecules (Fig. 5).

Complexes based on gellan and polycarboxylic acids are thermostable at the temperature interval 25-65°C due to additional stabilisation of IPC by hydrophobic interactions [4]. Stability of IPC on addition of dimethylformamide (DMF) was studied. It was shown that destruction of polycomplexes takes place at water/DMF = 80/20 vol.%.

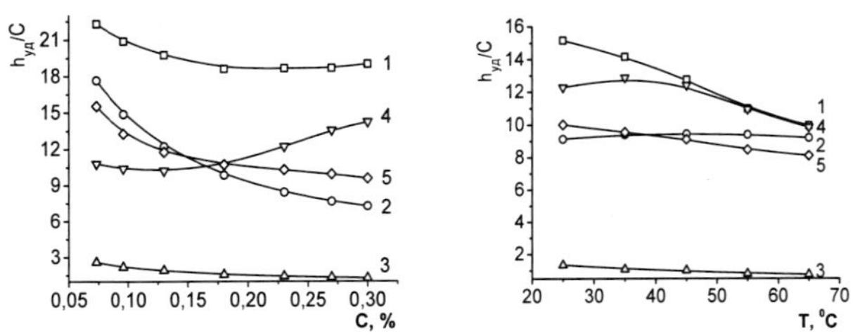

A B

Fig. 5: Influence of Dilution (A) and Temperature (B) on Reduced Viscosity of Solutions of Gellan (1) PAA (2), PMAA (3) and Polycomplexes Gellan-PAA (4), and Gellan-PMAA (5)

PRACTICAL SIGNIFICANCE

Soluble interpolymer complexes open a new range of promising application of this type of complexes in biomimetics, biotechnology and nanomedicine.

CONCLUSION

Soluble nonstoichiometric interpolymer complexes were obtained by interaction of natural polysaccharide gellan with polycarboxylic acids. Study of influence of complexation conditions on stability of polycomplexes allowed establishing the formation of stable polycomplex with participation of PAA as more complementary to gellan macromolecules.

REFERENCES

[1] Bimendina, L.A., Yashkarova, M.G., Kudaibergenov, S.E., Bekturov, E.A., Polymer complexes (preparation, properties and application), SGU, Semipalatinsk, 2003.

[2] Khutoryanskiy, V., Staikos, G., Hydrogen-bonded Interpolymer Complexes, Nova Sciences Publishers, New York, 2009.

[3] Desideri, P., Crescenzi, V., Yuguchi, Y., *Macromolecules*, 34, 1449, 2001.

[4] Koussathana, M., Lianos, P., Staikos ,G., *Macromolecules*. 30, 7798, 1997.

EFFECT OF Dy DOPING ON STRUCTURAL, ELECTRICAL AND MAGNETIC PROPERTIES OF Ni-Co SUBSTITUTIONAL SPINEL FERRITES

A.A. KADAM AND K.Y. RAJPURE

Electrochemical Materials Laboratory, Department of Physics, Shivaji University, Kolhapur
E-mail: rajpure@yahoo.com

ABSTRACT

In the present work nanocrystalline dysprosium doped $Ni_{0.8}Co_{0.2}Fe_{2-x}Dy_xO_4$ ferrite compositions with x = 0.00, 0.025, 0.050, 0.075 & 0.1 were synthesized by the solution combustion method. The influence of Dy doping on to the dielectric and electrical properties has been investigated. The dielectric properties such as dielectric constant, loss tangent and ac conductivity with frequency have been studied in the frequency range 20 Hz to 1 MHz. The variation of dc resistivity has been studied with temperature.

INTRODUCTION

Magnetic ferrite nanoparticles are widely used in permanent magnets, microwave devices and different electric and electronic devices. Ni-Co mixed ferrite has a high catalytic power than the other ferrites such as Ni-Zn ferrite, copper ferrite. In the Ni-Co ferrite, the Ni^{2+} and Co^{2+} cations occupy the octahedral sites, and the Fe^{3+} cations occupy the octahedral and tetrahedral sites equally [1]. Substituted nickel ferrite has been of great interest of extensive investigation because of their microwave applications. By introducing relatively small amount of rare earth ions, an important modification in both structural and dielectric properties can be obtained. Electric properties such as conductivity, resistivity is very much sensitive to cation distribution in substitutional ferrites. The valence of the rare earth dopants is influenced by the surrounding

V. Rajendran, P. Manivasakan and K.E. Geckeler (eds.)
Advanced Nano Materials for Industrial Applications, pp. 113-118 (2012).

environment. Most of the rare earth dopants are stable to +3 valency. In ferrites dysprosium was found to be B-site dopant [2]. By incorporating varius substitutions and different molar ratios properties of Ni-Co ferrite was reported by kadam and Patankar [3,4]. Rezlescu reported that substitution of Fe ions by small amount of rare earth ions may favorably influence the magnetic and electrical properties of Ni–Zn ferrites [5]. The effect of rare earth substitution on the dielectric properties and AC conductivity of Mg–Ti ferrites was reported by Ahmed [6]. Partial substitution of Fe^{+3} by Tb^{+3} and Dy^{+3} in the cobalt ferrite has been reported to cause an increase in the coercive field to about 2 kOe [7]. Effect on structural and dielectric properties of Nd^{3+} substituted Nickel-Zinc ferrite was reported by Shinde et al. [8]. The structural, ferromagnetic, ferroelectric, dielectric and transport properties of rare earth doped Ni ferrite were reported by Dwevedi et al. [9]. Among the rare earth elements, dysprosium has large anisotropy as well as large magnetostriction values [10]. Therefore it is of interest to investigate the effect of Dy^{3+} substitution on to the Ni-Co mixed ferrite. In the present work dielectric and electric properties of Dy doped Ni-Co ferrite has been investigated.

EXPERIMENTAL

Nanocrystalline ferrite samples with general formula $Ni_{0.8}Co_{0.2}Fe_{2-x}Dy_xO_4$ (where x = 0.00, 0.025, 0.050, 0.075 and 0.1) were synthesized by solution combustion method. The AR Grade nickel nitrate, cobalt nitrate, iron nitrate and dysprosium nitrate were used to prepare the ferrite compositions. Stoichiometric amount of corresponding metal nitrates (Ni, Co, Fe and Dy) acts as an oxidizing agent and fuel glycine acts as a reducing agent for the combustion process. 1M solution of respective metal nitrates such as Ni, Co, Fe, Dy and glycine were prepared separately and taken into the beaker. The solution was stirred slowly on magnetic stirrer to obtain homogeneous mixture and then stoichiometric solution of glycine was added into it to form the redox mixture for the combustion process. The resultant solution was then dehydrated slowly on a hot plate with continuous stirring until the viscous gel formed. On further heating, the temperature of the gel increased and at a certain temperature auto-ignition of the black and fluffy gel took place with evolution of gases. During ignition smoke is produced and the products left behind are voluminous and fluffy loose powder. The resulting powder from all the samples was ground for 30 minutes and presintered at 800°C for 5 hours at the rate of 20 min. in programmable furnace. The calcined powder then uniaxially pressed into the pellets by using 2% PVA binder and by applying pressure of about 7 tones for 10 minutes. These pellets were finally sintered at 1100°C for 11 hours to reduce porosity and increase density. Silver paste was applied on both the sides of the pellets for good ohmic contacts. The powders were characterized by X-ray diffractometer to confirm phase formation. The frequency dependence of dielectric permittivity (ε'), dissipation factor ($\tan \delta$) and ac conductivity in the range from 20Hz to 1MHz were studied using a precision LCR meter bridge (HP 4284 A). Resistivity measurement with respect to temperature was carried out using Keithly electrometer (6514).

RESULTS AND DISCUSSION

Dielectric Properties

Figure 1 shows variation of dielectric constant ε, with frequency for the $Ni_{0.8}Co_{0.2}Fe_{2-x}Dy_xO_4$ ferrite samples. Initially dielectric constant decreases with increase in frequency and finally at higher frequencies attains a constant value for all the samples. The decrease in dielectric constant is due to the dielectric dispersion in low frequency region. At high frequency the electron exchange between Fe^{2+}- Fe^{3+} do not follow the high alternating field [11]. It is observed that the relative dielectric constant increases with increasing in Dy content up to $x = 0.050$. Further increase in Dy content reduces the dielectric constant. This may be attributed to the change in cation distribution due to the substitution of iron by dysprosium. In the present ferrite, Ni^{2+} and Co^{2+} cations occupy the octahedral sites and Fe^{3+} cations occupy the octahedral and tetrahedral sites equally. When Dy is substituted for Fe, preferably it occupies B-Site because of larger ionic radius of Dy^{3+} (0.91 A^0) [12] than the Fe^{3+} (0.64 A^0), Ni^{2+}(0.74 A^0), and Co^{2+} (0.78 A^0) [13]. Incorporation of Dy^{3+} ions into the octahedral site of ferrite causes migration of Fe^{3+} ions from B- site to A-site. Also partial migration of Co^{2+} and Ni^{2+} ions from B-site to A-site causes an opposite transfer of Fe^{3+} ions from A-site to B-site in order to relax the strain [14, 15]. This results in increase in electron exchange mechanism between Fe^{3+} and Fe^{2+} ions. This causes increase in dielectric constant upto x = 0.050. Further increase in Dy^{3+} content reduces Fe^{3+} ion concentration on B-site. Consequently, dielectric constant decreases for higher concentration of dysprosium.

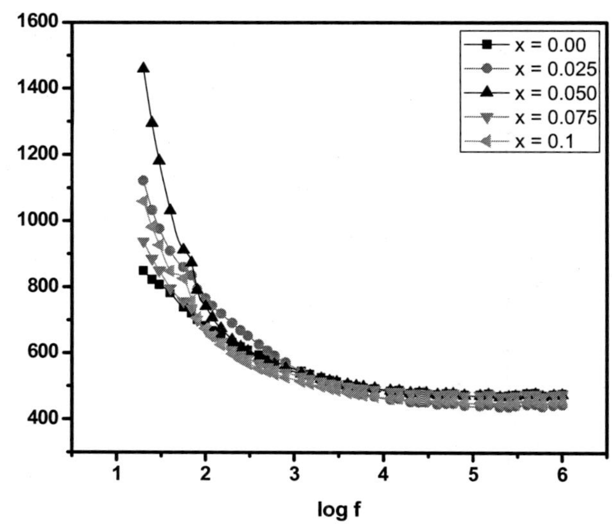

Fig. 1: Variation of Room Temperature Dielectric Constant with Frequency of $Ni_{0.8}Co_{0.2}Fe_{2-x}Dy_xO_4$ ferrite

The variation of tan δ with frequency given in Figure 2 shows a similar nature as the variation of ε′ with frequency. Dielectric losses are found to be low since domain wall motion is inhibited and magnetization is forced to change by rotation. The variation of AC conductivity with frequency is given in Fig. 3. It is well known that the electron and hole hopping between Fe^{2+}/Fe^{3+}, Ni^{3+}/Ni^{2+}, and Co^{3+}/Co^{2+} ions, is responsible for electrical conduction in mixed ferrites. In the present system net conduction due to the p-type and n-type of charge carriers increases [16]. The plots of AC conductivity are linear at low frequency, indicating that the conduction is due to small polarons. At higher frequencies, where conductivity increases greatly with frequency, the transport is dominated by contributions from hopping infinite clusters.

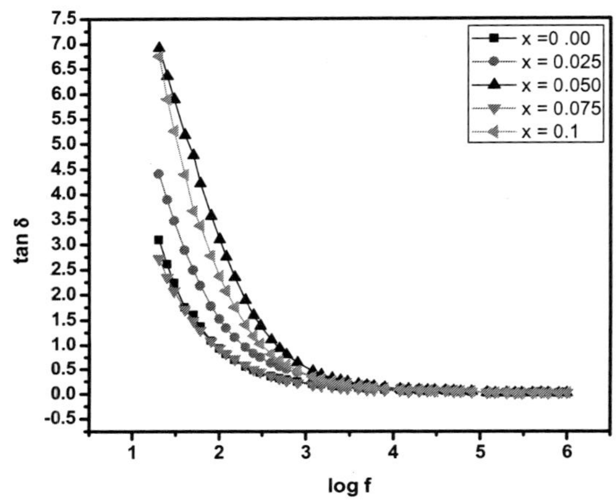

Fig. 2: Variation of Room Temperature Loss Tangent (Tan δ) with Frequency of $Ni_{0.8}Co_{0.2}Fe_{2-x}Dy_xO_4$ Ferrite

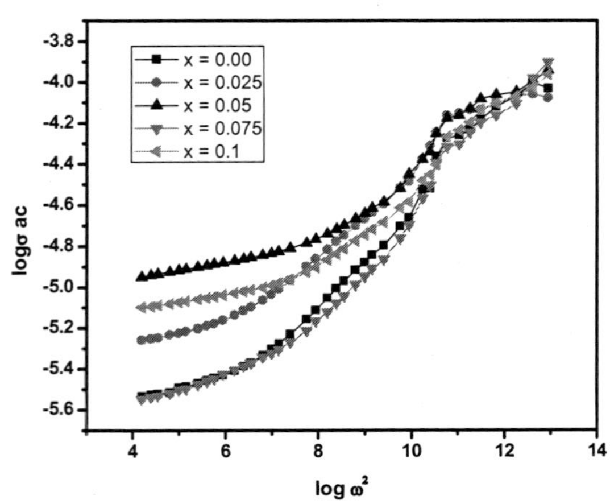

Fig. 3: Variation of AC Conductivity with Frequency of $Ni_{0.8}Co_{0.2}Fe_{2-x}Dy_xO_4$ Ferrite

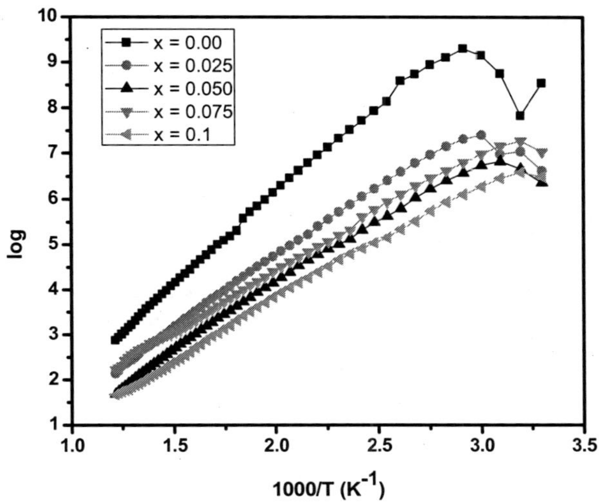

Fig. 4: Variation of DC Electrical Resistivity with Temperature of $Ni_{0.8}Co_{0.2}Fe_{2-x}Dy_xO_4$ Ferrite

Electrical Properties

Variation of dc resistivity with temperature for $Ni_{0.8}Co_{0.2}Fe_{2-x}Dy_xO_4$ ferrite is given in Fig. 4. The resistivity is found to decrease with increasing temperature due to the increase in the thermally activated drift mobility of charge carriers according to the hopping conduction mechanism. The electrical resistivity changes with the cation distribution in the ferrites. In case of ferrites electron hopping between $Fe^{2+}$and Fe^{3+} and hole hopping between Ni^{3+} to Ni^{2+} and Co^{3+} to Co^{2+} ions in B site decides electrical resistivity [17]. Resistivity decreases with increase in Dy^{3+} content except for x = 0.075. It is due to the fact that the ionic radius of Dy is larger than that of iron and it will occupy the octahedral sites displacing Ni^{2+} and Co^{2+} ions from B to A site which is accompanied by an opposite transfer of an equivalent number of Fe^{3+} ions from A to B sites in order to relax the strain [18]. This increase in the Fe^{3+} ions in the octahedral sites will increase the number of Fe^{2+} and Fe^{3+} pairs in octahedral sites, which gives increase in conductivity and decrease in resistivity of ferrites.

CONCLUSIONS

Fine nano particles of $Ni_{0.8}Co_{0.2}Fe_{2-x}Dy_xO_4$ [Dy (x = 0 to 0.1)] ferrites were successfully synthesized by solution combustion method. The dielectric constant and loss tangent (tan δ) increases with increase in Dy content up to x = 0.050 and then decreases for x = 0.075. The variation of AC conductivity with frequency suggests that the conduction is due to types of small polarons. The variation of dc resistivity with temperature shows semiconducting nature of the sample.

ACKNOWLEDGEMENTS

Authors are very much thankful to UGC and DST for the financial support through UGC-DSA-I, DST-PURSE and DIST-FST-II programmes.

REFERENCES

[1] G. Nabiyouni1, M. Jafari Fesharaki, M. Mozafari, J. Amighian, *Chin. Phys. Lett.* 27 (2010) 126401.

[2] Vasant Naidu, S. Gayathri Devi R. Legadevi, *Lakshmi Priya, International Journal of Computer Applications* 48 (2012) 15.

[3] K.K. Patankar, S.S. Joshi, B.K. Chougule, *Physics Letters* A 346 (2005) 337–341

[4] S. L. Kadam, K. K. Patankar, C. M. Kanamadi and B. K. Chougule, *Mater. Res. Bull.* 39 (2004) 2265-2272.

[5] N. Rezlescu, E. Rezlescu, *Solid State Comm.* 88 (1993) 139.

[6] M.A. Ahmed, E. Ateia, S.I. El-Del, F.M. Salem, *J. Mater. Sci. 38* (2003) 1087.

[7] F.Cheng, C. Liao, J.Kuang, Z.Xu, C. Yan, L.Chen, H. Zhao, and Z. Liu, *J. Appl. Phys*, 85 (1999) 2782-2784.

[8] T. J. Shinde, A. B. Gadkari , P. N. Vasambekar, *J. Mater. Sci: Mater Electron* 23 (2012) 697–705.

[9] S. Dwevedi, K. Kamala Bharathi and G. Markandeyulu, *IEEE Transactions on magnetics*, 45 (2009) 4253-4256.

[10] S. Legvold, J. Alstad, and *J. Rhyne Phys. Rev. Lett.* 10, 509 (1963).

[11] K.K. Patankar, C.M. Kanamadi, B.K. Chougule, *Phys. Lett.* A 346 (2005) 337–341.

[12] S. Cotton, *Lanthanide and Actinide Chemistry*, John Wiley & Sons, New York, 2006.

[13] K.J. Standley, *Oxide Magnetic Materials*, Clarendon, Oxford, 1972.

[14] F.X. Cheng, J.T. Jia, Z.G. Xu, B. Zhou, C.S. Liao, C.H. Yan, L.Y. Chen, and H.B. Zhao, *J. Appl. Phys.* 86 (1999) 2727

[15] W.C. Kim, S.W. Lee, S.J. Kim, S.H. Yoon, and C.S. Kim, *J. Magn. Magn. Mater.* 215 (2000) 217.

[16] R.S. Devan, Y.D. Kolekar and B.K. Chougule, *J. Phys.: Condens. Matter* 18 (2006), 9809–9821.

[17] A.D. Sheikh, V.L. Mathe, *J. Mater Sci.* 43 (2008) 2018-2025.

[18] R.C. Kambale, K.M. Song, Y.S. Koo, and N. Hur, *Journal of Applied Physics* 110(2011), 053910.

MgO-INCORPORATED NYLON 6 NANOFIBERS FOR PROTECTIVE APPLICATIONS

N.R. DHINESHBABU, G. KARUNAKARAN, R. YUVAKKUMAR, P. MANIVASAKAN, P. PRABU AND V. RAJENDRAN

Centre for Nanoscience and Technology, K.S. Rangasamy College of Technology, Tiruchengodu, Tamil Nadu
E-mail: veerajendran@gmail.com

ABSTRACT

The magnesia (MgO) nanoparticles and nylon 6 solutions were prepared using sol–gel method. The crystallinity and morphology of the prepared MgO nanoparticles were characterized by X-ray diffraction analysis (XRD) and transmission electron microscopy (TEM). The cubic-phase MgO nanoparticles with a crystalline size of 13.56 nm were examined by XRD and a flakelike structure was observed through TEM image. MgO nanoparticle-filled nylon solution was spun using electrospinning process to produce nanofiber mats. The prepared MgO/nylon 6 hybrid nanofiber mats were found to have excellent antibacterial activity against both gram-positive (Staphylococcus aureus) and gram-negative (Escherichia coli) bacteria. The mats were placed on cotton fabrics to measure the air permeability values of the fabrics. In sum, MgO/nylon 6 hybrid nanofiber mats prepared by the electrospinning process can be an excellent material for protective applications.

INTRODUCTION

Electrospinning is one of the effective methods used to produce polymer nanofiber with outstanding properties such as a high specific surface area and high porosity. Electrospun nanofibers are used for a wide variety of applications such as separation filters, wound-dressing

V. Rajendran, P. Manivasakan and K.E. Geckeler (eds.)
Advanced Nano Materials for Industrial Applications, pp. 119-124 (2012).

materials, tissue scaffold, sensors, protective clothing and catalysis reaction [1,2]. Electrospinning is one of the simplest and most versatile methods used to fabricate polymeric or composite fibers [3,4]. Recently, polymer/inorganic composite is used as a precursor to inorganic nanofibers using electrospun process [5,6]. The extensive research on magnesia (MgO) nanoparticles is undergoing due to their elastic properties such as thermal conductivity, heat resistance and good antibacterial activities [7]. Attempts have been made to use the nylon 6 and MgO nanoparticles in protective applications because of their antibacterial properties against *Escherichia coli* and *B. cereus* [8–10].

In this article, we attempted to prepare a novel MgO/nylon 6 hybrid nanofiber mat using electrospinning process. The prepared nanoparticles were characterized using X-ray diffraction (XRD) and transmission electron microscopy (TEM). The MgO nanoparticle-incorporated nylon 6 nanofibers were characterized using scanning electron microscopy (SEM). In addition, the functions of MgO/nylon 6 hybrid nanofiber mats against *Staphylococcus aureus* and *E. coli* bacteria were also studied.

EXPERIMENTAL

Materials

Magnesium nitrate ($Mg(NO_3)_2$; 99.9% assay; Merck) , sodium hydroxide (NaOH; 99.9% assay; Merck), nylon 6 (99% assay; Sigma), formic acid (99.9% assay; Merck), acetic acid (99.9% assay; Merck) and double-distilled water were used in this investigation.

Synthesis and Characterization of MgO Nanoparticles

In this experiment, MgO nanoparticles were prepared using simple sol–gel method. $Mg(NO_3)_2$ was used as a precursor and NaOH as a reducing agent. Magnesium nitrate (3 g) and sodium hydroxide (0.4 g) were dissolved separately in 100 and 50 ml double-distilled water in a beaker. Then, the NaOH solution was drop-wise added into $Mg(NO_3)_2$ solution with continuous stirring at 350 rpm at room temperature. Thereafter, the solution was centrifuged at 10,500 rpm at room temperature and the precipitate was washed several times with double-distilled water and then with ethanol to remove the ionic impurities. The obtained product was dried at 673 K under vacuum for 4 h. Finally, a white MgO nanopowder was obtained. The powder was characterized comprehensively using XRD and TEM.

Preparation and Characterization of Nylon 6 and Mgo/Nylon 6 Nanofiber Mats

The nylon 6 and MgO/nylon 6 hybrid functional nanofibers were prepared by an electrospinning process, as shown in Fig. 1. A nylon 6 pellet was dissolved in formic acid and acetic acid with a molar ratio of 3:1 at 10 wt% concentrations. On the other hand, MgO nanoparticles (5 wt%)

were added into the nylon 6 polymer solution with continuous stirring at 400 rpm at the room temperature. The nylon 6 solution was loaded into syringe and placed on a syringe pump. The parameter of feed rate was maintained at 0.09 ml h^{-1}. The distance between Taylor cone and collector was 18 cm. The syringe needle (diameter 0.07 mm) was positively charged using high-voltage power supply (18 kV) to spun the fiber mats. Simultaneously, MgO/nylon 6 hybrid nanofiber mats were formed under the similar conditions.

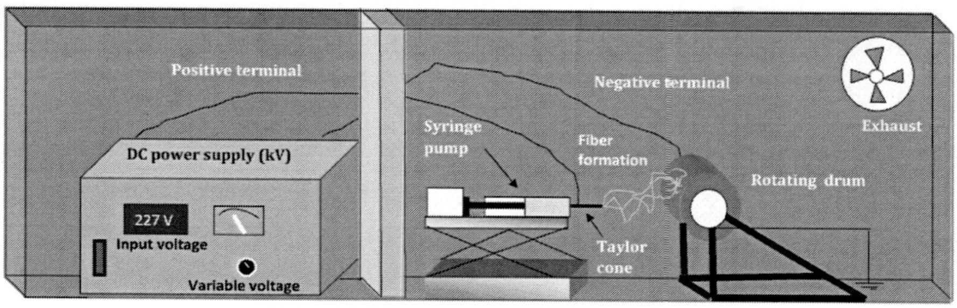

Fig. 1: Schematic of Electrospinning Unit

The viscous nylon 6 solution was measured using viscous meter (SNB-4; Hinotek, Korea). The ionic conductivity of the prepared nylon 6 and MgO/nylon 6 solution was measured using conductivity meter (Orion 5-Star; Thermo Scientific, USA). The surface morphology of nanofiber mats was examined through scanning electron microscopy (SEM) with energy-dispersive spectroscopy. The air permeability measurement of nanofiber mats was carried out according to the IS 11056-1984 and DIN 53887 standards.

Antibacterial Activity

The test bacteria *E. coli* and *S. aureus* were obtained from the National Collection of Industrial Microorganisms, National Chemical Laboratory (India) and maintained in nutrient agar (HiMedia, India) slants at 310 K for 24 h. Antimicrobial activity was analysed by Kirby–Bauer method (1966) using Mueller–Hinton Agar (MHA) medium (HiMedia). The MHA plates were prepared by pouring 15 ml molten media into sterile Petri plates. The plates were allowed to solidify for about 5 min and 0.1 ml inoculum was swabbed uniformly over the agar till it became invisible. MgO, nylon 6 and MgO/nylon 6 mats were loaded over the plate and incubated at 310 K for 24 h. After incubation, the plates were observed for inhibition zones around the mat and the zones were measured in millimeter using a transparent ruler. The experiment was performed in triplicates.

RESULTS AND DISCUSSION

Figure 2 shows the powder XRD pattern of the as-prepared sample obtained after calcinations at 673 K. The pattern is in good agreement with that of standard MgO nanoparticles (JCPDS

PDF no. 89-7746). To quantify the particle size effects, the mean grain sizes (D) of the powder sample is calculated with Scherer equation. The obtained mean grain size of sample is approximately 13.56 nm. The particle size, with an average diameter of 24–40 nm, and short flakelike morphology are clearly revealed from TEM image shown in Fig. 3.

Table 1 shows the ionic conductivity and viscosity of both nylon 6 and MgO-filled nylon 6 solutions measured at room temperature. During this process, the positive terminal is connected to the Taylor cone when the nylon 6 solution is ejected at a flow rate of 0.09 ml h^{-1} and the negative terminal is connected to the collector, which is a cylindrical drum, leading to the formation of long thread. The bunch of long thread is collected on the drum grounded.

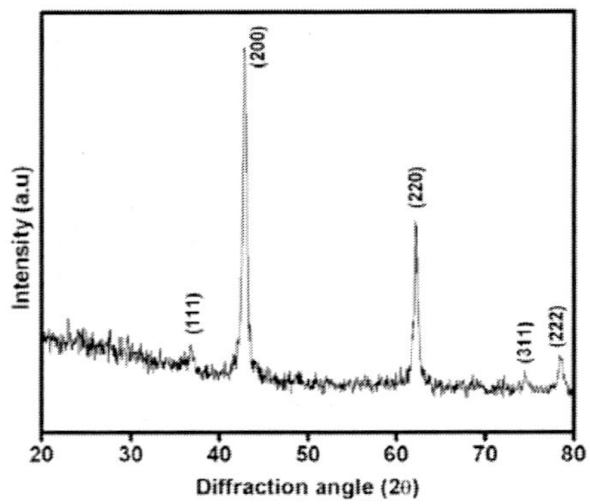

Fig. 2: XRD Analysis of MgO Nanoparticles

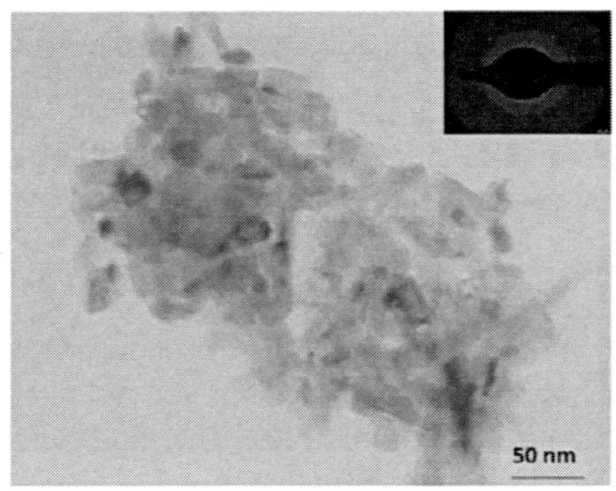

Fig. 3: TEM image of MgO Nanoparticles

On the other hand, MgO/nylon 6 hybrids are also spun onto the collector. Figure 4 shows the SEM image of nylon 6 and MgO/nylon 6 nanofibers obtained by electrospinning. It is observed that electrospun nylon 6 nanofibers have an average diameter of 210 nm, which is erratically placed to form a nonwoven mats. However, MgO/nylon 6 hybrid nanofibers reveal an average diameter (220 nm), which is higher when compared with that of the nylon 6 nanofibers.

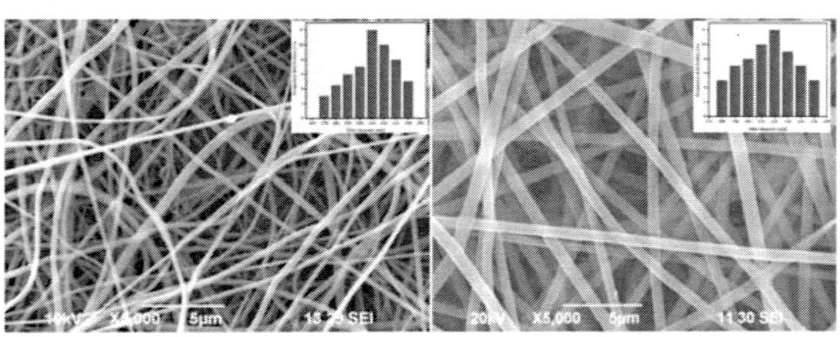

(a) Nylon 6 nanofiber (b) MgO/nylon 6 hybrid nanofibers

Fig. 4: SEM Image of Nanofibers Inside Fiber Diameters

The air permeability is the one of the important parameters to measure the spun nanofiber mats in protective applications. The spun nanofiber mats are placed onto the woven fabric surface. Table 2 compares the air permeability values obtained for the control woven fabric and the nylon 6, and MgO/nylon 6 nanofibers deposited on the fabrics. From the table, it is evident that the air permeability value is slightly reduced in the MgO/nylon fiber when compared with that of the nylon 6 nanofiber-coated and control woven fabrics.

Table 1: Properties of the Polymer Solutions

Measurements	*Nylon 6 Polymer Solution*	*MgO/nylon 6 Polymer Solution*
Viscosity (cP)	107±5	103±5
Ionic conductivity	3.5±0.1	3.8±0.1

Table 2: Air permeability Values of the Nylon 6 and MgO/nylon 6 Nanofibers Deposited on the Fabrics

Sample	*Nanofiber Density* *($g\ m^{-2}$)*	*Air Permeability** *($cc/s/cm^2$)*
Cotton fabric substrate	0	0.982±0.004
Nylon 6 nanofiber mat	3.51±0.02	0.641±0.004
MgO/nylon 6 hybrid nanofiber mat	3.95±0.18	0.588±0.004

*Air permeability error ±0.004

The antimicrobial activity (Fig. 5) reveals the zone of inhibition against *S. aureus* as 14, 6 and 18 mm zone, respectively, for MgO, nylon 6 and MgO/nylon 6 whereas that against *E. coli* is 12, 5 and 15 mm, respectively, for MgO, nylon 6 and MgO/nylon 6 nanofibers. Thus, from the results it is confirmed that the incorporation of Mg enhances the antimicrobial activity against both gram-positive (*S. aureus*) and gram-negative (*E. coli*) bacteria.

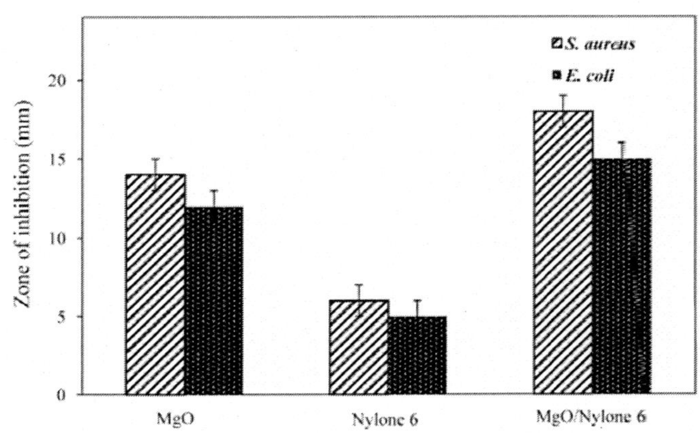

Fig. 5: Antimicrobial Activity of MgO, Nylon 6 and MgO/nylon 6 Fibers

CONCLUSION

MgO nanoparticles were prepared using sol–gel method. MgO/nylon 6 hybrid nanofiber mats were successfully prepared by electrospinning process. The antibacterial properties of MgO/nylon6 hybrid nanofiber mats were explored and compared. This study confirms a good antibacterial activity against gram-negative (*E. coli*) and gram-positive (*S. aureus*) pathogens. Therefore, an enhanced antibacterial activity and better air permeability values are most wanted characteristics in protective clothing.

REFERENCES

[1] Reneker D.H., Chun I., *Nanotechnology* 7, 216, 1996.

[2] Li D, Xia Y., *Adv. Mater.* 16, 1151, 2004.

[3] Park J.H., Karim, M.R., Kim, I.K, Cheong I.W., Kim J.W., Bae D.G., Cho J.W., Yeum J.H., *Colloid. Polym. Sci.*, 288, 115–121, 2010.

[4] Nagamine S., Ishimaru S., Taki K., and Ohshima M., *Materi. Lett.* 65, 3027–3029, 2011.

[5] Saho C.L., Kim, h. Y., Gong J., Lee d. R., *Nanotechnology* 13, 635, 2002.

[6] Dai H.Q.,, Gong J., Kim H., Lee D., *Nanotechnology* 13, 674, 2002.

[7] Lee J.D., *Concise Inorganic Chemistry*, 5th ed., Chapman and Hall, New York, 339, 1996.

[8] Stoimenov P.K., *Langmuir*, 18, 17, 6679–86, 2002.

[9] Huang L., *J. Inorg. Biochem.* 99, 5, 986–93, 2005.

[10] El-Newehy M.H., Al-Deyab S.S., Kenawy E.-R., Abdel-Megeed M., *J. Nanomater.* 8, 626589, 2011.

NATURAL RUTILE–DERIVED TITANIA NANOPARTICLES FOR PHOTOCATALYTIC APPLICATIONS

S. ARUNMETHA, A. KARTHIK, S.R. SRITHER, P. MANIVASAKAN AND V. RAJENDRAN

Center for Nanoscience and Technology, K.S. Rangasamy College of Technology, Tiruchengode, Tamil Nadu
E-mail: veerajendran@gmail.com

ABSTRACT

The titania (TiO_2) nanoparticles have been synthesised using spray pyrolysis techniques. The titanium sulphate, which is used as a precursor for production of TiO_2 nanoparticles, was obtained from natural rutile sand by acid extraction process. The crystallite size, structure, particle size distribution, morphology and purity of TiO_2 nanoparticles were identified by X-ray diffraction, Fourier transform infrared spectroscopy, particle size analyzer, energy-dispersive X-ray analysis and scanning electron microscope. As-produced TiO_2 nanoparticles were applied for textile effluent treatment.

INTRODUCTION

High surface area achieved due to small particle size is advantageous to many titania (TiO_2)-based devices such as beam splitters, optical and antireflection coatings, catalysis, gas sensors, ultraviolet absorbers, lithium batteries, and optical, electronic and electrochromic devices [1-5]. Owing to their high photocatalytic activity and chemical stability, TiO_2 nanoparticles have been used in clean technologies such as environmental remediation, self-cleaning glasses, pigments, paints, ceramics, cosmetics and solar energy conversions [6–8].

Many synthesis methods, such as thermal decomposition [9], precipitation and hydrolysis [10], sol–gel [11], hydrothermal [12], solvothermal [13], sonication [14], ball milling [15],

V. Rajendran, P. Manivasakan and K.E. Geckeler (eds.)
Advanced Nano Materials for Industrial Applications, pp. 125-130 (2012).

chemical vapour deposition [16] and spray pyrolysis [17], are being explored and developed for the production of TiO_2 nanoparticles. Spray pyrolysis is a powerful technique to synthesise a wide variety of highly pure, chemically homogeneous ceramic powders. The aim of this study was to investigate the process in detail and to understand the role that the chemistry of particle formation has in designing new techniques and developing new processes which can be used for the mass production of high-surface-area TiO_2 nanoparticles. The produced TiO_2 nanoparticles are applied for the textile effluent treatment because of their photocatalytic properties.

EXPERIMENTAL

The raw rutile sand, which consists of 96 wt% of TiO_2, was obtained from Kanyakumari region in coastal part of Tamil Nadu. TiO_2 nanoparticles were derived from the rutile sand using two methods: chloride process [7] and sulphur process [15]. The extraction of titanium sulphate from ilmenite sand was carried out using the established procedure [15]. The rutile sand was used as the starting material for synthesis of precursor. The sand was digested with concentrated H_2SO_4 in the temperature range of 423 K for 3 h in a muffle furnace. After the digestion, the mixture was turned into dry cake. The mixture containing water-soluble titanium sulphate was leached with double-distilled water.

$$\text{Rutile sand} + \text{Conc. } H_2SO_4 \rightarrow Ti(SO_4)_2 + \text{Residue waste}$$

Synthesis of TiO_2 Nanoparticle Using Spray Pyrolysis Method

The precursor extracted from rutile sand was hydrolysed using aqueous ammonia solution that was used as the starting phase for spray pyrolysis. The well-developed automated spray pyrolysis experimental setup available in the authors' lab was used in this study [15].

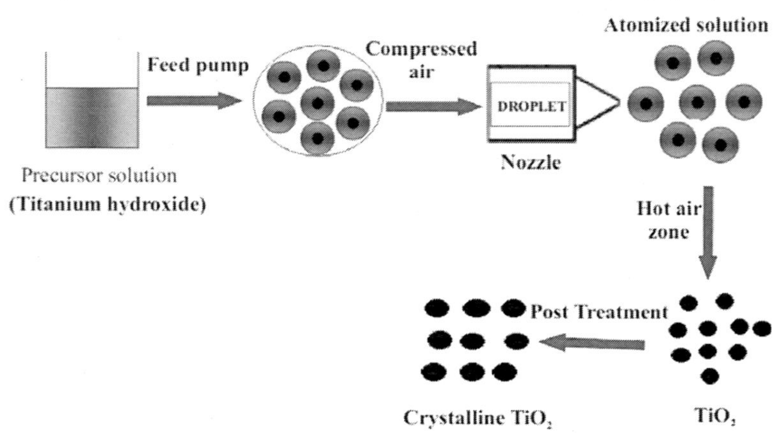

Fig. 1: Spray Pyrolysis Atomisation Process

The titanium hydroxide colloidal solution was diluted with double-distilled water and was fed through a nozzle into a furnace with compressed air to obtain transparent solution. In hot-air zone, the atomised droplets were decomposed to form TiO_2 nanoparticles. Atomised solution was dried in a chamber. TiO_2 nanoparticles were collected from the cyclone where particles were stored, as shown in Fig. 1. The powders were subjected to the post-synthesis treatment sintered at 673 K for 3 h to form crystalline nanoparticles.

$$Ti(OH)_4 \rightarrow TiO_2 + 2H_2O\uparrow$$

RESULT AND DISCUSSION

Figure 2 shows the XRD patterns of TiO_2 nanoparticles. It represents the phase and crystallite dimensions of as-synthesised and calcined samples. All the experimentally observed reflections have been assigned on the basis of the reported anatase phase of titania (JCPDS, file no. 21-1272, 1999). The crystallite size is obtained through XRD peak using Scherer equation [18]. The average crystallite size of TiO_2 nanoparticles with anatase phase is 7 nm.

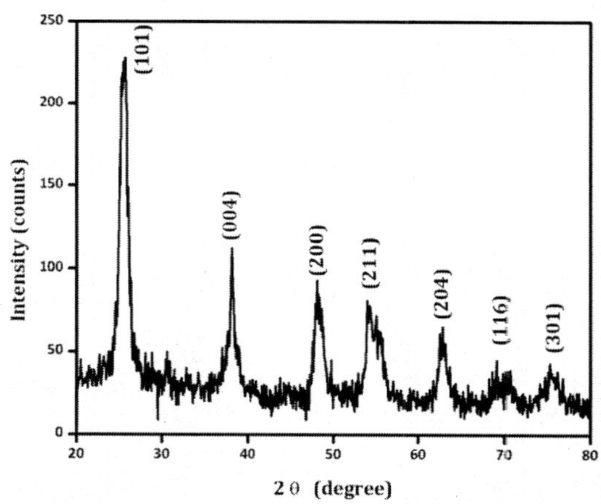

Fig. 2: XRD Patterns of TiO_2 Nanoparticles

The Fourier transform infrared (FTIR) spectra of TiO_2 nanoparticles, shown in Fig. 3, show broad absorbance peaks between 500 and 700 cm^{-1} that correspond to stretching modes of Ti–O bonds. The peaks observed at 1634 cm^{-1} and 985 cm^{-1} are due to existence of hydroxyl groups of surface-adsorbed water and Ti–OH respectively [15,19]. Similarly, the peak observed at 1136 cm^{-1} is ascribed to $-CH_3$ bonds [15].

The particle size distribution of TiO_2 nanoparticles dispersed in water is shown in Fig. 4, which is in the range between 25 and 55 nm. However, the maximum distribution (d_{50}) of the particles is 38 ± 3 nm. Spray pyrolysis method yields a reduced particle size with narrow distribution.

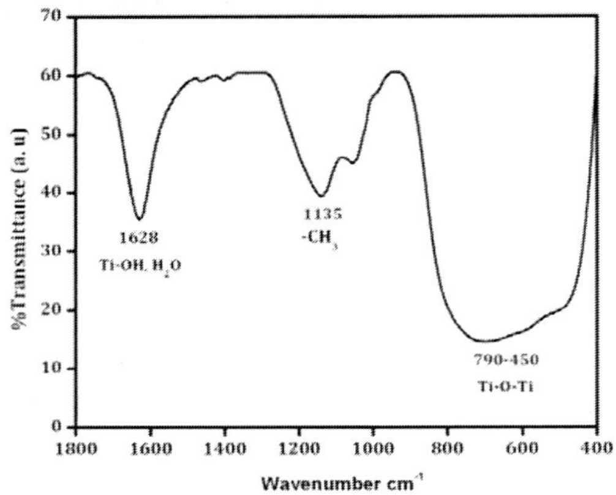

Fig. 3: FTIR Patterns of TiO_2 Nanoparticles

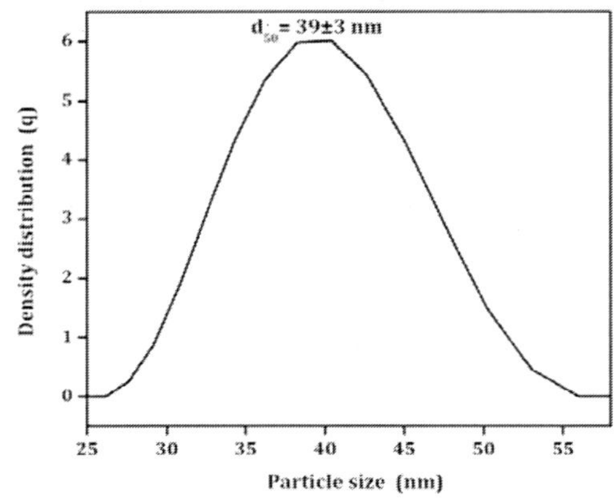

Fig. 4: Particle Size Distribution of TiO_2 Nanoparticles Dispersed in Water

The surface area of TiO_2 nanoparticles is 102 m^2g^{-1}. The surface-to-volume ratio is increased with decreased particle size, leading to high-surface-area particles with free-flowing structure. The topographic features are characterized through scanning electron microscopy (SEM) images [Fig. 5]. It is evident from the figure that the particles are weakly agglomerated with spherical morphology. In the nanoparticles titanium and oxygen elements were present in large quantity, and sulphur impurity elements were present in less quantity. The experiment was performed to know the element distribution in the sample. It confirmed that purity of the sample was 99%.

Fig. 5: SEM Micrographs of Powder TiO_2

APPLICATION

The oxidative degradation of an actual textile dyehouse wastewater was investigated by photocatalysis in the presence of TiO_2. The UV-A-induced photocatalytic oxidation over TiO_2 suspensions was capable of decolourising the effluent completely, as well as reducing chemical oxygen demand (COD) sufficiently. The COD reduction generally varied between about 40% and 90%, depending on the operating conditions. The degradation enhanced for base, neutral and acidic conditions. Photocatalytic degradation was to be tested in varied UV-A-induced time and visible region (40 W; Osram, HQL, MBF-U). Photocatalytic experiments with artificial irradiation were performed as follows: 100 mL textile effluent was introduced in the reaction vessel and an appropriate amount of TiO_2 was added to achieve the desirable concentration. In all the cases where runs were carried out at near-neutral or an acidic condition, the initial pH was adjusted adding the appropriate amount of HCl. From Table 1, it is clear that the extent of photocatalytic degradation was found to increase with increasing TiO_2 concentration up to 1.0 g L^{-1} TiO_2, above which degradation remained practically constant. The COD values decreased as such from the base, neutral to the acidic condition. More organic compounds removal was performed in the acidic condition. The COD values decreased with increasing treatment time, which also indicates an increased percentage of organic compounds removal.

Table 1: COD Range and Percentage of Organic Compounds Removal

Photocatalytic Process		*Before Treatment (COD mg L^{-1})*	*After Treatment (COD mg L^{-1})*	*Percentage of Organic Compounds Removal*
Concentration of TiO_2	0.3 g L^{-1}	260	128	51
	0.5 g L^{-1}	260	116	55
	1.0 g L^{-1}	260	112	57
pH variation	9	260	124	52
	7	260	110	58
	3	260	40	85
UV treatment time	2 h	260	120	54
	4 h	260	116	55
	6 h	260	100	62

CONCLUSION

Nano-sized TiO_2 powder with narrow particle size distribution and pure spherical morphology was successfully prepared from rutile sand using spray pyrolysis. The obtained nanoparticles were tested for photocatalytic treatment of the textile wastewater and organic pollutant degradation. The photocatalytic oxidation over TiO_2 suspension led to reduction in the COD. The COD reduction generally varied between about 40% and 80% depending on the operating conditions. The extent of photocatalytic degradation was found to increase with increasing

TiO_2 concentration up to 1.0 g L^{-1} TiO_2, above which degradation remained practically constant. Furthermore, textile effluent degradation is enhanced at acidic conditions. The produced TiO_2 nanoparticles have enhanced photocatalytic properties. They can be used for solar panel and semi-conducting nanoparticles application.

REFERENCES

[1] Lee H.Y., Kale G.M., *Int. J. Appl. Ceramic. Technol.* 2008, 5, 657–665.

[2] Garzella C., Comini E., Tempesti E., Frigeri C., Sberveglieri G. *Sens. Actuators B* 2000, 68, 189–196.

[3] Apatiga L.M., Rivera E., Castano V.M. *J. Am. Ceram. Soc.* 2007, 90, 932–935.

[4] Zhang Y., li G., Luo Y., Zhang L. *J. Phys. Chem.* 2005, 109, 5478–5481.

[5] Choi S.Y., Mamak M., Coombs N., Chopra N., Ozin G.A. *Nano Lett.* 2004, 4, 1231–1235.

[6] Lam S.W., Gan W.Y, Chiang K., Amal R. *J. Aus. Ceramic Soc.* 2004, 44 6–11.

[7] Suzuki Y., Pavasupree S., Yoshikawa S., Kawahata R. *Phys. Stat. Sol.*2007, 204, 1751–1756.

[8] Vargas W.E., Greenwood P., Otterstedt J.E., Niklasson G.A., *Sol. Energy*, 2000, 68 553–561.

[9] Banerjee I., Karmakar E.S., Kulkarni N.V., Nawale A.B., Mathe V.L., Das A.K., Bhoraskar S.V., *J. Nanopart. Res.* 2010, 12, 581–590.

[10] Cassaignon S., Koelsch M., Jolivet J.-P. *J. Phys. Chem. Solids* 2007, 68 695–700.

[11] Sahni S., Reddy S.B., Murty B.S., *Mater. Sci. Eng. A*, 2007, 452–453, 758–762.

[12] Chae S.Y., Park M.K., Lee S.K., Kim T.Y., Kim S.K., Lee W.I. *Chem. Mater.* 2003, 15, 3326–3331.

[13] Kang M., Lee S.-Y., Chung C.-H., Cho S.M., Han G.Y., Kim B.-W., Yoon K.J. *J. Photochem. Photobiol. A: Chemistry*, 2001, 144, 185–191.

[14] Xia X.H., Luo Y.S., Wang Z., Liang Y., Fan J., Jia Z.J., Chen Z. H. *Mater. Lett.* 2007, 61 2571–2574.

[15] Manivasakan P., Rajendiran V., Rauta P.R., Sahu B.B., Sahu P., Panda B.K., *J. Am. Ceram. Soc.* 2010, 93, 2236–2243.

[16] Pradhan S.K., Reucroft P.J., Yang F., Dozier A., *J. Cryst. Growth* 2003, 256 83–88.

[17] Feng L., Xu D., Lei A. *J. Thermal Spray Tech.* 2008, 17, 473–477.

[18] Suprabha T., Roy H.G., Thomas J., Kumar, K.P., Mathew S., *Nanoscale Res. Lett.* 2009, 4, 144–152.

[19] Pradhan S.K., Reucroft P.J., Yang F., Dozier A., *J. Cryst. Growth* 2003, 256 83–88.

CORE-SHELL NANOPARTICLES: SYNTHESIS AND CHARACTERIZATION OF GOLD COATED THERMOSENSITIVE MICROGELS

LUHLUH JAHAN K., R.A. SABARISH AND SURESH BHAT

Polymer Science and Engineering Division, National Chemical Laboratory, Pune
E-mail: sk.bhat@ncl.res.in

ABSTRACT

Our work describes the synthesis and characterization of gold coated thermoresponsive microgels. These particles are of nanometer scale and are being developed to serve as unique drug-delivery vehicles since they have the ability to respond to ambient changes in temperature. In this work we synthesize citrate stabilized gold nanoparticles(~13 nm) and surface modify it with amine terminated linear pNIPAM. These surface modified nanoparticles were then used as seeds for the microgel synthesis with a desire to obtain pNIPAm coated Au nanoparticles. The TEM results seem to show that instead of pNIPAm forming the shell, they form as core around which gold nanoparticles form the shell.

INTRODUCTION

The synthesis and application of small polymeric nanoparticles for drug delivery and protein/ enzyme encapsulation is an expanding area of research.[1-4] In this application area, the nanoparticles should have high control over size on nanometer scale. The ease of conjugation of these nanoparticles to biological macromolecules and their assembly into higher order structures are also required.[5]With advances in the synthesis of polymeric nanoparticles, researchers have designed various interesting architectures, one being the core-shell structure which can provide a robust platform for incorporating multiple orthogonal functionalities into nanoparticles.[6]

V. Rajendran, P. Manivasakan and K.E. Geckeler (eds.)
Advanced Nano Materials for Industrial Applications, pp. 131-136 (2012).

INTRODUCTION

Polymer coated nanoparticles are very interesting nowadays because of their stimuli-sensitive phase transition behavior. Poly(N-isopropylacrylamide)(PNIPAm) is one of the most commonly studied thermosensitive polymer which undergoes a phase transition from a hydrophilic water swollen state to hydrophobic globular state when heated to above its lower critical solution temperature (LCST) which is about 31°C in water. It is also sensitive to some other stimuli such as pH, ionic strength and light[7]. Many research groups have studied the synthesis and characterization and applications of such stimuli-responsive materials.[8-10]

Many of the researchers synthesized hydrogel coated Au nanoparticles using metal templating approach. They used Au nanoparticles having larger diameter such as 60nm.[11] Some others used silica as the metal template. But they yielded polymer coated nanoparticles having radii in the range of 100-200nm[12]. This size range is typically regarded as not appropriate for many applications in targeted gene and drug delivery.

In this report, we describe the method of synthesis of gold nanoparticles coated with pNIPAM. This method have been previously used by other groups also to synthesize hydrogel nanoparticles.[13] Our method involves the initial formation of hydrophobic polymer nuclei, which act as seeds for the growing polymer chain. When the polymerization is maintained above the LCST of the polymer, the growing polymer chains get phase separated and collapse onto the hydrophobic nuclei to form a cross linked polymer shell around the core nanoparticles (AuNPs). When we followed this method we have seen that instead of pNIPAM forming the shell, they form as core around which the gold nanoparticles form the shell. We are expecting that, by controlling the molecular weight of the pNIPAM and making them shorter can lead to pNIPAM forming the shell around each gold nanoparticles.

EXPERIMENTAL

Materials

Gold tetrachloroauratetrihydrate($HAuCl_4$) and sodium citrate were purchased from SRL. N-isopropylacrylamide (NIPAM), ammonium persulfate (APS), sodium chloride (NaCl), N,N methylene acrylamide(BIS) and sodium dodecyl sulfate (SDS) were purchased from Sigma aldrich. The NIPAM was purified by recrystallization from 2:1 solvent mixture of n-hexane and toluene. All other chemicals were used as-received.

Au nanoparticle Synthesis

Poured 20mL of 1.0mM $HAuCl_4$ into a 50mL beaker. Added a magnetic stir bar and heated the solution to boiling on stir/hot plate while stirring with the magnetic bar. After the solution began to boil, we added 2mL of 38.8mM $Na_3C_6H_5O_7$. Continued to boil and stirred the solution until it is gets a deep red colour. As the solution boils, we added distilled water as needed to keep the total solution volume near 22mL. The sodium citrate reduces the Au metal. The excess citrate anions in the solution sticked to the Au metal surface giving an overall negative charge to each Au nanoparticles. When the solution attains a deep red colour, the hot plate was turned off and brought to the room temperature.

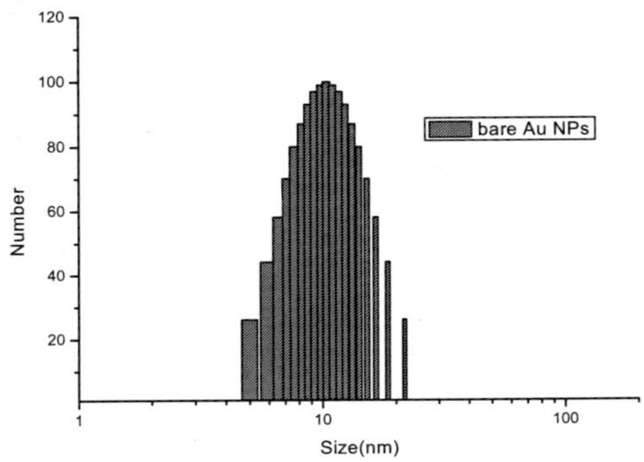

Fig. 1: Size Distribution of Bare Gold Nanoparticles

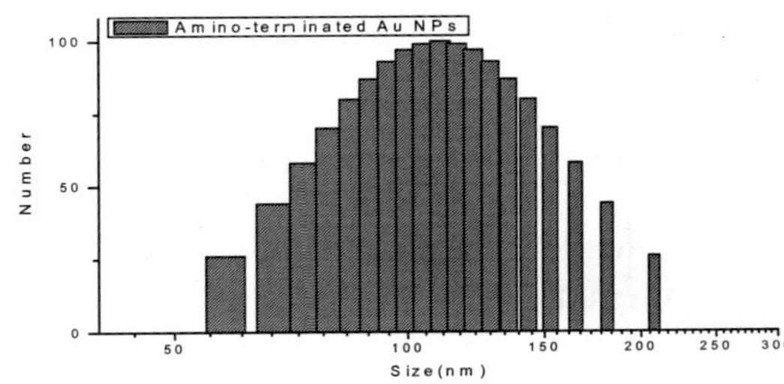

Fig. 2: Size Distribution of Amino-terminated Gold Nanoparticles

pNIPAM synthesis

The NIPAM is re-crystallized using the standard procedure by dissolving it in toluene and stirring at 60°C. The solution was kept at room temperature for 24 hours. The toluene was removed using rotary evaporator.

Thermoresponsive pNIPAM microgel particles were synthesized by emulsion polymerization. The surfactant used was SDS and KPS was used as the free radical initiator. The NIPAm monomer,cross-linker and surfactant were dissolved in 500 mL of distilled, deionized water, filtered through a 0.2 μ m nylon membrane filter to remove any large particulate matter and then continuously stirred in a three-neck, 250mL round bottom flask. This solution was heated to 70°C while being purged with nitrogen gas. Approximately 1 hour later, the temperature of the solution was stable at 70 °C. Fifteen minutes later, the reaction was initiated by adding a hot (70°C) solution of KPS. The solution turned turbid within 10min, indicating successful initiation. The reaction proceeded for 4 h under a constant stream of nitrogen. Following synthesis, the particle dispersions were dialyzed (using 10000 MWCO) for 2 weeks against distilled, deionized water with a daily exchange with fresh water). This polymer solution was then taken in a 1000ml round bottom flask and the flask was dipped in liquid N_2 to freeze the water. Finally this flask is attached to a freeze dryer for around 8 hours to remove the solvent completely and get the polymer in a powdered form.

Amino-terminated pNIPAM Synthesis

Amino terminated pNIPAM was prepared by radical polymerization in aqueous solution by using the redox couple APS/AET as the initiator[13]. The monomer was dissolved in water (0.1

mol in 80 ml of water). The temperature was adjusted at 29°C by using the water bath. The initiators (1.1 mmol of APS and 2.2 mmol of AET.HCl) were dissolved separately in 10 mL of water and added after 1hour to monomer solution. The reaction time was 3 hours. The product was purified by dialysis against water through a membrane (10000 MWCO) and freeze dried.

Amino-terminated pNIPAM required to cover the surface of gold nanoparticles is as follows: 500μL of 1M NaCl was added to 500μL of Au colloid solution after the Au colloid had been incubated overnight with 0.1 mM amino terminated pNIPAM.[5]

Au-pNIPAMcore/Shell Synthesis

The procedure for the synthesis of Au-pNIPAM core/shell is described by Andrew Lyon group[5]. The polymer adsorbed Au nanoparticles were heated to 70°C under nitrogen atmosphere. The aqueous solution of monomer, cross-linker (BIS) and 65μL of surfactant (SDS) were added to the reaction. After 1 hour the reaction was initiated by adding 130μL of 0.1M ammonium persulfate (APS) and the reaction was kept at 70°C for 4 hours. The monomer acrylic acid was added just before the reaction initiation to synthesize pNIPAM-co-AAC nanogels. A molar composition of 85% monomer, 5% cross-linker, and 10% co-monomer was used.

Dynamic Light Scattering

The particle size of the bare and the surface modified gold nanoparticles were determined using Brookhaven 90 plus particle size analyzer. The hydrodynamic radii of the particles were calculated from the diffusion coefficients using the Stokes-Einstein equation.

UV-Vis Spectrometer

The optical measurements of the samples are carried out in a Perkin Elmer precisely Lambda 950 UV/VIS Spectrometer.

Transmission Electron Microscopy(TEM)

The synthesized nanoparticles were characterized using FEI TECNAI G2 TF Transmission Electron Microscopy(TEM). All the TEM samples were prepared by casting a drop of nanoparticle solution (diluted 10 times) on a Cu TEM grid. The grid was then dried overnight at ambient temperature.

RESULTS AND DISCUSSION

Citrate stabilized Au nanoparticles synthesized by the reduction of $HAuCl_4$ were used for the formation of polymerization seeds. The surface modification was done by using amino-terminated pNIPAM. We then tried to coat the pNIPAM on the surface of the modified gold nanoparticles to make Au-pNIPAM core-shell.

Dynamic Light Scattering

The particle size of the synthesized nanoparticles was measured using Brookhaven 950 plus particle size analyzer. It shows that the gold nanoparticles have an effective diameter around 15nm and polydispersity 0.238. The particle size distribution of bare and surface modified gold nanoparticles are shown in the Figures 1 and 2 respectively. It shows that the particles are not that much monodisperse. The particle size of the amino-terminated pNIPAM is greater than that of bare gold nanoparticles show that the particles are successively surface modified. This result says that the amino-terminated pNIPAM can be used for the surface modification of the gold nanoparticles and it gives a positive charge to the surface of Au nanoparticles. The coating of the pNIPAM on this surface modified Au particle is now easy as pNIPAM have the negative charge.

UV-Vis Spectra

The UV-Vis spectra for the synthesized colloid showed an absorption band with a maximum around 518 nm due to the surface plasmon resonance of spherical Au nanoparticles. Seeds compatible with the precipitation polymerization of pNIPAm were formed by the adsorption of a layer of amino-terminated pNIPAm onto the Au nanoparticles. When heated to 70°C (above the LCST of pNIPAm), the adsorbed pNIPAm layer collapses onto the Au nanoparticle surface, thus serving as a hydrophobic nucleus for the subsequent precipitation polymerization step. Figure 3 shows the UV-Vis spectrum of bare and amino terminated pNIPAm adsorbed Au nanoparticles.

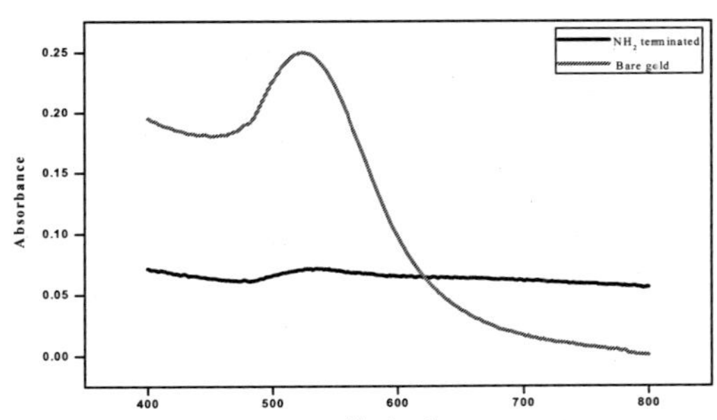

Fig. 3: UV-Vis Spectra of Bare and Amino-Terminated Gold Nanoparticles

Transmission Electron Microscopy

To characterize the architecture and shape of the core-shell, TEM imaging was performed. Figure 4 shows representative TEM images of Au nanoparticles before (Figure 4a) and after the adsorption of polymers (Figure 4b). As seen from Figure 4a, the Au Nanoparticles have the diameter less than 20nm. But the figure 4b shows a surprising result that, the pNIPAm is coated with the Au nanoparticles. This may happen due to the higher molecular weight of the pNIPAM hydrogel. A controlled polymerization of pNIPAM to reduce the molecular weight seems to be the key for forming the shell on the gold nanoparticles. The future work will also aim to synthesize core-shell nanoparticles with different size and shapes.

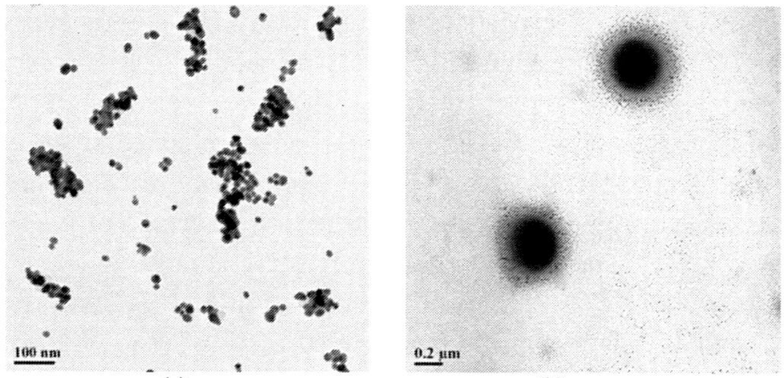

(a) (b)

Fig. 4: TEM Images of (a) Bare Gold and (b) Gold Coated pNIPAMmicrogel

CONCLUSION

In this work we synthesized colloidal gold nanoparticles encapsulated with amino terminated pNIPAM and characterized them using dynamic light scattering, UV-Vis spectroscopy and transmission electron microscopy. The TEM results seem to show that instead of pNIPAM forming the shell, they form as core around which gold nanoparticles form the shell. By controlling the molecular weight of the pNIPAM and making them shorter can lead to pNIPAM forming the shell around each gold nanoparticle.

REFERENCES

[1] Meier, W. *Chem. Soc. Re*V.2000, *29*, 295.
[2] Langer, R. *Science* 1990, *249*, 1527.
[3] Peppas, N.A.; Langer, R. *Science* 1994, *263*, 1715.
[4] Marinakos, S. M.; Anderson, M.F.; Ryan, J.A.; Martin, L.D.; Feldheim, D.L. *J. Phys. Chem. B* 2001, 105, 8872.
[5] Singh, N., Lyon, L.A., *Chem. Mater.* 2007, 19, 719-726.
[6] Nayak, S., Lee, H., Chmielewski, J., Lyon, L.A. *J. Am. Chem. Soc.* 2004, 126, 10258.
[7] Qui, Y. Park K., *Adv. Drug Delv. Rev.* 53 (2001) 321-339
[8] Nayak, S., Lyon, L.A., *Angew. Chem., Int. Ed. Engl.,* 2005, *44*, 7686.
[9] Pelton, R., *Ad*V. *Colloid. Interface Sci.* 2000, 85, 1.
[10] Wang, C., Flynn, N.T., Langer, R., *Ad*V. *Mater.* 2004, 16, 1074.
[11] Kim, J.-H., Lee, T.R., *Chem. Mater.* 2004, 16, 3647.
[12] Zha, L., Zhang, Y., Yang, W., Fu, S. *Ad*V. *Mater.* 2002, 14, 1090.
[13] Durand, A., Hourdet, *D. Polymer* 1999, 40, 4941.

SILAR METHOD POTENTIAL IN MORPHOLOGICAL TAILORING

R. CHANDRAMOHAN, M. KARUNAKARAN, J. THIRUMALAI[1], M. BASHEER AHMED[1], S. BALAMURALI[2] AND T. MAHALINGAM[3]

Department of Physics, Sree Sevugan-Annamalai College, Devakottai
[1]Department of Physics, B.S. Abdur Rahman University, Chennai
[2]Department of Physics, PARK college of Eng.& Technology, Coimbatore
[3] Department of Physics, Karunya University, Coimbatore
E-mail: rathinam.chandramohan@gmail.com; phy_karuna@yahoo.co.in

ABSTRACT

The growth of highly textured Zinc oxide (ZnO) thin films with a preferred (101) orientation has been prepared by employing Successive ionic layer growth by Adsorption Reaction (SILAR) using a sodium zincate bath on glass substrates has been reported. Morphology of the films were found to be varying well from stacked nanorods to shoots of nanotubes and clustered spherical and hilly morphology with alteration of deposition conditions like pH, thickness and subsequent treatment The study shows that morphological tailoring could be easily achieved by SILAR method.

INTRODUCTION

Zinc oxide (ZnO) is a versatile material of compound semiconductors with excellent properties and extensive applications in electronics, optoelectronics, sensors, and catalyses [1]. ZnO thin films have attracted considerable attention because they can be tailored to possess high electrical conductivity, high infrared reflectance and high visible transmittance by different coating technique [2-5]. Some of the remarkable properties of ZnO are its wide direct-band gap of 3.37 eV, the binding energy of the exciton of ZnO (60 meV) which makes it an excellent material for excitonic devices [6-8]. Other favorable aspects of ZnO include its broad chemistry

V. Rajendran, P. Manivasakan and K.E. Geckeler (eds.)
Advanced Nano Materials for Industrial Applications, pp. 137-142 (2012).

leading to many opportunities for wet chemical etching, low power threshold for optical pumping, radiation hardness and biocompatibility[9-12]. Together, these properties of ZnO make it an ideal candidate for a variety of devices ranging from sensors through to ultra-violet laser devices and nanotechnology based devices such as displays. As fervent research into ZnO continues, difficulties such as the fabrication of p-type ZnO that have so far stated that the development of devices had overcome[13-20]. Mitra et al [11] has prepared Zinc Oxide thin films using chemical deposition technique. The structural, morphological properties of the prepared films are characterized using X-ray diffraction and scanning electron microscope. They have used Zn salts as precursor and successfully synthesized ZnO films. The growth of highly textured Zinc oxide (ZnO) thin films with a preferred (101) orientation has been prepared by employing chemical bath deposition using a sodium zincate bath on glass substrates has been reported by [2]. In this study the innumerable ways by which morphology of films may be modified are presented using Mn doped ZnO thin films.

SYNTHESIS, STRUCTURE AND MORPHOLOGY OF THE ZnO THIN FILMS

ZnO thin films were grown using a two-step chemical bath deposition technique using a solution comprising 0.1 M Zinc Sulphate (99% emerk) , 0.2 M sodium hydroxide with a pH value of 9±0.2 deposited at bath temperature of 90 °C under optimized condition. For Fe doped ZnO (FZO) thin films $Fe(SO_4)_3$ was used at a concentration of 0.1mM. Before deposition, the glass substrates were cleaned by chromic acid followed by cleaning with acetone. The well-cleaned substrates were immersed in the chemical bath for a known standardized time followed by immersion in hot water for the same time for hydrogenation. The process of solution dip (step 1) followed by hot water dipping (step 2) is repeated for known number of times. The cleaned substrate was alternatively dipped for a predetermined period in sodium zincate bath and water bath kept at room temperature and near boiling point, respectively. According to the following equation, the complex layer deposited on the substrate during the dipping in sodium zincate bath will be decomposed to ZnO due to dipping in hot water. The proposed reaction mechanism is according to the following equations [2,4]

$$ZnSO_4 + 2\ NaOH \rightarrow Na_2ZnO_2 + H_2SO_4 \uparrow$$

$$Na_2ZnO_2 + H_2O \rightarrow ZnO + 2\ NaOH$$

Part of the ZnO so formed was deposited onto the substrate as a strongly adherent film and the remainder formed as a precipitate. The addition of MSO_4 in the ratio of Zn:M as 100:1 in the first dip solution leads to the formation of M doped zinc oxide nano thin films where M stands for metallic dopants like Fe, Ni, Mn, etc. (Here it is Mn). The structural, optical and transport properties of ZnO thin films grown are reported in our papers elsewhere[3-5]. Here the extent of morphological variations that can be achieved using this simple SILAR method are presented. As a representative system Mn doped ZnO thin films are taken and the interesting morphological variations obtained are presented Possible Morphological Modifications.

RESULTS AND DISCUSSION

Figures 1(a)–(d) show the morphology as obtained from the SEM for various thicknesses of SILAR grown thin films of ZnO undoped and Mn doped grown over glass substrates. Different magnifications were used to study the morphology of the Mn doped ZnO thin films with 40 number of dippings employing the concentration of Zn; Mn as 100: 10 or 10:1. Typical magnifications around 1–20K were employed for the samples. At low magnification, samples appear very homogeneous and compact. However, at high magnifications, the differences observed in terms of chemical composition are reflected also in the morphology of films. It is also possible to observe that these large grains are made up of many smaller crystallites within them. It is observed that the films are continuously formed and consists of some spherical and needle like particulates embedded inside the continuous region connected by a spiky nano fiber like shoots. The EDX obtained at various points like needles, blocky features. It is observed that Mn has entered the system to little atomic percentages. Also this is possible because the ionic radii of Mn is larger than the Zn and hence the space available.

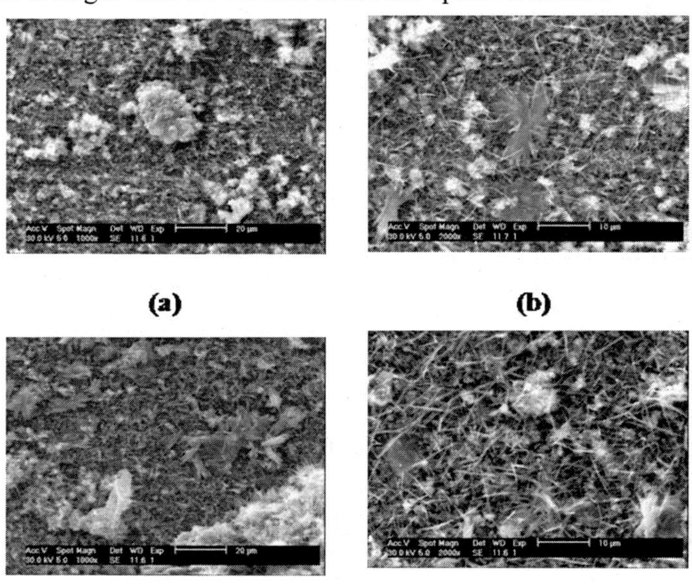

Fig. 1 (a)–(d): Shows the Morphology as Obtained from the SEM for Various Thicknesses

The Mn doped ZnO have some monoclinic shaped grains with uniform distribution of grains. The average surface grain size increases with increasing Mn content. The MZO have the flower grains with agglomeration. Some hexagonal shaped grains without voids are also clearly visible. The particle size is around 0.5 micrometer. The film had a blockish white grain agglomerated at some region and the SEM micrograph obtained at these regions is shown in

Fig. 2. (a) and (b). The EDX elemental mapping shows the uniform distribution of Mn and ZnO which confirms the formation of MZO film. In this system no carbon contamination was present. However, the blocky particles present in the film surface are also found to be made of Mn and Zn with different composition as is evinced by the variation in Mn intensity seen in EDX. At higher thicknesses, these spherical grains tend to agglomerate and due to thermal energy associated become a layer adhering to the adjacent layers adhered to the substrate which before being crystallized becomes sites for the molecules which once again try to agglomerate. The difference in energy used in the film formation process required adhering to the substrate and the cohesive bonding required for subsequent layer group might have supplied the thermal energy needed for the agglomerates. Such process lead to a morphology as revealed in Fig. 2(a) and (b) respectively. These islands of grain that can be used for various applications including solar cells. The thickness improvement via additional dippings changes the morphology considerably as shown in Figure 3 (a-d). This is understandable as the ionic and atomic radii of Mn is considerably larger and also addition of layers to the surfaces would alter the these state considerably leading to different morphologies. The various morphologies like needle, rod and flower structure. The dopant atoms occupancy is one of the reasons for various morphology formation. This can be explained by the following kinetics. The voids in the structure may invite more atoms to make them as sites. Different number of dippings in SILAR not only produces more layers but also adds more atoms to these vacant sites and inertial. Due to the nucleation in these sites and subsequent growth of the layer initially the one dimensional growth is favored. This leads to a needle like grains and their spreading to entire surface. The width of the needle throughout the film is nearly uniform in diameter (530 nm) and having lengths up to several hundreds of micro-meters. These needle like grains tend to agglomerate. When the thermal energy is not sufficient further agglomeration and the neighbouring layers cohesive force towards the substrate dominates during which the film being before crystallized leads to the needle like morphology. The continuous growth of the film in the material developed on the substrate on more number of dipping may alter this condition resulting in more clustering and agglomeration conditions. The rod like morphology is due to the favorable conditions of one dimensional extension of these needle morphologies to two dimensions along length and diameter.

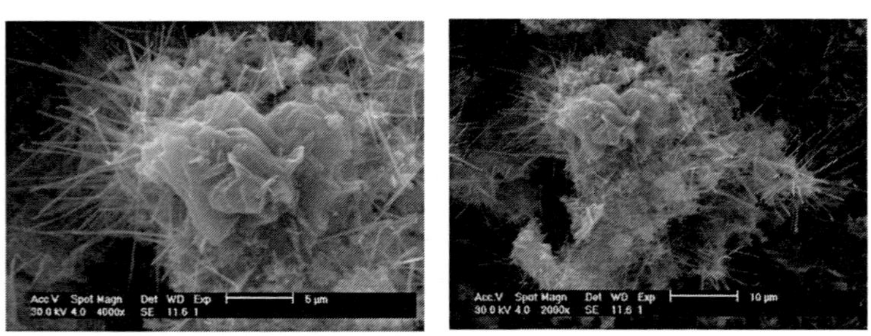

(a) (b)

Fig. 2 (a and b): The Morphology as Obtained from the SEM for Needle with Klower Structure

This is two dimension growths on the substrate. The continuous growth of these grains and agglomerations leads to a flower like morphology. Many rods may be joined together forming a typical spinel structure. With the increase of Mn content in the film, strain in the film also increases which tends to resist this growth. It is observed that proper choice of number of dipping, pH and concentration of the dopant the shape and size of the grains constituting the film surface may be tailored by SILAR method. Hence Morphology tailoring can be achieved quite easily using SILAR method.

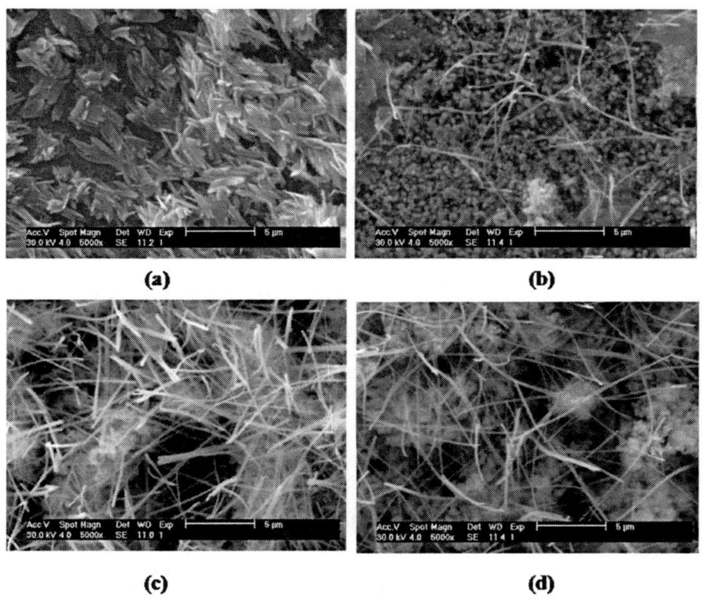

Fig 3 (a)–(d): SEM Micrographs of Mn Doped Films with Various Thickness

CONCLUSIONS

Transparent conducting Mn doped ZnO thin films have been prepared by SILAR method. The deposition conditions have been optimized based on their structure and on the formation of smoothness, adherence and stoichiometry. The results of studies by X-ray diffraction, Scanning Electron Microscopy (SEM), reveal the varieties of structural and morphological modifications feasible with SILAR method. The interesting morphological variations with bath conditions are observed and discussed. The films quality is comparable with those grown with physical methods and is suitable for spintronic and gas sensing applications. It is observed that proper choice of number of dipping, pH and concentration of the dopant the shape and size of the grains constituting the film surface may be tailored by SILAR method. Hence Morphology tailoring can be achieved quite easily using SILAR method.

REFERENCES

[1] Das, S.; Chaudhuri, S. (2007). Mg^{2+} substitutions in ZnO–Al_2O_3 thin films and its effect on the optical absorption spectra of the nanocomposite. *Appl. Surf. Sci.*, 253., 21., 8661-8668, ISSN: 0169-4332.

[2] Ramamoorthy, K.; Arivanandhan, M.; Sankaranarayanan, K. & Sanjeeviraja, C. (2004). *Mater. Chem. Phys.*, 85, 2-3 (June & 2004) 257-262, ISSN: 0254-0584.

[3] Vijayan, T. A.; Chandramohan, R.; Valanarasu, S.; Thirumalai, J.; Venkateswaran, S.; Mahalingam, T. & Srikumar, S.R. (2008). Optimization of growth conditions of ZnO nano thin films by chemical double dip technique. *Sci. Tech. Adv. Mater.*, 9 (April & 2008) 035007, ISSN: 1468-6996 .

[4] Vijayan, T. A.; Chandramohan, R.; Valanarasu, S.; Thirumalai, J.; Subramanian, S. P. (2008). Comparative investigation on nanocrystal structure, optical, and electrical properties of ZnO and Sr-doped ZnO thin films using chemical bath deposition method. *J. Mater. Sci.*, 43, 6 (March & 2008) 1776–1782, ISSN: 0022-2461.

[5] Chandramohan, R.; Thirumalai, J.; Vijayan, T. A.; ElhilVizhian, S.; Srikanth, S.; Valanarasu, S. & Swaminathan, V. (2010). Nanocrystalline Mg Doped ZnO Dilute Magnetic Semiconductor Prepared by Chemical Route. *Adv. Sci. Lett.* 3, 3 (September & 2010) 319-322, ISSN: 1936-6612.

[6] Saeed, T. & Brien, P. O'. (1995). Deposition and characterisation of ZnO thin films grown by chemical bath deposition. *Thin Solid Films*, 271., 1-2 (December & 1995) 35-38, ISSN: 0040-6090.

[7] Natsume, Y. & Sakata, H. (2000). Zinc oxide films prepared by sol-gel spin-coating. *Thin Solid Films.,* 372., 1-2., (September & 2000) 30-36, ISSN: 0040-6090.

[8] Ryu, Y. R.; Zhu, S.; Budai, J. D.; Chandrasekhar, H. R.; Miceli, P. F.; White, H.W. (2000). Optical and structural properties of ZnO films deposited on GaAs by pulsed laser deposition. *J. Appl. Phys.* 88., 1., 201-204, ISSN: 0021-4922.

[9] Wang, L.; Giles, N. C. (2003). Temperature dependence of the free-exciton transition energy in zinc oxide by photoluminescence excitation spectroscopy. *J. Appl. Phys.* 94., 2., 973-978., ISSN: 0021-4922.

[10] Yang, Y.; Tay, B.K.; Sun, X.W.; Han, Z.J.; Shen, Z.X.; Lincoln, C.; & Smith, T, (2008). Nanoelectronics Conference, INEC 2008 2 nd IEEE International, Nanyang Technical University, Singapore 24-27 March 2008.

[11] Mitra, P.; Chatterjee, A.P. & Maiti, H.S. (1998). Chemical deposition of ZnO films for gas sensors. *J. Mater. Sci: Mat in Electronics.,* 9, 6 (December & 1998) 441-445, ISSN: 0957-4522.

[12] Chapparro, M.; Maffiotte, C.; Gutierrez, M.T. & Herrero, J. (2003). Study of the spontaneous growth of ZnO thin films from aqueous solutions. *Thin Solid Films.*, 431, 1 (May & 2003) 373-377, ISSN: 0040-6090.

DEVELOPMENT OF NANOSTRUCTURED MULTILAYER COATINGS OF Zn-Ni-Al_2O_3 AND THEIR CHARACTERIZATION

YATHISH ULLAL AND A. CHITHARANJAN HEGDE

Electrochemistry Laboratory, Department of Chemistry, National Institute of Technology Karnataka, Srinivasnagar
E-mail: hegdeac@rediffmail.com

ABSTRACT

Nanocrystalline composite coatings of Zn-Ni-Al_2O_3 has been developed on mild steel from chloride bath containing gelatin and 3-aminophenol as additives and using direct current (DC). The deposition was made from conventional Zn-Ni bath containing Al_2O_3 nano-particulates to be co-deposited. The influence of additives on the nucleation process was investigated by cyclic voltammetry (CV) method. The effect of cathode current density, phase structure and nickel content of the coatings on corrosion behaviors of the coated film have been identified. The inert Al_2O_3 particles were found to be responsible for enhancing the Ni content and hardness of the coating. The corrosion performance of all electrodeposited coatings was studied by electrochemical AC and DC methods in 5% NaCl medium. The deposition conditions and bath compositions were optimized for peak performance of the coatings against corrosion. The corrosion stability of the coatings was further improved by cyclic multilayer coating technique. The superiority of the nanostructured multilayer coating developed using pulsed DC over the monolayer coating developed have been discussed. The corrosion tendency of the coatings was found in the order: monolayer alloy > monolayer composite > nano-structured multilayered > nano-structured multilayered composite coating. The electrodeposited layers and their layered degradation after corrosion study were confirmed by Scanning Electron Microscopy (SEM) analysis. Improved corrosion resistance multilayer coating was found to be more a function of increased number of interfaces due to layering, rather than composition of the individual layers.

V. Rajendran, P. Manivasakan and K.E. Geckeler (eds.)
Advanced Nano Materials for Industrial Applications, pp. 143-148 (2012).

INTRODUCTION

A new type of coating system well known as nanostructured multilayer coatings (NMC) is gaining prominence in materials research because these layered-structure coatings possess improved properties or novel phenomenon such as increased mechanical strength, micro hardness, giant magneto resistance and corrosion resistance. [1-4] The NMC's consist of alternate layered deposits of metals/alloys and each film is characterized by distinct properties. [5]

Zinc and its alloys are widely used as corrosion resistant coatings in various industries due to their decent cathodic and barrier protection. A noteworthy improvement in the corrosion characteristics of NMC is established relative to pure Zn coatings, this is because of the finer structure in each thin layer and the multiple layer effect. Electrochemical stripping methods have been proposed to characterize the alloy deposition process and product. Stripping studies are an accelerated corrosion test of the deposits [4]. Studies on NMC coatings of Zn with Ni, Fe and Co have been carried out by many researchers using both single bath and dual bath technique. [5-9] Many investigators have proposed use of composite coating technique to improve the corrosion protection ability and hardness of Zn and its alloys using metal oxides, ceramic particles and polymer particles of micro and nano meter range. [10-13] In the present study, we have attempted to merge the two approaches to produce novel NMC's containing composite particles.

EXPERIMENTAL

Electrodeposition was carried out from chloride bath with following composition: $ZnCl_2$ 25g/L, $NiCl_2.6H_2O$ 75g/L, H_3BO_3 20g/L, sodium gluconate 100g/L, m-aminophenol 2g/L and gelatin 0.2g/L. For composite coatings, 2g/L Al_2O_3 nanopowder was added to the electrolyte and mechanically stirred for 24h for uniform dispersion. The cyclic voltammetry (CV) was used to analyze influence of additives. Monolayer and multilayered deposits were obtained at different current densities (c.d's) on polished mild steel plates with exposed area of 7.5 cm^2 using high precision DC power analyzer AGILENT 6705A. The current patterns generated and used in the present work are shown in figure 1.

The coated samples with 1cm^2 exposed coating areas were subjected to electrochemical investigations in a three electrode system with platinum electrode as counter and saturated calomel electrode as reference electrode. A luggins capillary with agar in sat.KCl was used to minimize ohmic drop. All electrochemical studies were made using Potentiostat/Galvanostat (VersaSTAT3, Princeton Applied Research) in a three-electrode configuration cell using saturated calomel electrode as reference electrode. The 5% NaCl solution was used as corrosion medium. Potentiodynamic polarization study was carried out in a potential ramp of ± 250 mV around open circuit potential (OCP), at scan rate of 1mVs^{-1}. The formation of multilayers and corrosion mechanism were examined using scanning electron microscopy (SEM, Model JSM-6380 LA from JEOL, Japan).

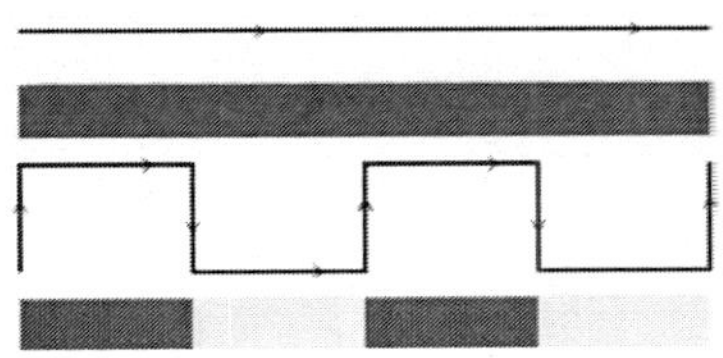

Fig. 1: Current Patterns used for Producing Monolayer and NMC

RESULTS AND DISCUSSIONS

Cyclic Voltammetric Studies

The CV was performed on platinum electrodes to understand the deposition process. The obtained voltammograms obtained with and without Al_2O_3 nanopowders is Shown in Figure 2. The composite deposits, i.e., with nanopowder are less susceptible to dissolution and hence render more corrosion protection.

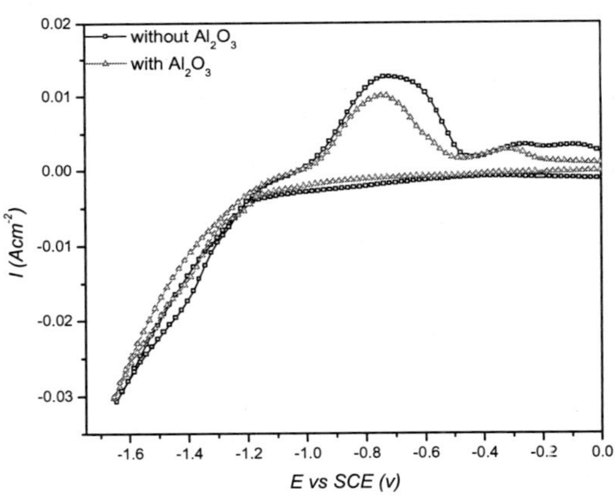

Fig. 2: Voltammogram of Electrolytes Used to Produce Alloy and Composite Coatings

Powder X-Ray Diffraction Studies

The Figure 3 shows the XRD patterns of Zn-Ni alloys obtained at two different current densities from the same bath. The necessary criteria to electrodeposit multilayers using single bath technique is that phase structure should vary with switching currents. In present study, the deposits obtained at 2.0 and 4.0 A/dm^2 current densities exhibit a marked variation in the major phase content. Accordingly the multilayers were developed switching the current densities between these two values. In case of composite coatings, the deposit showed no marked variation in XRD patterns.

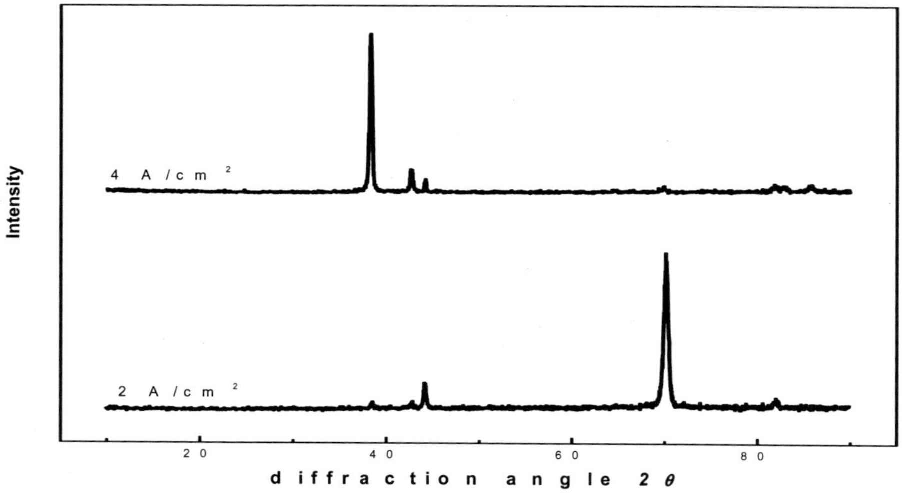

Fig. 3: Powder XRD Patterns of Coatings at Different C.D

Development of Monolithic Zn-Ni Alloy and Zn-Ni-Al_2O_3 Composite Coatings

The monolithic Zn-Ni alloy and Zn-Ni-Al_2O_3 composite coatings were galvanostatically developed at different current densities. The effect of c.d. on deposit thickness, corrosion rate and appearance is reported in Table 1. The c.d. has marked impact on Ni percentage and thus the corrosion resistance of the alloys. In case of composite coatings, inclusion of nano-Al_2O_3 particles improved the hardness of the deposits but no change in Ni percentage was seen. The current density was the key factor harnessed to tailor the properties and decide switching cathode current densities to develop multilayered coatings successfully. The monolayer alloy and composite coatings developed at 4.0 A/cm^2 was found to exhibit least corrosion rate.

Development of Nanostructured Multilayered Coatings of Alloy and Composites

The NMCs were deposited from single bath technique with square current pulse with 2.0 Adm^{-2} as base and 4.0 Adm^{-2} as peak current density. The multilayers with 10, 20, 60, 120, 300 and 600 layers were deposited. The total time was 600 seconds for a cycle and time was divided with required number of layers. In Figure 4, the potentiodynamic polarization curves of NMCs are shown. The i_{corr} shifts to lower values as the number of layers increases upto 300 layers and then increases. The decrease in corrosion rate is mainly due to the increased number of barriers in the path of corrosive medium. A similar trend is observed for composite multilayers with 120 layered coating showing least corrosion resistance. The fitment results are shown in Table 2.

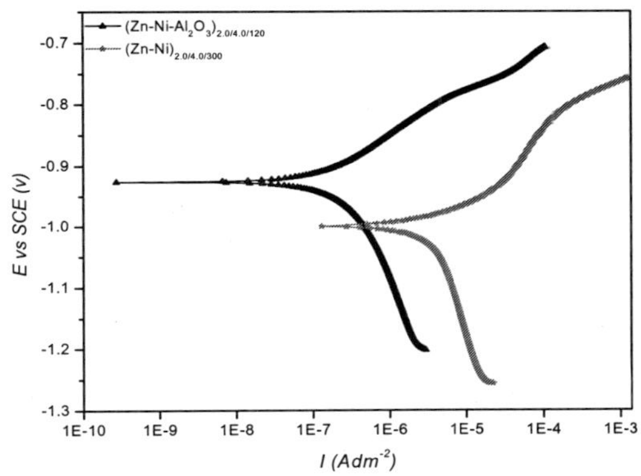

Fig. 4: Potentiodynamic Polarization Curves of Multilayers Having Highest Corrosion Resistance

Table 1: Corrosion Datas of the Optimum Coating Configuration

Coating Configuration	*Thickness*	*-E_{corr} vs SCE (v)*	*$I_{corr} \times 10^{-6}$ A/cm^2*	*CR × 10^{-2} mmpY*	*Appearance*
Zn-Ni Alloy					
$(Zn\text{-}Ni)_{4.0}$	20	1.03	18.09	26.4	Bright
300	24	0.91	0.26	0.40	Semi bright
Zn-Ni-Al_2O_3 Composite					
$(Zn\text{-}Ni\text{-}Al_2O_3)_{4.0}$	23	0.98	13.0	19.3	Semi bright
120	21	0.93	0.14	0.21	Dull

SEM Analysis

The scanning electron microscope images are shown in Figure 5 below. In Fig. 5 (a) and 5 (b), the variation in surface morphology without and with composite coatings is shown respectively. The surface is faded and composite particles are distributed all over the coating which was confirmed by EDAX. In fig 5 (c), the cross section of NMC with 10 layers is shown. The clear demarcation between layers is due to the change in phase composition. In fig 5(d), image of NMC with 10 layers after electrochemical corrosion tests is shown and we see only 4 layers are disrupted. The underlying layers are intact providing protection to the substrate. This confirms the preferential degradation of top layers providing improved corrosion resistance.

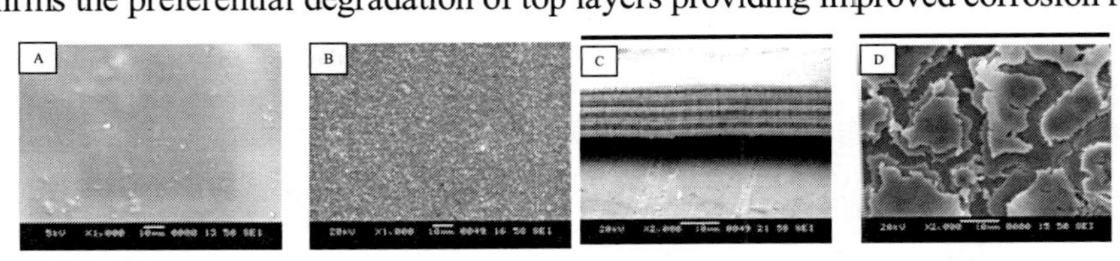

Fig. 5: SEM Images of Monolithic and Multilayered Alloy and Composite Coatings

CONCLUSIONS

- The monolithic alloy and composite coatings were obtained successfully on mild steel panes.
- The composite coatings were harder and less prone to dissolution as seen from CV and OCP values.
- NMC's were successfully deposited from single bath technique.
- Alloy NMC with 300 layers was found to be ~66 times more corrosion resistant than monolayer coating.
- Composite NMC with 120 layers was ~92 times more corrosion resistant compared to monolithic composite coating.

REFERENCES

[1] Milan Paunovic and Mordechay Schlesinger, (2006). "Fundamentals of Electrochemical Deposition, Second Edition". John Wiley & Sons, Inc., Hoboken, New Jersey.

[2] Gabe, D.R. (1994). "Protective layered electrodeposits." *Electrochim. Acta.*, 39, 1115-1121.

[3] Fratesi, R., and Roventi, G. (1996). "Corrosion resistance of Zn-Ni alloy coatings in industrial production." *Surf. Coat. Technol.*, 82, 158-164.

[4] Gavril, M., Millet, J.P., Mazille, H., Marchandise, D. and Cuntz, J.M. (2000). "Corrosion behavior of zinc-nickel coatings, electrodeposited on steel." *Surf. Coat. Technol.*, 123(2-3), 164-172.

[5] Conde, A., Arenas, M. A., & De Damborenea, J. J. (2011). "Electrodeposition of Zn–Ni coatings as Cd replacement for corrosion protection of high strength steel." *Corrosion Science,* 53(4), 1489-1497.

[6] Short, N.R., Abibsi, A. and Dennis, J.K. (1984). "Corrosion Resistance of Electroplated Zinc Alloy Coatings." *Trans. Inst. Met. Fin.*, 67(1), 73-77.

[7] Fei, J., & Wilcox, G. (2006). "Electrodeposition of zinc–nickel compositionally modulated multilayer coatings and their corrosion behaviours." *Surf. Coat. Technol*, 200, 3533-3539.

[8] Yogesha, S., and A. Chitharanjan Hegde. (2011). "Optimization of deposition conditions for development of high corrosion resistant Zn–Fe multilayer coatings." *J. Mater Proces Technol*, 211, 1409-1415.

[9] Low, C.T.J.; Willis, R.G.A. & Walsh, F.C. (2006). "Electrodeposition of composite coatings containing nanoparticles in a metal deposit." *Surf. Coat. Technol.*, 201, 371-383.

[10] Bahrololoom, M.E. & Sani, R. (2005). "The influence of pulse plating parameters on the hardness and wear resistance of nickel–alumina composite coatings." *Surf. Coat. Technol.*, 192, 154-163.

[11] Jung, A.; Natter, H.; Hempelmann, R. & Lach E. (2009). "Nanocrystalline alumina dispersed in nanocrystalline nickel: enhanced mechanical properties." *J. Mater. Sci.*, 44, 2725–2735.

[12] Krishnaveni, K. Sankara Narayanan, T.S.N. & Seshadr, S.K. (2008). "Electrodeposited Ni–B–Si_3N_4 composite coating: Preparation and evaluation of its characteristic properties." *J. Alloys Comp.*, 466, 412-420.

[13] Gay, P.A.; Bercot, P. & Pagetti, J. (2001). "Electrodeposition and characterization of Ag–ZrO_2 electroplated coatings." *Surf. Coat. Technol.*, 140, 147-154

EFFECT OF ACETALDEHYDE ON THE PROPERTIES OF Zn-Ni ALLOY ELECTRODEPOSITS FROM EDTA BATH

K. Juliet Gnana Sundari, C. Joseph Kennady[1] and S. Rajendran[1]

Department of Chemistry, Karunya University, Coimbatore, Tamil Nadu.
[1]RVS College of Engineering and Technology, Dindigul, Tamil Nadu
E-mail: juliemoses@yahoo.com

INTRODUCTION

Zinc is one of the most widely used metal and its electrochemical properties are considered in the production and application. Due to high corrosion resistance of zinc in atmospheric and other environments, zinc coatings are used for corrosion protection of steel structures. For the prevention of corrosion of steel fabrications exposed to water, zinc metal coatings are widely used [1]. For meeting the demands for products with high quality coatings, alloys were used between the last two decades [2]. As zinc alloy deposits exhibit better corrosion resistance and physical properties than pure zinc layer, zinc can be alloyed with metals like Ni, Co, Cu, Al, etc. [3-6]. Since Zn-Ni alloy finds better corrosion protection applications, it is electrodeposited on mild steel from alkaline bath containing EDTA and acetaldehyde. The effect of acetaldehyde on the properties of Zn-Ni alloy electrodeposits is studied by X-ray diffraction studies, scanning electron microscopy (SEM) analysis and electrochemical methods like Tafel polarization studies and electrochemical impedance spectroscopy.

EXPERIMENT

Galvanostatic Electrodeposition was carried out using mild steel plate as cathode and zinc as anode. Anode and cathode were cleaned with 5% H_2SO_4 and then washed in deionized water and dried. Electrolytic bath solution was prepared using ZnO, NaOH, $NiSO_4$ and disodium

V. Rajendran, P. Manivasakan and K.E. Geckeler (eds.)
Advanced Nano Materials for Industrial Applications, pp. 149-154 (2012).

salt of EDTA. The concentration of zinc and nickel in the bath solution was 5 g/l and 0.5 g/l respectively. The area of the mild steel plate exposed during electrodeposition was 6.25cm^2.

Preparation of the Electrodeposit

Zn-Ni alloy electrodeposit was prepared with the electrolytic bath solution and the electrodeposition was carried out using DC power supply unit [7] at 1 A/dm^2 and 30°C and the duration of electrodeposition was 30 min. Nature of the electrodeposit obtained was analysed. To get more bright and high corrosion resistant deposit, electrodeposition was carried out with 0.1 to 1.0g/l acetaldehyde additive.

Alloy Composition and Surface Structure

To know the percentage of zinc and nickel in the electrodeposits obtained from optimized bath solution with acetaldehyde and without acetaldehyde Varian Spectra 200 Atomic Absorption Spectrometer was used. Surface structure of the Zn-Ni alloy electrodeposit was characterized using Schimadsu -6000 series X-ray diffractometer. The results were compared with the standard values obtained from JCPDS data card.

Corrosion Studies

Corrosion behavior of the electrodeposits obtained with and without acetaldehyde additive was studied by the electrochemical techniques like Tafel polarization studies [8] and electrochemical impedance spectroscopy [9] using CH1660C electrochemical work station model. From the electrochemical techniques, corrosion rate and corrosion resistance of the electrodeposits were determined.

Tafel polarization studies were performed using a conventional three electrode cell in which the coated mild steel plate was the working electrode, platinum wire was the counter electrode and saturated calomel electrode was the counter electrode. The studies was carried out in 0.05M H_2SO_4 at the sweep rate of 0.01v/sec. Corrosion current density I_{corr} was determined from the anodic and cathodic slopes. Using the electrochemical impedance spectra obtained from electrochemical impedance measurements, Nyquist plots were drawn from which the charge transfer resistance was determined. Impedance measurements were taken in 0.05M H_2SO_4 in the frequency range from 1 Hz to 1, 00,000 Hz

RESULTS AND DISCUSSION

Zn-Ni alloy is electrodeposited on mild steel in alkaline solution with and without acetaldehyde additive. The results of the characterization of the electrodeposits are discussed below.

Nature of the Electrodeposit

Zn-Ni alloy electrodeposit prepared without acetaldehyde additive is uniform, dark grey and semi bright. Addition of acetaldehyde has a great influence on the surface brightness of the electrpodeposit.The electrodeposits prepared with acetaldehyde in the concentration range of 0.1 to 1.0g/l from the alkaline bath solution are bright, uniform and well adherent. The significant effect of the additive on the brightness of the surface is due to the easy adsorption of the groups in the additive on the cathode surface (10).

Alloy Composition and Surface Structure

Percentage of Zn and Ni in the alloy electrodeposit is analysed and it is found that the electrodeposit contains 89% Zn and 10.83% Ni .The electrodeposit prepared with acetaldehyde contains 89.5% Zn and 10.5% Ni. The high zinc content of the alloy is due to the inhibition of Ni^{2+} discharge by the Zn $(OH)_2$ on the surface which increases the surface pH with the evolution of hydrogen.

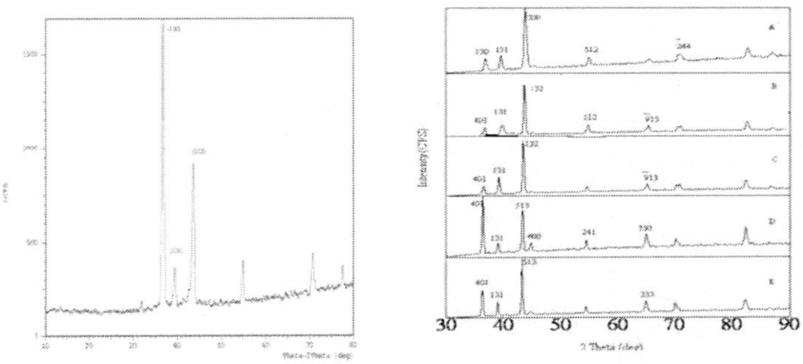

(a) (b)

Fig. 1: XRD Pattern of the Zn-Ni Alloy Electrodeposit (a) Without Acetaldehyde and (b) With Acetaldehyde

XRD patterns of the electrodeposit with and without acetaldehyde are given in Figure 1.Crystallite size calculated from XRD pattern obtained without acetaldehyde is 165.79 nm and with acetaldehyde it is found to be 140.66 nm. The results are compared with the standard values obtained from JCPDS and the structure is found to be monoclinic with acetaldehyde and face centered monoclinic without acetaldehyde.

Corrosion Studies

Tafel Polarization Studies

From the Tafel polarization studies in 0.05M H_2SO_4, the corrosion rate of Zn-Ni alloy electrodeposit prepared without acetaldehyde is found to be 0.3690 mm/yr. Corrosion rates of the electrodeposits obtained with 0.1g to 1g/l acetaldehyde are given in the Table 1.The tafel curves are shown in Figures 2 and 3. It is observed that corrosion rate decreases with an increase in the concentration of aldehyde.

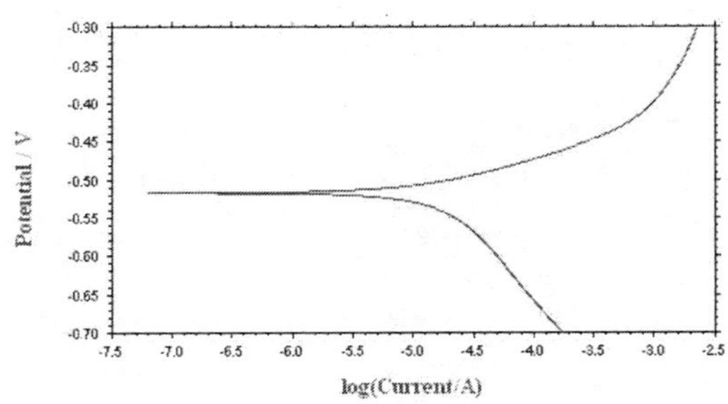

Fig. 2: Tafel Polarization Curve of Zn-Ni Alloy Electrodeposit in 0.05M H_2SO_4

The acetaldehyde added is getting adsorbed on the electrode surface resulting in the blocking of high energy sites for the crystallization of metal. The possibility of electrodeposited alloy undergoing corrosion is prevented due to the adsorption of acetaldehyde on the surface [11].

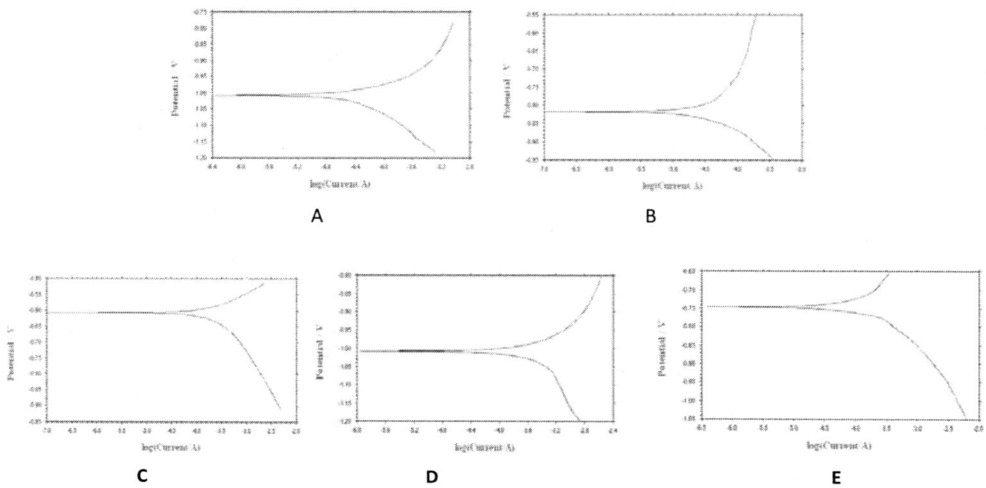

Fig. 3: A to E. Tafel Polarization Curves of the Electrodeposit with 0.1, 0.25 0.50, 0.75 and 1.0g/l Acetaldehyde in 0.05 M.H_2SO_4

Table 1: Corrosion Rate of Zn-Ni Alloy Electrodeposit with Acetaldehyde in 0.05M H_2SO_4

Concentration of Acetaldehyde g/l	*E_{corr} mV Vs SCE*	*Corrosion Current Density I_{corr} mA/cm^2*	*Tafel Slopes*		*Corrosion Rate mm/yr*
			Anodic mV/decade	*Cathodic mV/decade*	
0.10	-927	0.5349	245.94	227.32	0.4865
0.25	-799	0.4081	294.46	141.12	0.3728
0.50	-543	0.3784	114.57	179.15	0.3458
0.75	-1015	0.2275	160.54	314.37	0.2080
1.00	-548	0.1467	147.51	185.91	0.1343

Electrochemical Impedance Spectra

Zn-Ni alloy electrodeposited plates with and without acetaldehyde are subjected to electrochemical impedance studies and the charge transfer resistance [R_{ct}] of the electrodeposit in 0.05M H_2SO_4 without aldehydes is 0.8354 kΩ.The R_{ct} values of the electrodeposits with 0.1g to 1g/l acetaldehyde are given Table 2. The Nyquist plots obtained from impedance studies are shown in Figures 4 and 5. It is observed that R_{ct} values increase with increase in the concentration of aldehyde.The corrosion of the alloy is prevented by the incorporation of acetaldehyde molecules and the carbonyl group in acetaldehyde is acting as efficient corrosion inhibitor for mild steel [12].

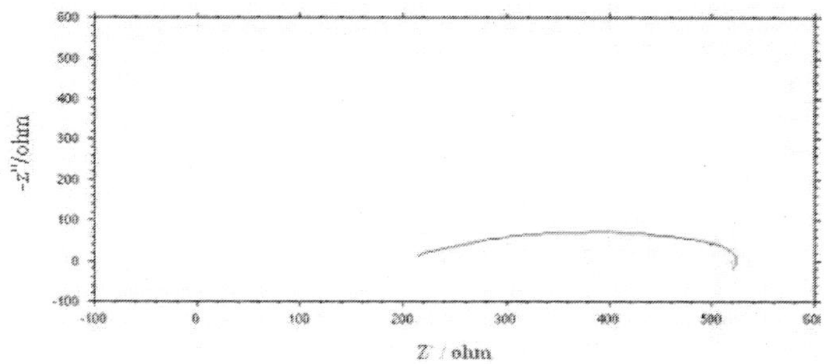

Fig. 4: Impedance Spectra of the Zn-Ni alloy Electrodeposit in 0.01M and 0.05M H_2SO_4

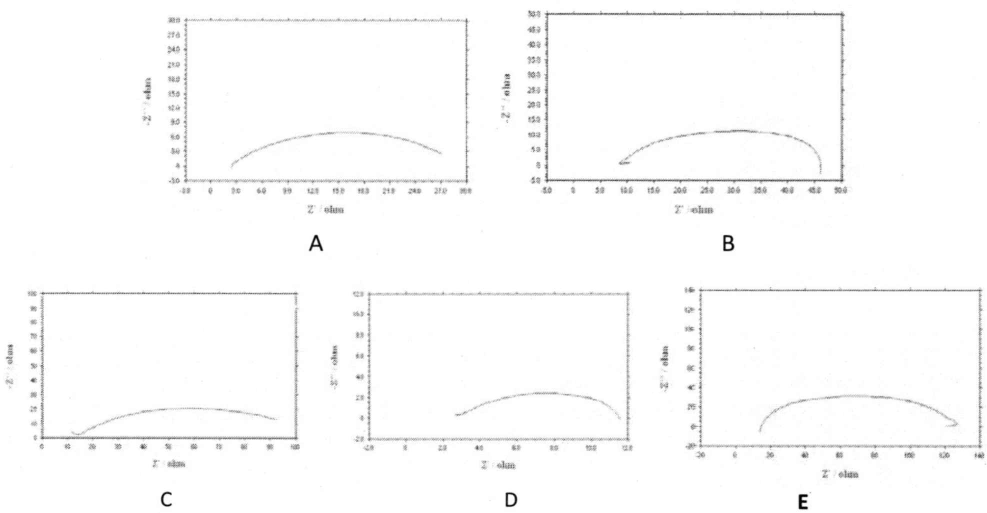

Fig. 5: A to E Impedance Spectra of the Electrodeposit with 0.1 0.25, 0.50, 0.75 and 1.0g/l Acetaldehyde in 0.01M H_2SO_4

Table 2: Effect of Acetaldehyde on Charge Transfer Resistance of Zn-Ni Alloy Electrodeposit in 0.05M H_2SO_4

Concentration of Acetaldehyde g/l	*Charge Transfer Resistance R_{ct} kΩ (In 0.05M H_2SO_4)*
0.10	0.3040
0.25	0.6141
0.50	0.6520
0.75	0.7200
1.00	0.7560

CONCLUSION

Electrodeposition of Zn-Ni alloy on mild steel was carried out from bath solution containing acetaldehyde as additive. The surface of the electrodeposit without aldehyde is uniform and semi bright and with acetaldehyde it is uniform and brighter. Percentage composition of Zn and Ni in the electrodeposit is not significantly changed by the addition of aldehyde. From XRD studies it is shown that the structure is face centered monoclinic in the absence of aldehyde and monoclinic in the presence of aldehyde.SEM photographs show the presence of crystallites of small and large size covering the surface. Corrosion studies of the electrodeposit reveal that the corrosion resistance of Zn-Ni alloy with acetaldehyde is greater than the corrosion resistance of electrodeposit without acetaldehyde.

REFERENCES

[1] Bowden, C.Matthews.A., Study of the corrosion properties of PVD Zn-Ni Coatings, *Surf.Coat.Technol.*, 76/77,508-515 ,1995.

[2} Hsin-Yi Hsu and Chao-Cher Yang.Z., Electrodeposition and Corrosion Resistance of Zn-Pd Alloy from the $ZnCl_2$-EMIC(1-ethyl-3-methylimidazolium chloride)-$PdCl_2$ Room Temperature Molten Salt, *Naturfor.* 58b1055-1062, 2003.

[3] Beltowska-Lehman E P.Ozga, Z.Swaiatek, C.Lupi, Electrodeposition of Zn-Ni protective coatings from sulfate–acetate baths, *Surf. Coat. Technol.*, 444,151-152,2002.

[4] Koura. N.Endo T. Edemoto Y., *J. Non.Cryst solids*, 650, 1996.

[5] Shih H.C.Hsu. J.W.Sun C.N. Chung, S.C.,Morphology and corrosion resistance of Cr(III)-based conversion treatments for electrogalvanized steel, *Surf.Coat.Technol.*, 150. 70,2002.

[6} Guzman, L. Adami, M.gissler. W.Klose S. DeRossi. S. *Surf.Coat.Technol.* 125, 218, 2002.

[7] Lin Y.P, Selman J.R., *Electrodeposition of NiZn Alloy, J. Electrochem. Soc.*, 140,1304-1311, 1993.

[8] Mansfeld F et al.,Polarization resistance techniques for measuring corrosion currents ,in M.G.Fontanaand R.W.Staehle (1976).Second Edition.Advances in corrosion science and technology, Plenum press, New York

[9] Geary A.L,The Mechanism of Passivating type Inhibitors, J. Electrochem. Soc., 105, 638, 1958.

[10] G.Y. Li,J. S. Lian, L.Y. Niu and Z.H. Jian., Investigation of nanocrystalline zinc-nickel alloy coatings in an alkaline zincate bath, *Surface and coatings Technology.*, 191, 59-67, 2005.

[11] Visalakshi Ravindran and Muralidharan V.S.,Characterisation of Zinc–Nickel alloy electrodeposits from sulphamate bath containing substituted aldehydes, *Bull. Mater. Sci.*, 29(3), 293-301, 2006.

[12] Silvans Sundersingh., Evaluation of antibiotics as corrosion inhibitors, Ph.D. Thesis, Bharathiar University, Coimbatore, India, 2009.

SYNTHESIS AND STRUCTURAL CHARACTERIZATION OF GADOLINIUM DOPED CERIA ELECTROLYTE MATERIAL FOR SOLID OXIDE FUEL CELL APPLICATION

KOTESWARARAO P., M.B. SURESH[1], B.N. WANI[2] AND P.V. BHASKARA RAO[3]

Centre for Nano Science and Technology, Jawaharlal Nehru Technological University, Hyderabad
[1]Center for Ceramic Processing, International Advanced Research Centre for Powder Metallurgy and New Materials (ARCI), Hyderabad
[2]Chemistry Division, Bhabha Atomic Research Centre, Trombay, Mumbai
[3]Physics Department, St. Mary's Engineering College, Deshmukhi, R.R. (Dt.)
E-mail: koteswarphysics@gmail.com

ABSTRACT

Gadolinium doped Ceria solid solution was prepared by sol gel processing method and its structural characterization was carried out by X-ray diffraction (XRD), Particle size measurement and Scanning electron microscopy. XRD analysis reveals the cubic fluorite structure. Crystallite size and lattice parameter were evaluated from XRD pattern. Particles are spread over from 70nm to 600nm with an average particle size of 220 nm. Density of greater than 95% is obtained at a sintering temperature of 1350°C for 2h. Microstructure of the sintered pellet shows an average grain size of 1-2micron. AC impedance analysis is used to evaluate the ionic conductivity of the material in the temperature range of 500-800°C and frequency range of 10MHz to 0.1Hz.

INTRODUCTION

Ceria based solid solutions have drawn attention from a decade as a candidate material for solid oxide fuel cell electrolyte due to their high ionic conductivity at intermediate temperatures

V. Rajendran, P. Manivasakan and K.E. Geckeler (eds.)
Advanced Nano Materials for Industrial Applications, pp. 155-158 (2012).

[1-3]. Doping trivalent rare-earth oxides like Samarium or Gadolinium oxide or divalent oxides like magnesium oxide into ceria creates oxygen vacancies due to charge compensation. In principle, high temperature sintering above 1400°C is required for gadolinium doped ceria to achieve a density more than 90%. GDC ceramics have good electrochemical properties even at 800°C. However, it's sintering behavior and mechanical properties have not been fully investigated. Sammes et al [4] reported 20 mol% Gd-doped ceria ceramics by sintering at 1600°C for 16hr and obtained 96% of relative density.

In this paper, GDC20 ceramics were prepared by sintering $Ce_{0.8}Gd_{0.2}O_{1.9}$ powder compacts at 1350°C, 1400°C, 1550°C and 1600°C for 2hr. The true density and apparent density were measured to calculate the relative density. In addition, structural analysis was done on the sintered samples to correlate the density with grain size.

EXPERIMENTAL

Commercially available Gadolinium doped Ceria powder (20GDC, Cottor International, India) with a particle size of 200 nm was used for the processing of the specimens. The powder was mixed with 2wt% of methyl cellulose as a binder and deionized water as solvent. The slurry is allowed for ball billing using zirconia balls as grinding media for 12 hrs. Thus obtained slurry is dried in an oven and the resultant powder is granulated. The granulated powder is used to compact into cylindrical green samples of 10mm diameter. The green samples were sintered at different temperatures at constant soaking time for optimum densification.

A phase of the sintered sample was analyzed by XRD (Bruker's D8 advance XRD). Micro structural analysis of polished and thermally etched specimens was carried out using SEM. Average Grain sizes of the specimens were measured using linear intercept method.

RESULTS AND DISCUSSION

Figure 1 shows the DLS particle size distribution curve and X-ray diffraction pattern of as received 20GDC powder. DLS measurement revealed a mean particle of size of 20GDC powder of around 200nm. All the peaks in the XRD pattern correspond to the single phase polycrystalline cubic CeO_2 phase. No phases other than cubic CeO_2 were found in the XRD diffraction pattern of the sample.

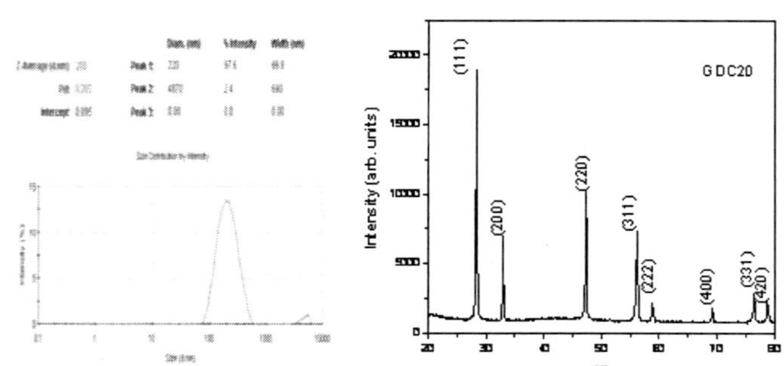

Fig. 1: (a) Particle size Analysis of 20 GDC Powder, and (b) XRD of 20 GDC Powder

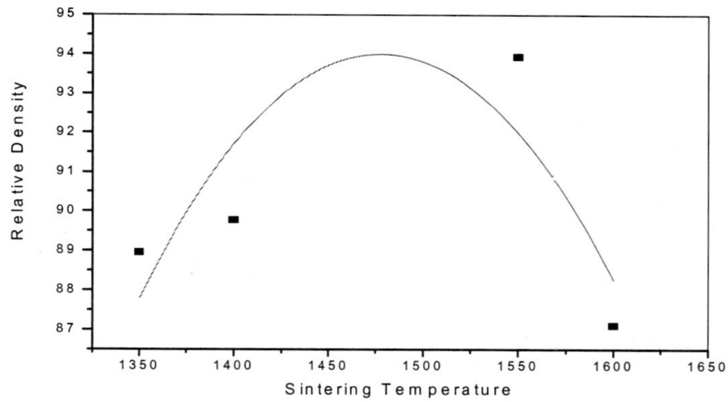

Fig. 2: Variation of Relative Density with Sintering Temperature of 20 GDC Pellets

Figure 2 shows the variation of relative density against the sintering temperature. The data at each temperature is an average of measurements on three different samples. Densification of GDC samples increases with increasing sintering temperature. However, it's started decreasing at higher temperatures after 1550°C due to pore formation at high temperatures. At 1350°C, the density is 92% and it reaches to its maximum value at 1550°C, equal to 96%. This clearly shows that more the sintering temperature more is the densification but at higher temperature such as at 1600°C the density observed is low due to the formation of non uniform grains which happened because of more than one type of grain formation example bimodal formation. Generally, a density greater than 95% of the theoretical is considered to be acceptable for electrolyte functionality.

Fig. 3: SEM Images of 20GDC Samples Sintered at Different Temperatures

Figure 3 shows the microstructures of air sintered 20GDC samples at different temperatures. The SEM image reveals dense microstructure at all temperature with an increasing trend in the densification up to 1550°C. The average grain size of the sample sintered at 1350°C is around 1.5 micron and growth is marginal up to 1400°C. On increasing the temperature to 1550°C, a significant increase in the grain size to 10 micron is observed.

Figure 4 shows the variation in grain size as sintering temperature. From graph we can interpret that in the intermediate temperature range as sintering temperature is increased grain size also increases, SEM images (shown later in this section) proves this fact. It reaches it maximum value at 1550°C which equal to 10.42 μm. According to the graph as temperature is a further increased grain size decrease which tells about the less uniformity in grain distribution and bimodal grain formation. Further study will be carried out to characterize the developed electrolyte for solid oxide fuel cell application in the future

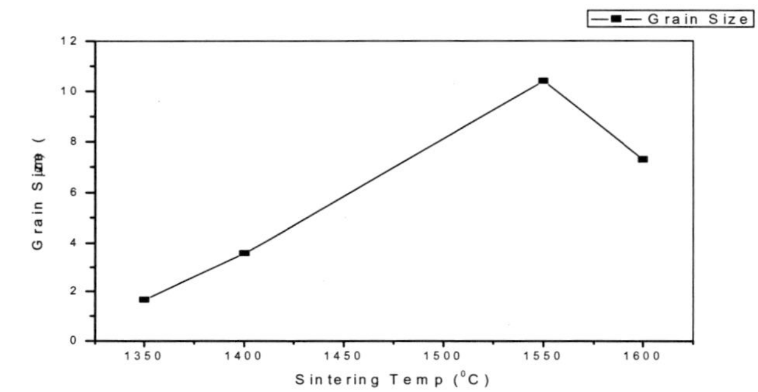

Fig. 4: Variation of Grain Size with Sintering Temperature

CONCLUSIONS

The effect of sintering temperature on the structural and morphological properties of 20GDC was investigated. The samples sintered at 1550oC, showed highest density. Surface morphology of the samples showed decreased porosity with a grain size in the range of 1-10 microns. Further study will be carried out to characterize the developed electrolyte for solid oxide fuel cell application in the future.

REFERENCES

[1] H. Inaba and H. Tagawa, *Solid State Ionics*, Vol. 83, no. 1-2, pp. 1-16, 1996.

[2] D.L. Maricle, T.E. Swarr and Karovalis, *Solid State Ionics*, Vol. 52, no. 1-3, pp. 173-182, 1992.

[3] H.Yahiro, T.Ohuchi, K.Eguchi, and H.Arai, *Journal of Materials Science*, Vol. 23, no. 3, pp. 1036-1041, 1998.

[4] Leng Y.J., Chan S.H., jiang S.P. and Khor K.A. 2004. *Solid State Ionics,* 170 9.

[5] Laukaitis G., Dudonis J. and Milcius D. 2007, *Materials Science*, 13, 23.

MAGNETIC PROPERTIES OF Ni-Mn-Ga FOR VIBRATION DAMPING APPLICATIONS

C. MAHALAKSHMI[1,2], S.VINODHKUMAR[1], S.SEENITHURAI[1], S. REKHA[1,3], A. SAJITHA BANU[1,4] AND M. MAHENDRAN[1]

[1]*Smart Materials Lab, Dept. of Physics, Thiagarajar College of Engg., Madurai*
[2]*Department of Physics, V.V. Vanniaperumal College for Women, Virudhunagar*
[3]*Department of Physics, SACS MAVMM Engineering College, Madurai*
[4]*Department of Physics, Sri G.V.G. Visalakshi College for Women, Udumalpet*
E-mail: cmaha27@gmail.com

ABSTRACT

Ferromagnetic Shape Memory Alloy (FSMA) is a class of smart material which can be magnetically controlled to produce motion and force in sensors and actuators. The large magnetic field induced strain exhibited in Ni-Mn-Ga is due to the rearrangement of martensitic twins. The $Ni_{52.4}Mn_{23.4}Ga_{24.2}$ sample was prepared using arc melt technique. The alloy was annealed at different temperatures. The structural properties of the sample were studied using XRD and the magnetic properties were carried out with Vibrational Sample Magnetometer (VSM), which shows the ferromagnetic nature of the sample. The alloy has been tested to check the feasibility to use it for vibration damping applications.

INTRODUCTION

The physical properties of ferromagnetic shape memory alloys (FSMAs) for potential applications have been studied by many researchers [1-4]. The FSMAs show a large magnetic field-induced strain (MFIS) by the rearrangement of twin variants in the martensitic phase [5-10]. Unlike the case of conventional shape memory alloys, these Ni-Mn-Ga FSMAs are potential candidate materials for magnetic actuators because, in this mechanism the speed of

V. Rajendran, P. Manivasakan and K.E. Geckeler (eds.)
Advanced Nano Materials for Industrial Applications, pp. 159-162 (2012).

shape memory is not limited. The Heusler Ni-Mn-Ga alloys are characterized by a wide variety of T_M from 200 K to 330 K by changing the compositions [13, 14]. As of now, data reported on Ni-Mn-Ga are mostly related to single crystals and thin films. It is observed that that in single crystals, the maximum strain is observed when the magnetic field orients along the [001] crystal axis [14]. But in polycrystalline sample, the orientation direction will vary from grain to grain. As a result small strain can be observed in polycrystalline Ni-Mn-Ga. In the present paper, we reported the structural and magnetic properties of $Ni_{52.4}Mn_{23.4}Ga_{24.2}$ polycrystalline alloy to determine the feasibility of using it in vibration damping applications.

EXPERIMENTAL PROCEDURE

The $Ni_{52.4}Mn_{23.4}Ga_{24.2}$ alloys were prepared by repeated arc melting of pure Ni (99.99%), pure Mn (99%) and pure Ga (98%) in an argon atmosphere. To obtain the homogenized samples, the reaction products were sealed in the double evacuated silica tubes annealed at 950°C for 15 hours and then quenched into water. Then the samples were exposed to X-ray powder diffraction (Rigaku RINT 2500) measurement with Cu Kα (λ = 1.5416 Å) radiation. The XRD profile was determined from 20° to 70 ° in 2θ using Cu-Kα radiation. A scanning rate of 4° min^{-1} was used. The magnetic properties of $Ni_{52.4}Mn_{23.4}Ga_{24.2}$ FSMA were investigated using a vibrating sample magnetometer (VSM-5, TOEI industries). The magnetization (emu/g) was studied at room temperature by varying the magnetic field from +10 to –20 kOe.

RESULTS AND DISCUSSION

Many of the physical properties of Ni-Mn-Ga alloys are mainly depend on their crystal structure of martensite [14]. X-ray diffraction pattern is recorded from the powder specimen of the alloy. The diffracted lines observed are identified as martensite superlattice reflections and indexed on the tetragonal phases (not shown). The BCC structure is stable at high temperature above 800 K can be retained at a lower temperature by fast cooling. The reflection peaks are appeared with slightly broadening features indicating disordered crystal structure [14]. The width of the XRD peaks is attributed to the size of the powder particles. From the present studies, it is noted that the (220) peak is the highest peak and inferred that the intensities are function of ageing temperature.

The magnetization curves of the $Ni_{52.4}Mn_{23.4}Ga_{24.2}$ FSMAs are plotted in Fig. 1. The saturation magnetization, another important parameter increases when particle size is decreased. When the magnetic field is increased, the magnetization increases slowly and reaches saturation nearly at 8 kOe. A large coercive force is obtained when the magnetization is decreased. The retentivity of 0.025 emu has been observed. It is studied that the magnetization depends on the state of the alloy. The composition of the $Ni_{52.4}Mn_{23.4}Ga_{24.2}$ alloy changes are expected to affect tremendously the ferromagnetic and structural transformation temperatures. At one point of time the magnetization is increased rapidly to exhibit the saturation which is slowly reached. A new hysteresis loop is obtained suggesting that a new magnetic easy axis in the direction of the magnetic field has been developed.

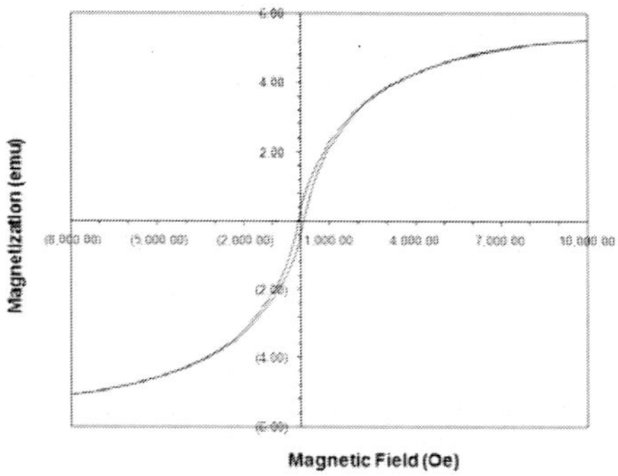

Fig. 1: Orientation of Twin Boundries Under the Infleunce of Magnetic Field

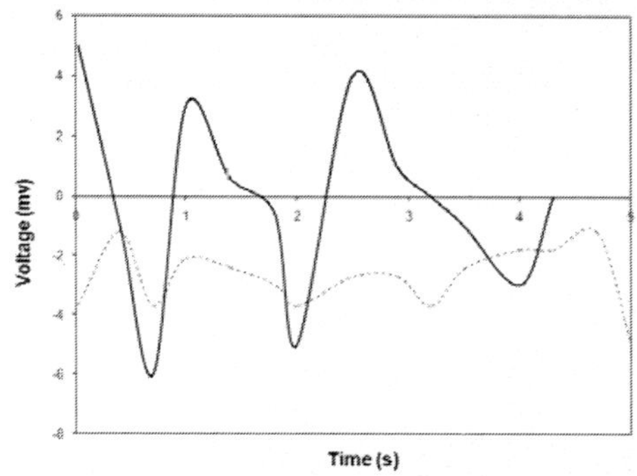

Fig. 2: Decrease of Amplitude Stress in $Ni_{52.4}$ $Mn_{23.4}$ $Ga_{24.2}$ as a Funcition of Time

A new experimental set up (not shown) has been designed which consists of measuring the acoustic intensity at the piezo sensor and noting the attenuation, for different drive frequencies (Fig. 2). It is assumed that the back of the piezo stack actuator to be stationary. The first interface is that between stack and sample under test (Ni-Mn-Ga). From this interface the wave is reflected back to piezo stack and transmitted into sample under test. The wave transmitted into the sample under test gets reflected and transmitted at the second interface. Again these two waves must be treated at the next interface they see, the reflected wave sees the sample/stack interface where it splits, and the transmitted wave sees the free end of the large mass, where it is totally reflected. At an instant, if the wave goes from the interface (stack and sample) another wave is expected to come from another interface (sample and PZT). The transit time of the acoustic wave depends on the nature of the materials. The reflected wave and transmitted wave produce more vibration of the particle where the crests or troughs of these waves collided. Therefore, even at higher frequencies, particle vibration is controlled, yet waves are passed through with low intensity. Now, another resonance is occurred due to the presence of residual waves. Finally, the oscillations are damped due to the occurrence of more stress amplitude.

CONCLUSION

We have fabricated $Ni_{52.4}$ $Mn_{23.4}$ $Ga_{24.2}$ FSMA and studied the structural and magnetic properties of the alloy. The saturation magnetization measurements reveal the coercive force and the remanence of the FSMAs. It is concluded that the annealing an off-stoichiometric compound will increase the thickness of the martensitic phase. An increase of the martensitic transformation temperature and the appearance of the martensitic structure at room temperature are observed in $Ni_{52.4}$ $Mn_{23.4}$ $Ga_{24.2}$. Our new experimental set up suggests that this active material could be used for vibration damping applications.

ACKNOWLEDGEMENTS

One of the authors (MM) would like to thank Prof.Dr.Robert C.O'Handley for introducing him to the subject.

REFERENCES

[1] Vallal Peruman, K., and Mahendran, M., Ball Milling Effect on Structural and Magnetic Properties of Ni–Mn–Ga Ferromagnetic Nanoparticles, *Pure Appl.Chem.*, 83, 2071, 2011.

[2] Pushpanathan, K., Vallalperuman, K., Seenithurai, S., Kodipandyan, R., and Mahendran, M., Effect of Mn Substance on Matrensitic Tansformation Temperature in Ni-Mn-Ga Shape Memory Alloys, *Mod. Phy. Lett.*, 25, 1577, 2011.

[3] Rama Rao, N.V., Gopalan, R., Manivel Raja, M., Arout Chelvane, J., Majumdar, B., and Chandrasekaran, V., Magneto-Structural Transformation Studies in Melt-Spun Ni–Mn–Ga ribbons, *Scrip.Mater.*, 56, 405,2007.

[4] Ullakko, K., Huang, J.K., Kantner, C., O'Handley, R.C., and Kokorin, V.V., Large MagneticFieldInduced Strains in Ni_2MnGa Single Crystals, *Appl. Phys. Lett.*, 69, 1966, 1996.

[5] Jame, R.D., and Wuttig, M., Temperature Variation of Structure and Magnetic Properties of NiMnGa Alloy, *Phil.Mag.* A., 77, 1273, 1998.

[6] Wu, G.H., Yu, C.H., Mng, L.Q., Chen, J.L., Yang, F.M., Qi, S.R., Zhan, W.S., Wang, Z., Zeng, Y.F., Zao, L.C., Thermal Stability of NiMnGa Heusler Alloy with High Temperature Transformation, *Appl. Phy. Lett.*, 75, 2990,1999.

[7] Tickle, R., James, R.D., Shield, T., Wuttig, M., Kokorin, V.V., Ferromagnetic Shape Memory in Ni MnGa System, *IEEE Trans. Magn.*, 35, 4301, 1999.

[8] Murray, S.J., Marioni, M., Allen, S.M., 0'Handley, R.C., Lograsso, TA., Magnetic-Field-Induced Strain by Twin-Boundary Motion in Ferromagnetic Ni–Mn–Ga, *Appl. Phys. Lett., 77,* 886, 2000.

[9] Ullakko, K., Ezer, Y., Sozinov, A., Kimmel, G., Yakovenko, P., Lindroos, V.K., Magnetic-Field-Induced Strains in Polycrystalline Ni-Mn-Ga at Room Temperature, *Scripta Mater.*, 44, 475, 2001.

[10] Webster, P.J., Ziebeck, K.R.A., Town, S.L., Peak, M.S., Magnetic Order and Phase Transformations in NiMnGa , *Phil. Mag.*, B 49, 295,1984.

[11] Chemenko, V.A., Compositional Instability of β-Phase in Ni-Mn-Ga Alloys., *Script. Mater.*, 40, 523, 1999.

[12] Jin, X., Marioni, M., Bono, D., Allen, S.M., O'Handley, R.C., Hsu, T.Y., Empirical Mapping of Ni–Mn–Ga Properties with Composition and Valence Electron Concentration, *J. Appl. Phys.*, 91, 8222, 2002.

[13] Ullakko, K., Huang, J.K., Kokorin, V.V., and O'Handley, R.C., Magnetically Controlled Shape Memory Effect in Ni_2MnGa Intermatallics, *Script. Mater.*, 36, 1133, 1997.

[14] Manickam Mahendran, Microstructural Analysis and Phase Transformation in Ni–Mn–Ga Ferromagnetic Shape Memory Alloys, *Smart Materials and Structures.*, 14, 1403-1409, 2005.

MAGNETO-MECHANICAL PROPERTIES OF DISORDERED Ni-Mn-Ga NANO MATERIALS

A. Sajitha Banu[1,2], S. Vinodhkumar[1], S. Seenithurai[1], S. Rekha[1,3], C. Mahalakshmi[1,4] and M. Mahendran[1]

[1]*Smart Materials Lab, Department of Physics, Thiagarajar College of Engineering, Madurai*
[2]*Department of Physics, Sri G.V.G. Visalakshi College for Women, Udumalpet*
[3]*Department of Physics, SACS MAVMM Engineering College, Madurai*
[4]*Department of Physics, V.V. Vanniaperumal College for women, Virudhunagar*
E-mail: manickam-mahendran@tce.edu

ABSTRACT

Ni–Mn–Ga Ferromagnetic Shape Memory Alloys (FSMAs) are emerging as active materials due to their large Magnetic Field Induced Strain [MFIS]. Although strains offered by FSMA are similar to conventional temperature induced SMA, the faster response due to magnetic control makes them more promising for a variety of applications. The nano structured $Ni_{52.8}Mn_{24.2}Ga_{23.1}$ alloy was prepared using the ball milling technique. The distortions in X-ray diffraction pattern of the milled powder shows that the alloy need further annealing. Magnetic measurement has been performed using VSM. It is concluded that Ni-Mn-Ga in appropriately chosen ratios might be useful for vibration damping applications.

INTRODUCTION

FSMAs exhibits large changes in shape and size under strong magnetic field. The key factor behind this phenomenon is the martensitic phase transformation of the crystal lattice below a certain temperature. They are acknowledged for their giant Magnetic-Field Induced Strain (MFIS), which occurs due to the reorientation of martensitic variants under an applied magnetic field. Because of their large MFIS, more consideration has been shown to utilize these materials

V. Rajendran, P. Manivasakan and K.E. Geckeler (eds.)
Advanced Nano Materials for Industrial Applications, pp. 163-166 (2012).

for practical applications such as actuators and sensors [1-3]. The FSMA differs from the conventional SMA by its twin boundary motion in the martensitic phase rather than the structural change [4-9].

Ni -Mn-Ga alloy is currently the best and the most well known FSMA compound as they show significant deformation when they are exposed to magnetic field. The switching process is faster in Ni-Mn-Ga alloy and it is better controlled when compared to conventional shape memory alloys.

Fine Ni–Mn–Ga nano particles were prepared by ball milling technique and effects of milling and annealing on their structural and magnetic properties were investigated. The present study revealed that the ball milling is a simple and effective technique for producing fine nano particles.

EXPERIMENTAL PROCEDURE

Ni-Mn-Ga alloy was prepared using the conventional arc melting technique in argon atmosphere. To remove air inside the arc melting chamber, as much as pure argon was pumped into the chamber. Argon atmosphere inside the chamber would help the arc to travel and reach the sample. After the first melt the ingots were reverted in their same place and melted once again in the same atmosphere. This melting procedure was repeated upto six times by exposing the ingots in all the dimensions, in order to achieve homogeneity in the sample.

Finally the samples were taken out of the chamber and sealed in a quartz tube with argon atmosphere for thermal annealing. The ingots were crushed mechanically followed by planetary ball mill to achieve nano size particles. Ball mass-to-powder mass ratio of 10:1 was chosen to produce the nano powders. In the presence of toluene, the wet ball milling was performed with a rotation speed of 400 rpm. Finally, the ball milled sample was annealed for 10 hours at 900°C using high-temperature furnace.

The composition of the resultant powder was found to be $Ni_{52.8}Mn_{24.2}Ga_{23.1}$ using Inductively Coupled Plasma Optical Emission Spectroscopy (ICP-OES) technique. The powder X-ray diffraction measurement has been carried out to study the crystal structures using Cu-Kα radiation at room temperature. Magnetization measurements were performed using a Vibrating Sample Magnetometer (VSM-5, TOEI Industries) at room temperature.

RESULTS AND DISCUSSION

Many of the physical properties of ferromagnetic Ni-Mn-Ga alloys are mainly depend on their crystal structure of martensite [10-13]. The X-ray diffraction pattern of the powdered specimen is recorded, in order to determine its crystal structure. The XRD pattern of the $Ni_{52.8}Mn_{24.2}Ga_{23.1}$ is as shown in Figure 1.

For ball milled sample, only one prominent peak appears with broadening features indicating disordered/amorphous state [14]. Peaks in the XRD pattern always represent the martensite phase of the alloy. So, the presence of the martensite phase in the room temperature is also confirmed by the pattern.

Peak broadening is due to strain with decrease in particle size. This is common to all metallic systems prepared by ball milling technique. The broadening feature indicates the disordered crystal structure. The distortion shows that the sample requires further annealing. Moreover it shows that the sample did not attain homogeneity. Magnetic properties are measured using VSM. Figure 2 shows the M–H curve of the milled and annealed Ni–Mn–Ga alloy.

There is a gradual increase of magnetization when the magnetic field reaches about –5.427 Oe. There is a sudden rise of magnetization when the field reaches about –1.391 Oe and it increases up to 1.684 Oe. The alloy attains nearly saturation region when is the field is 1.684 Oe and the complete saturation is obtained when the field is 15.615 Oe.

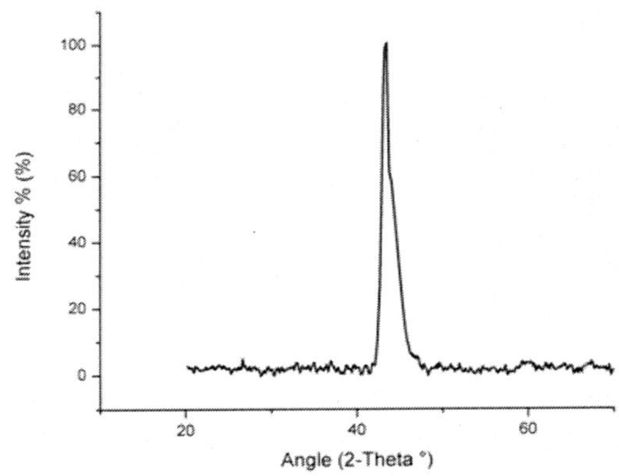

Fig. 1: X-ray Powder Diffraction Patterns of $Ni_{52.8}Mn_{24.2}Ga_{23.1}$ Nano Powder

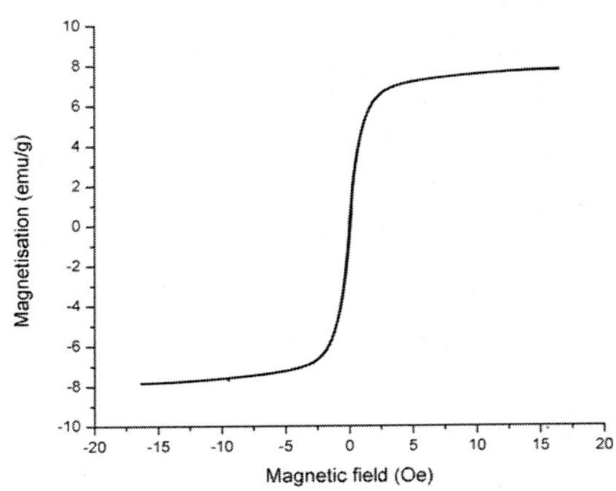

Fig. 2: M–H curves of $Ni_{52.8}Mn_{24.2}Ga_{23.1}$ Alloy

CONCLUSION

From these results, we can conclude that the nano $Ni_{52.8}Mn_{24.2}Ga_{23.1}$ alloy produced by ball milling present a ferromagnetic behaviour. XRD peak broadening indicates that reduction in particle size is accompanied with an increase of lattice strain due to milling process.VSM study reveals that annealed sample shows the improved ferromagnetic nature. Therefore, the annealed Ni–Mn–Ga fine particles can be used in polymer composites of various shapes and moreover these composites could be useful for vibration damping applications.

ACKNOWLEDGEMENT

The author (MM) thanks Prof. Dr. Robert C. O'Handley for introducing him to the subject.

REFERENCES

[1] Manickam Mahendran, Jorge Feuchtwanger, Ratchatee Techapiesancharoenkij, David Bono, Robert C.O' Handley, *J. Magn.Magn. Mater.*, 323, 1098 (2011).

[2] Ratchatee Techapiesancharoenkij, Jari Kostamo, Samuel M. Allen, Robert C. O'Handley, *J. Appl. Phy.*, 105, 093923 (2009).

[3] K.Vallal Peruman, M.Mahendran, S.Seenithurai, R.Chokkalingam, R.K.Singh, V.Chandrasekaran, *J. Phy. Chem. Solid*, 71, 1540 (2010).

[4] K.Ullakko, J.K Huang, C. Kantner, and R.C.O'Handley, *Appl. Phy. Lett.*, 69, 1968 (1996).

[5] V.A.Chernenko, S.Besseghini, P.Müllner, G.Kostorz, *Sensor Lett.*, 5, 233 (2007).

[6] S.J. Murray, M. Marioni, S.M. Allen, R.C. O'Handley, and T.A. Lograsso, *Appl. Phys. Lett.*, 77, 886 (2000).

[7] A. Sozinov, A.A. Likhachev, N. Lanska, and K. Ullakko, *Appl. Phys. Lett.*, 80, 1746 (2002).

[8] P. Mullner, V.A.Chernenko, M.Wollgarten, and G. Kostorz, J. Appl. Phy., 92, 6708 (2002)

[9] P. Mullner, and K. Ullakko, *Phy. Stat. Solidi (b)*, 208, R1-R2 (1998).

[10] I. Takeuchi, O.O. Famodu, J. C. Read, M. A. Aronova, K-S.Chang, C. Craciunescu, S.E. Lofland, M. Wuttig, F.C. Wellstood, L. Knauss, and A. Orozco, *Nature Mater*, 180, (2003).

[11] C. Biswas and S. R. Barman, *Appl. Surf. Sci.*, 252, 3380 (2006).

[12] P. Mullner, V.A.Chernenko, and D. Mukherji, *MRS Proc.*, D122, 785 (2004).

[13] Z.H. Liu, M. Zhang, Y. T. Cui, Y.Q. Zhou, W. H. Wang, G.H. Wu, X. X. Zhang and Gang Xiao, App. Phys. Lett., 82, 424 (2003).

[14] J. Liu, T.G. Woodcock, N. Scheerbaum, and O. Gutfleisch, *Acta Mater.*, 57, 4911 (2009).

NANOTECHNOLOGY ENABLED ANTENNA DEVELOPMENT: A REVIEW

Rajendra R. Patil, Vani R.M.[1] and P.V. Hunagund[2]

E&CE Dept, Appa Institute of Engineering & Technology, Gulbarga, Karnataka
[1]Gulbarga University, Gulbarga, Karnataka
[2]Department of Applied Electronics, Gulbarga University, Gulbarga, Karnataka
E-mail: rajurpatil@yahoo.com

ABSTRACT

This paper presents wide and profound impact of nanotechnology on antennas. Antennas are devices used for radiating and receiving electromagnetic waves, commonly used in a large number of telecommunication applications like cellular, Wi-max, system-on-chip, RFID, wireless ad-hoc networks, and space applications etc. Microstrip patch antennas (MSA) are becoming increasingly popular, as size and weight reduction are important factors in a market driven by device miniaturization consideration. Large bandwidth, frequency tunability and mechanical flexibility are desirable properties of antennas. Nanotechnology is enabling the development of antennas, where they will be able to radiate electromagnetic signals from few GHz to THz. Apart from electromagnetic communication in the THz band or optical frequencies; nanotechnology enables research in antennas in the development and discovery of new nano materials, which present potential application in antenna miniaturization. This paper reviews nanotechnology enabled novel techniques and approaches used for antenna miniaturization and study of other antenna parameters for tailor made applications.

INTRODUCTION

A radio antenna is still considered the largest barrier for radio frequency system because its size reduction is limited by the physics of electromagnetic radiation. Several strategies have

---

V. Rajendran, P. Manivasakan and K.E. Geckeler (eds.)
Advanced Nano Materials for Industrial Applications, pp. 167-172 (2012).

been used for size reduction of the antenna, and each affects the antenna performance differently. Numerical solutions can optimize antenna structure, but enhancement of the performance could come from the nanotechnology. Nanotechnology also offers new possibilities for antenna development. In this paper, conventional techniques followed by employing nanotechnology are reviewed for antenna miniaturizations, which are reported in various literatures.

(a) Geometric technique (b) Dielectric loading (c) Fractal technique (d) Pin shorting

Fig. 1: Antenna Miniaturization by Conventional Techniques

ANTENNA MINIATURIZATION BASED ON ANTENNA TOPOLOGY

Geometrical Modification

This strategy tries to reduce the overall dimension of the antenna to lengthen the current path in the patch. Figure 1(a) presents geometrical modification using slots in conducting patch. Drawback with geometrical modification is reduction in radiation resistance.

Dielectric Loading

This technique use high permittivity substrate, which allows reduction in electrical size of the antenna through reduction in effective wavelength. Antennas that use this technique are patch antenna and PIFA antennas, given in Fig. 1(b), where $\varepsilon_2 > \varepsilon_1$. But use of higher permittivity reduces the existing bandwidth and efficiency of the antenna [1].

Fractal Geometries

A fractal antenna illustrated in Fig. 1(c), created using fractal geometry, a self-similar pattern built from the repetition of a simple shape. Fractal structures have jagged shapes with many corners that these discontinuities increase bandwidth and the effective radiation of antennas. Disadvantages: gain loss and complexity numerical limitations.

Shorting Plane

This technique use shorting pins and planes make part of the antenna becomes redundant and can be removed. Monopole and PIFA antennas make use of this approach; see Fig. 1(d). But there is a reduction in gain and efficiency of the antenna.

NANOTECHNOLOGY ENABLED ANTENNA DEVELOPMENT

In the previous discussion, we have reviewed many well-established techniques and approaches based on novel architectures that are used for miniaturizing antennas. In spite of this, there are plenty of researches looking for innovative ways of antenna miniaturization. One of the fastest growing field is nanotechnology (synthesis, production, and characterization of nanomaterials and carbon nanotubes for various applications) where development and discovery of new materials which present potential for antenna miniaturization. This section will devote how nanotechnology [2] could be used to obtain miniaturized radio frequency and optical/infrared antennas using novel materials. For antenna miniaturization, nanotechnology focuses on optimizing parts of antenna. A part of them is optimizing the employed antenna substrate. On the other hand, there have been researches on optimizing conductive patch of the antenna. Optimizing parts of antenna using nanotechnology that miniaturize antennas as well increase the antenna performance are reviewed.

Optimizing Substrate of MSA: Employing Engineered Magneto-Dielectric Material

Antenna miniaturization using high permittivity materials as substrates have been attempted in the past [1]. Drawbacks with dielectric loading effects are discussed in introduction. Magneto-dielectric can also miniaturize the antenna by the same factor however using moderate values of permittivity and permeability. Thus, the issue of high Q-factor is minimized and the medium is far less capacitive when compared to the dielectric with only high permittivity material. The characteristic impedance of magneto-dielectric is close to that of the surrounding medium, it allows for ease of impedance matching over a much wider bandwidth.

Reducing bulk material in to nanoscale size leads to novel magnetic coupling phenomena resulting in higher permeability. Magnetic nanocomposites show permeability value much higher than that obtainable from the conventional material due to the reduced eddy current losses and antiparticles exchange interaction, which leads to magnetic ordering within a grain, extends out to neighboring environments.

Nano structured magneto-dielectric materials are synthesized by using variety of techniques [2] such as: gas condensation, chemical co-precipitation technique, hydrothermal synthesis, glass ceramic technique, sol-gel technique, and precursor technique. These techniques yield nano particles with controlled size, uniform morphology, magnetic/thermal stability, and defined preparative composition and stoichiometry. Nanoparticle characterization is necessary to understand and control of particle size, information about magnetic ordering existing in nanocrystalline materials etc. T.D. Xiao et al. [3] reported Nickel-Zink-Ferrite magneto-dielectric nanocomposite of size 5 to 15nm. The permeability of this material varies from 2.5-3 at GHz frequency range, depending on the volume of ferrite loading in epoxy, the tangent losses were less than 0.001 dB, and Q in excess of 100, making it a good candidate for miniaturized antennas. Jungyub Lee et al. [4] presented miniaturized antennas for mobile handsets using magneto-dielectric material. They synthesized magneto-dielectric material from BaO, Fe_2O_3 and MeO able to get permittivity and permeability around 13 and 2.2. The tangent loss is negligible in the measured frequency band. The antenna is miniaturized with more improved bandwidth compared to the one of the dielectric chip antenna at the same resonance frequency. Deepak Ram Prasath S et. al [5] demonstrated miniaturization of aperture coupled antenna with the loaded magneto-dielectric material. They synthesized Nickel-Zink-Ferrite material and obtained $\mu_r = 2.3$ and $\varepsilon_r = 10.3$. An aperture coupled patch antenna, illustrated in Fig. 2(a) loaded with the said synthesized material above the slot of the feeding structure validates the miniaturization capability of the magneto-dielectric substrate. The loading of the synthesized material increases the effective permeability and permeability of the combinations and hence shifts the resonant frequency, which in turn reduces the size of the antenna and increases the gain.

Optimizing Conductive Patch: Employing Nanofilm

An important aspect of fabrication involving nanomaterials for MSA is reliability of contacts between nanomaterial conducting patch and metallic feeding line. In order to overcome this challenge Fabio Urbani et. al [9] have fabricated a MSA with ultra thin iron patch of thickness ~10 nm, where there is no physical contact with bulk metallic (copper) feed line. They used a feeding technique known as aperture-coupled configuration is illustrated in Fig. 2(a). This technique significantly minimizes practical manufacturing difficulties such as ensuring a reliable electrical contact. The fabricated MSA showed excellent performance, and an ultra-wide bandwidth of 1GHz.

Optimizing Conductive Patch: Employing Carbon Nanotubes (CNTs)

In this section, we review optimizing antenna conductive parts instead of the substrate pattern. While copper is typically used in applications where high conductivity is required, it does not maintain its bulk conductivity when scaled to nanometer dimensions. In contrast, nanotubes have better conductivity than copper when scaled to their diameter.

Carbon Nanotubes (CNTs) and Graphene Nanoribbons (GNRs) emerged as potential candidate for replacement of conventional conducting metal patch. Quite recently, application of new CNT and conductive polymer ink instead of metallic conductors has led to the development of low cost, flexible, conformal and lightweight antennas using different fabrics as a substrate that can be cotton, fleece, nylon etc [6]. Single Walled Carbon Nanotubes (SWCNTs), see Fig. 2(b), have a current density of $\sim 4 \times 10^9$ A/cm^2 [7], which are three orders of magnitude higher than a typical metal such as copper or aluminium. Thus using CNT instead of traditional metal, one can build antenna that will lead to efficient ways of electromagnetic radiation. Controlled deposition and patterning of CNTs on flexible substrates opened doors to a number of exciting applications in the field of flexible/wearable electronics in wireless communication such as RFID devices and wearable devices. P.K. Patra et al [6], have successfully developed CNT based electronic ink that has enough threshold conductivity used to make wearable MSA on flexible substrate, illustrated in Fig. 3(a). To realize flexible antenna they used a highly automated process involving inkjet printing of CNTs on various flexible substrates, such as fabric, paper, and transparency.

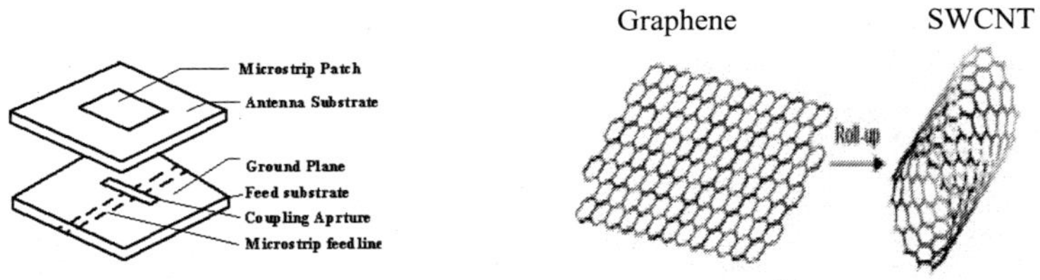

(a) (b)

Fig. 2: (a) Aperture Coupled Microstrip Antenna, (b) SWCNT

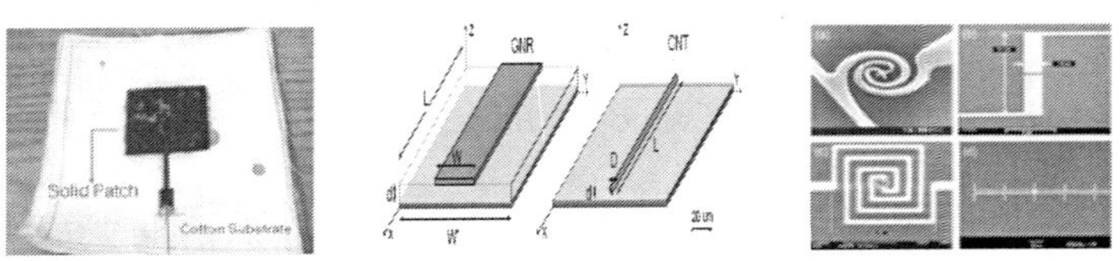

(a) (b) (c)

Fig. 3: (a) Wearable Textile Antenna using CNT, (b) A Nano-patch Antenna based on GNR and a Nano-dipole Based on a CNT, (c) Optical Antennas

Recently, J.M. Jornet and I.F. Akyildiz [8] modeled a nano-dipole and a nano patch antenna using transmission line model of CNT. The results show that for a maximum antenna size in the order of several hundred nanometers, both a nano-dipole and a nano patch antenna will be able to radiate electromagnetic waves in the terahertz band (0.1-10.0 THz), illustrated in Fig. 3(b).[8].

Optical and Infrared antennas: Employing CNT's

Optical antennas are defined in [10] as antenna-coupled optical detectors, also named as optical antennas, are being developed as detection devices with micro- and nano-scale features for their use in the millimeter, infrared, and visible spectral range. They are optical components that couple the electromagnetic radiation in the visible and infrared wavelengths in the same way that radio electric antennas do at the corresponding wavelengths. Optical antennas show polarization dependence, tuneability, and rapid time of response. In [11] optical antenna is defined as devices designed to convert efficiently free propagating optical radiation to localized energy and vice versa. Different forms of fabricated optical antennas based on nanotubes shown in Fig. 3(c) [11]. Emerging applications of optical/IR antennas are nano-antenna probes for nano-imaging, signal conversion and processing, etc.

CONCLUSION

This paper reviewed miniaturization of antennas using conventional techniques and nanotechnology. The research has shown that indeed physical, electrical, and electronic properties of nanomaterials and nanotubes are remarkable. The CNT based radio frequency antenna and optical/IR antenna design are ideal for space, defence and wearable applications where lightweight and robust system are crucial. The authors' feels that in coming days, if proven to be effective, nanomaterials/CNTs will eventually replace the current bulk materials being used in antennas.

REFERENCES

[1] Chun-On H.; Hwang, Y. ; Lam, E.K.W. ; Lee, B., Miniature Aperture Coupled Microstrip Antenna of very high Permittivity, *Electron. Lett*, 33, 9-10, 1997.

[2] Sharma A., Bellare J., and Sharma A., Advances in nanoscience & nanotechnology, 1st ed., National Institute of Science Communication and Information Resources, New Delhi 2004.

[3] Xiao T. D., et al., Magnetic Nanocomposite Paste: An Ideal High-ì, k and Q Nanomaterial for Embedded Inductors in High Frequency Electronic Applications, International Symposium on International Symposium on *Advanced Packaging Materials: Processes, Properties and Interfaces,* pp. 272-275, March 2005.

[4] Lee J., Heo J., and Lee J., Design of Small Antennas for Mobile Handsets using Magneto-Dielectric Material, *IEEE Trans. Antennas Propag*., 60, 2080-2084, 2012.

[5] Prasath S.D.R., et al., Miniaturization of patch antennas using magneto-dielectric materials, *A Workshop on Advanced Antenna Technology, Indian Antenna Week*., pp. 1-4, May 31-June 4, 2010.

[6] Patra P.K. et al., Textile based carbon nanostructured flexible antenna, National textile center annual report, NTC Project no: M06-MD01, November 2007.

[7] Bajwa H., Patra P., Ikram A.A. and Mirza J., Nanostructured conformable patch antenna array, International conference on emerging technologies ICIET-2010, pp. 1-4, 14-16 June 2010.

[8] Jornet J.M., and Akyildiz I.F., Graphene-based Nano-antennas for Electromagnetic Nano Communications in the Terahertz *Band, European conference on antennas and propagation EuCAP-2010*, pp. 1-5, 12-16 April 2010.

[9] Urbani F., Stollberg D.W., and Verma A., Outstanding Performance of a Nanofilm Microstrip Antenna, *IEEE nanotechnology materials and devices conference*, USA, pp. 160-163, October 12-15, 2010.

[10] Cui X., Erni D., Dong L., and Zhang W., Optical Antennas based on Segmented Metal-Nanotube Structures, Second European Topical meeting on Nanophotonics and metamaterials NANOMETA-2009 Australia, 5-8, January 2009.

[11] Bharadwaj P., Deutsch B., and Novotny L., Optical Antenna, *Adv. Opt. Photon*., vol. 1, no. 3, pp. 438-483, November 2009.

BULK PRODUCTION OF CARBON NANOTUBES (CNT) – A LITERATURE REVIEW

Prasanna Kumar M.K., Vidyashankar S. and P. Martin Jebaraj[1]

Dept. of Mech. Engg. Bangalore Institute of Technology, K.R.Road, Bangalore
[1]City Engg. College, Doddakallsandra, Kanakapura Main Road, Bangalore
E-mail: prasannabit@gmail.com

ABSTRACT

In this paper an overview of the continuous production of carbon Nanotube as well as methods to carry out its bulk production is presented. The mass production interest of nanotubes is as a result of the demand from several applications in the material field, such as fiber components, electronic industries as an additive to improve mechanical properties or electrical properties, or as a component in the electronic and optoelectronic product integration. In this paper the production methods adopted and reported by investigators are reviewed and presented. The outcome of the literature review it is inferred that there are three distinct routes in the literature for high volume production of CNT. They work on three basic concepts: (i) Multiple reactor usage; (ii) Moving bed and rotate plate reaction; (iii) Catalyst injection and feed step gas blowing method; They are presented in the paper.

INTRODUCTION

A carbon nanotube (CNT) is a tubular structure made of carbon atoms, having diameter of nanometer order but length in micrometers. Although this kind of structures was synthesized, studied and reported by several researchers during 1952–1989 [1], Since Ijima proved to be a discovered CNT in 1991 [2]. The significant interest in the production, research and development of carbon nanotubes stems from the unique chemical, mechanical, and physical properties inherent in these materials. These desired properties include high tensile strength, high electric and thermal conductivity, lightweight, high surface area per gram [3]. Because of these

V. Rajendran, P. Manivasakan and K.E. Geckeler (eds.)
Advanced Nano Materials for Industrial Applications, pp. 173-178 (2012).

properties, many researchers and product developers have been attracted to carbon nanotubes for a broad array of potential applications including in the preparation of Molecular probes, Pipes, Wires, Bearings, springs, Gears, Pump in addition to that future application, i.e. Molecular transistors, Field emitters, Building blocks for bottom-up electronics, Smaller & lighter weight components for next generation spacecraft, Enable large quantities of hydrogen to be stored in small low pressure tanks, Space elevator, etc. The demand for CNT is even increasing but with the existing system of preparation of CNT bulk production of CNT is difficult. Hence in this paper the focus is on the information available for the mass production of CNT.

BACKGROUND OF ALL PROCESS

The synthesizing CNT have been introduced such as, Chemical Vapor Deposition (CVD) [4], Arc discharge [5] and Laser ablation of carbon [6]. Each of the process has the following characteristic feature and is presented in Table 1.

Table 1: Comparison of Different CNT Production Methods [7]

Sl. No	*Process* / *Properties*	*Arc Discharge*	*Laser Ablation*	*Chemical Vapor Deposition*
1	Process control	Difficult	Difficult	Easy and it can be automated easily
2	Reactor Design	Difficult	Difficult	Design for large scale process
3	Production rate	Low	Low	High
4	Yield process	Moderate (70 %)	High (80–85%)	High (95–99%)
5	Post treatment requirement	Required Refining	Required Refining	No extensive refining requirement
6	Purity of product	High	High	High
7	Process nature	Batch type	Batch type	Continuous
8	Per unit cost	High	High	low

Study of the information availability in the literature, patents etc., it appears that CVD method is more conducive of CNT as shown in the Table 1.These process as are suggested and developed for mass production of CNT and hence the present literature review of bulk production of CNT.

CNT SYNTHESIS

Figure 1 shows a schematic line diagram of the experimental setup used for CNT growth by Chemical vapor deposition techniques is well known as CVD. It consists of a furnace reaction chamber maintained at 700 to 1200°C under inert atmosphere. The catalyst in a ceramic boat is placed inside the reaction chamber. The organic gas is blown over the catalyst. The gas decomposes leaving CNT on the catalyst surface. Depending on the catalyst, reaction temperature and reaction gas, it is possible to produce single wall or multiwall CNT. Some of the catalyst and the reaction gases are listed in Table 2.

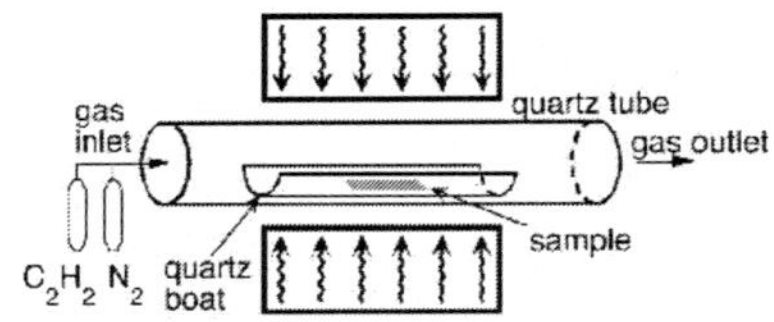

Fig.1: Line Diagram Showing the Facility for Making CNT

Table 2: Catalyst and Gases Used for Synthesizing CNT

Catalyst	*Hydrocarbon Gas*	*Inert Gas*
• The frequently used catalysts are Fe, Co, Ni. • Other metals such as Sc, Ti, V, Cr, Mn, Y, Zr, Nb, Mo, Hf, Ta, and W, Re or a combination of them.	Methane, ethane, ethylene, acetylene, or other higher hydrocarbon is used without dilution.	Argon, helium, etc.

BULK PRODUCTION METHODS FOR CNT BASED ON CVD METHODS

The bulk productions of CNT based on CVD method adopted are listed in the Table 3.

Table 3: Bulk Production Methods for CNT based on CVD Technique

Sl. No.	*Method Adaptation*
i	Multiple reactor usage where sequentially CNT is produced in each reactor unit- Published Literature [8]
ii	Moving bed reactor using a conveyor carrying the catalyst into the reaction zone- Published Literature [9]
iii	Multistage rotating plate reaction technique – Patent [10]
iv	Catalyst injection into reaction gas - Patent [11]

EXISTING OF ALL PROCESS

Producing CNT Using Multiple Reactor

This apparatus has been constructed for continuous mass production in continuous supply of the catalyst to the inside of the reaction chamber, temperature control and the reaction gas supply. Figure 2 is a schematic diagram of the multiple reactor apparatus usage for continuous mass production for CNT.

The major parameter involved in this process for capable of achieving mass production of CNT include a number of reaction chambers, a heater and a drive motor the continuous mass synthesis system carries out a preheating process in which the catalyst is melted stably in a temperature range 200 to 700°C under an argon atmosphere, the reaction process of synthesizing CNTs via reaction of the reaction gases with the catalyst by supplying a carbonization gases such as C_2H_2, CH_4 at a flow rate of 10 to 1000 sccm in the temperature range of 600 to 1000°C, by decomposing the gas at the high-temperature where CNT are synthesized and a cooling process is sharply lowering the temperature of the reaction chambers to a temperature of 700 to 200°C under an argon atmosphere, after completion of the reaction.

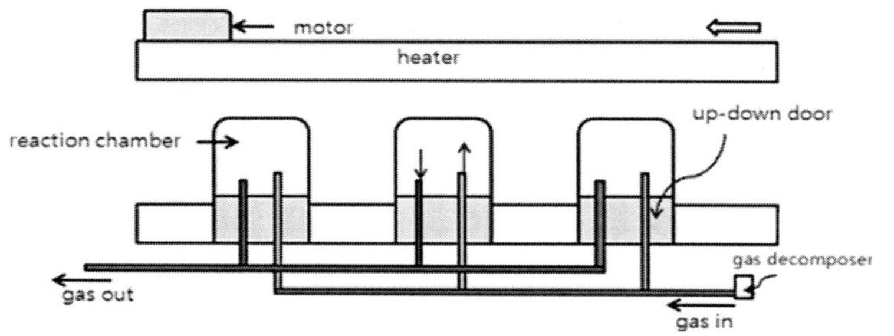

Fig. 2: Schematic Diagram of the Multiple Reactor Apparatus usage for Continuous Mass Production for CNT

Moving bed Reactor using a Conveyor Carrying the Catalyst into the Reaction Zone

In this method the reaction can run in a continuous fashion and produced a carbon nanotubes are in high quality and at high production rates. The moving bed reactor is illustrated schematically in Figure 3.

In this method the prereduced catalyst was packed into the solid sampler in an inert environment, followed by heating the reactor in a flow of purified N_2 from room temperature to the desired reaction temperature. Then speed regulator was operated for prereduced catalyst dropped down automatically and covered the conveyor belt evenly, and the belt moved forward automatically. Subsequently, CH_4 was introduced at a flow rate of 50 sccm, and carbon nanotubes were formed on the belt by deposition of carbon atoms obtained from decomposition of CH_4. Finally, the sample of a black power dropped into the collector.

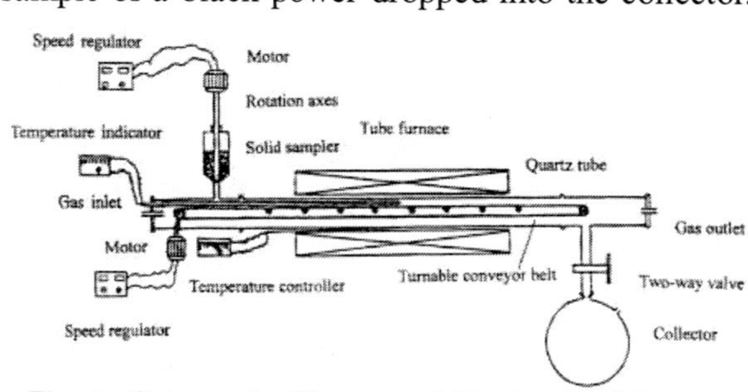

Fig. 3: Schematic Diagram of Moving Bed Reactor

Multistage Rotating Plate Reaction Technique

In rotating plate reaction apparatus for synthesizing CNT consists vertical reaction chamber includes dispersion plates that partition the chamber into a plurality of section, in which a mixture of catalyst causes free fall through a dropping hole of each of the dispersion plates and in which CNTs are continuously produced by using free falling catalyst. In this apparatus capable of reducing the time required to supply a catalyst and synthesize the CNTs continuously. The Figure 4 represents the schematic diagram of the rotate plate reaction.

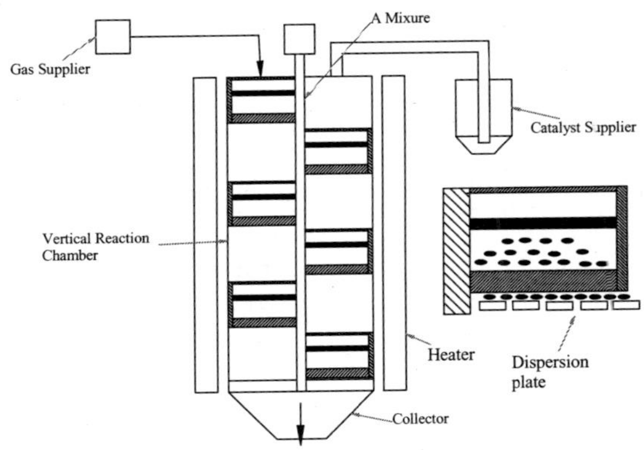

Fig. 4: Schematic Diagram of Rotate Plate Reaction

Catalyst Injection into Reaction Gas

The catalyst injection unit synthesis the CNTs continuously by uniform length and infusion on substrate. Referring to Figure 5 schematic diagram of a reactor for synthesis of a CNT infused substrate, catalyst source 1, Carbon feedstock source 2, and carrier gas source 3 are introduce at the top of the CNT growth zone 4 through an inlet device 5. A heating element 6 can he used to raise the temperature of the mixture to promote the formation of CNTs. As the CNTs grow, they cart pass through a dispersion hood 7 to cool prior to entering the infusion chamber 8 containing the substrate 9. The synthesized CNTs can infuse to substrate 9 to produce a CNT infused substrate before passing out of the reactor.

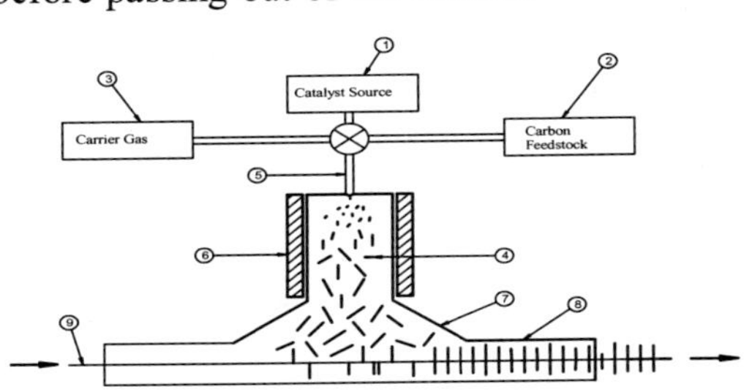

Fig. 5: The Schematic Diagram for Catalyst Injection and Feed Step Gas Blowing Method

CONCLUSIONS

Conclusions drawn from the current attempted mass production systems. In these methods the net outcomes are shown in Table 4.

Table 4: Outcome Information

Method No.	*Method*	*Summary*
1	Producing CNT Using Multiple Reactor	Using Fe/Mo catalyst a high yield can be grown by the continuous mass production apparatus.
2	Moving bed reactor	High qualities CNT were continuously produced by using moving bed reactor.
3	Multistage rotating plate	CNTs are continuously synthesized by using the free falling catalyst & an apparatus capable of reducing the time required and thus enhancing productivity
4	Catalyst injection into reaction gas.	The continuous production of CNT of uniform length and distribution along spoolable lengths of tow, tapes, fabrics and other 3D woven structure is successful.

REFERENCES

[1] L. V. Radushkevich and V. M. Lukyanovich, *Zh. Fizich. Khim.* 26, 88 (1952).

[2] S. Iijima, *Nature* 354, 56 (1991).

[3] Reid Lifset, Industrial Environmental Management Program, Analysis of Nanotechnology from an Industrial Ecology Perspective Part II Revised Draft - November 2005.

[4] A.M. Benito, Y. Maniette, E. Munoz and M.T. Martinez, *Carbon* 36[5-6] (1998) 681-683.

[5] H. Zeng, L. Zhu, G.M. Hao and R.S. Sheng, *Carbon* 36[3], (1998) 259–26.

[6] R.Z. Ma, B.Q. Wei, C.L. Xu, J. Liang and D.H. Wu, *Carbon* 38[4] (2000) 636-638.

[7] Muhammad Musaddique Ali Rafique, , Javed Iqbal, Production of Carbon Nanotubes by Different Routes—A Review, *Journal of Encapsulation and Adsorption Sciences*, 2011, 1, 29-34.

[8] Dae-Sup Soa, B, Heon Hamc, Woo Sik Kima, Sin Young Bangd, Bong Geun Choid, Kwang Bo Shima,D and Haiwon Leea,E, Continuous mass production system of carbon nanotubes synthesis and processing, *Journal of Ceramic Processing Research*. Vol. 10, No. 1, pp. 105-108 (2009).

[9] Bao Chun L.I.U., Mei Zhen Q.U., Zuo Long Y.U., Continuous Production of Carbon Nanotubes by Using Moving Bed Reactor, *Chinese Chemical Letters* Vol. 12, No. 12, pp. 1135 – 1138, 2001

[10] United States Patent, Patent no. US 7,972,571B2, July 5, 2011.

[11] United States Patent Application Publication, Patent no. US 2010/0260931 A1, October 14, 2010.

INVESTIGATION ON THE ADDITION OF PLASTICIZERS IN PVCAC BASED POLYMER ELECTROLYTE SYSTEM

D. Vinoth Pandi and S. Rajendran

School of Physics, Alagappa University, Karaikudi, Tamil Nadu
E-mail: sraj54@yahoo.com

ABSTRACT

Polymer electrolyte films are prepared from poly(vinyl chloride-co-vinyl acetate) and $LiClO_4$ with different plasticizers such as EC, PC, GBL and DMC by using solution casting technique. Ionic conductivity of the polymer electrolyte has been measured in the temperature range 303-343 K. The maximum conductivity value 3.52×10^{-6} S/cm has been obtained for EC-based complex. The structure and complex formation have been confirmed by XRD and FTIR spectroscopic studies.

INTRODUCTION

In recent years, there has been a tremendous interest in the preparation of polymer electrolytes with high ionic conductivity, good mechanical strength and thermal stabilities because these polymer electrolytes play a major role in solid-state batteries, electrochromic windows, sensors, fuel cells etc [1]. Solid polymer electrolytes (SPEs) have several advantages over the liquid counterpart such as desirable shape mouldability, free from leakage, mechanical strength and flexibility of design, thereby permitting miniaturization [2]. In order to enhance the performance of SPE in electrochemical utilities, several modifications are explored such as addition of plasticizers [3], incorporation of fillers etc in polymer electrolytes. The major issue in these modifications is focused on improving ionic conductivity, which promises simultaneous mechanical properties and stable electrode/electrolyte interface. PVCAC copolymer often has

V. Rajendran, P. Manivasakan and K.E. Geckeler (eds.)
Advanced Nano Materials for Industrial Applications, pp. 179-184 (2012).

superior physical properties to either of the related homopolymers. The thermal stability is often impaired. Polymer electrolyte films are prepared from poly(vinyl chloride-co-vinyl acetate) and $LiClO_4$ with different plasticizers such as EC, PC, GBL and DMC by using solution casting technique.

EXPERIMENTAL

Poly(vinyl chloride-co-vinyl acetate) (PVCAC; Aldrich, USA) and Lithium perchlorate ($LiClO_4$; Aldrich, USA) were used as received. All the solid polymer electrolyte films were prepared by solution casting technique. Appropriate weight of polymer and salt were dissolved in Tetrahydrofuran (THF; Merck, India) separately at room temperature. After the complete dissolution of polymer and salt they were mixed together and then the measured plasticizer added to it. This mixture was stirred for 24 hours and then it was allowed to evaporate slowly until a homogeneous solution was obtained. The obtained highly viscous solutions were cast onto clean glass plates to obtain the desired thin film membranes. The films were further dried for 5 hours in vacuum to remove the residual solvent THF.

The X-ray diffraction patterns were recorded with X'pert PRO PANalytical diffractometer in the scanning range 2θ = 10 to 80° using Cu-Kα (wavelength, λ = 1.541Å). The Fourier Transform Infrared (FT-IR) spectra of the samples were taken with Perkin Elmer 577-IR spectrophotometer in the range of 400-4000 cm^{-1} in transmittance mode at ambient conditions.

The ionic conductivity of the polymer electrolytes was carried out with the help of stainless steel blocking electrodes by using a computer controlled μ-Autolab type-III Potentiostat/ Galvanostat in the frequency range 40Hz–300 KHz. The conductivity (σ) was calculated using the formula $\sigma = l/R_bA$, where l is the thickness of the polymer electrolyte film, R_b is the bulk resistance, and A is the surface area of the polymer electrolyte film.

RESULTS AND DISCUSSION

XRD Analysis

In order to investigate the effect of various plasticizers in PVCAC/$LiClO_4$ system, XRD analysis has been performed and their respective diffraction pattern of pure PVCAC, $LiClO_4$ and EC, PC, GBL, DMC based complexes are shown in Fig. 1. The high intense diffraction peaks observed in Fig. 1(a) of $LiClO_4$ reveals the crystalline nature of Lithium salt. No peak is observed in Fig. 1(b), which indicates the amorphous nature of host polymer PVCAC. The diffraction peaks observed in $LiClO_4$ were completely absent in all polymer-salt complexes. This reveals the complete dissolution of the inorganic salt in the polymer in presence the of plasticizer. This amorphous nature gives greater ionic diffusivity.

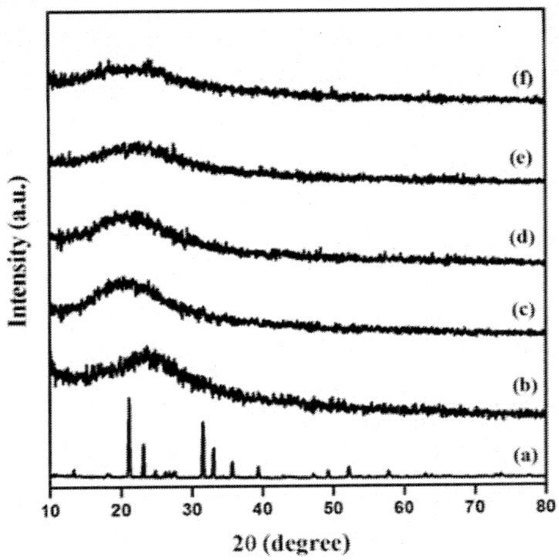

Fig. 1: XRD Patterns of (a) $LiClO_4$, (b) Pure Polymer PVCAC and their Complexes PVCAC/$LiClO_4$/X[X= (c) EC, (d) PC, (e) GBL and (f) DMC]

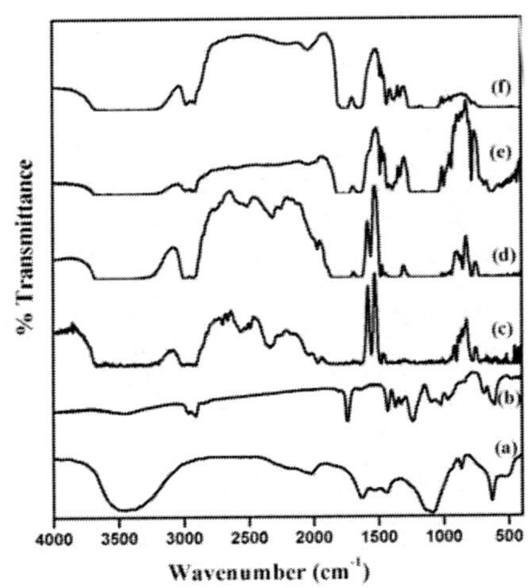

Fig. 2: FT-IR Spectra of (a) $LiClO_4$, (b) Pure Polymer PVCAC and their Complexes PVCAC/$LiClO_4$/ X[X= (c) EC, (d) PC, (e) GBL and (f) DMC]

FTIR Analysis

Figure 2 depicts the FT-IR transmittance spectra of PVCAC/$LiClO_4$ with different plasticizers such as EC, PC, GBL and DMC. The spectral pattern of copolymer PVCAC in Fig. 2(b) consists of bands at 2976 and 2911 cm^{-1} (-C-H stretches), 1737 cm^{-1} (C=O stretch), 1426 and 1434 cm^{-1} (CH_3 rocking), 1376, 1329, 1235, 1103 and 1022 cm^{-1} (C-O stretch coupled with C-C vibrations), 962 cm^{-1} (skeletal vibrations) and 748 and 706 cm^{-1} (C-Cl stretches and skeletal vibrations)[4,5].

The absorption peaks corresponding to EC (2934, 1807, 1390, 1137, 849 and 774 cm^{-1}), PC (1340 and 849 cm^{-1}), GBL (1648, 1280 and 676 cm^{-1}) and DMC (2963, 1894, 1606 and 1282 cm^{-1}) were found to be shifted in all the complexes. The change in vibrational peaks indicates the strong interaction taking place between the plasticizer and the polymer-salt complex. The vibrational peaks at 1620, 1320 and 940 cm^{-1} are ascribed to pure $LiClO_4$ [6]. These results show that some of the significant vibrational peaks in PVCAC are absent and some are shifted from the actual value.

The vibrational peaks at 2911, 1426, 962, 748 cm^{-1} of pure PVCAC are shifted in EC-based complex as 2932, 1478, 979, 779 and in PC-based complex as 2920, 1480, 991, 774 cm^{-1} and in GBL-based complex as 2921, 1425, 932, 778cm^{-1} and in DMC-based complex as 2915, 1427, 980, 790 cm^{-1} respectively. The characteristic weak peak at 932 cm^{-1} of ClO_4^- anion is observed in the polymer complexes which has been used as an indicator for the dissociation of $LiClO_4$ [7]. In addition some new peaks are present and some of them disappeared in the electrolytes. Thus the spectral analysis confirms the complexation of these polymer and lithium salt.

Conductivity Measurements

The ac impedance plot of the EC based complex at room temperature is shown in Fig.3. The ionic conductivities of the polymer PVCAC with different plasticizers (EC, PC, GBL and DMC) at different temperatures are exemplified in Fig. 4.

From the Fig. 4, it is concluded that the electrolyte containing EC as plasticizer exhibits maximum ionic conductivity of the order of 3.5148×10^{-6} Scm^{-1} which is mainly due to the specific properties of the plasticizer such as higher dielectric constant and lower viscosity. In addition, the higher ionic conductivity is mainly attributed to the higher amorphous nature of PVCAC. The polymer electrolytes were also subjected to conductivity studies in the temperature range (303-343 K). It is formed that as temperature increases, the conductivity values are also found to increase. This behavior is in agreement with the free volume model theory.

At high temperature, the polymer can expand easily and produce free volume. This leads to an increase in ion mobility that will assist ion transport and virtually compensate the retarding effect of ion clouds. The temperature dependence plot shows non-linear relation of temperature with conductivity. It is a well established fact that the addition of plasticizers increases the overall conductivity of the polymer electrolyte.

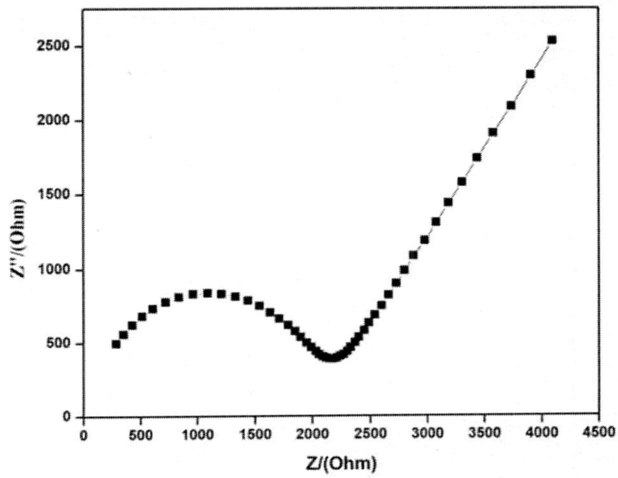

Fig. 3: Impedance Diagram of PVCAC/$LiClO_4$/ EC at Room Temperature

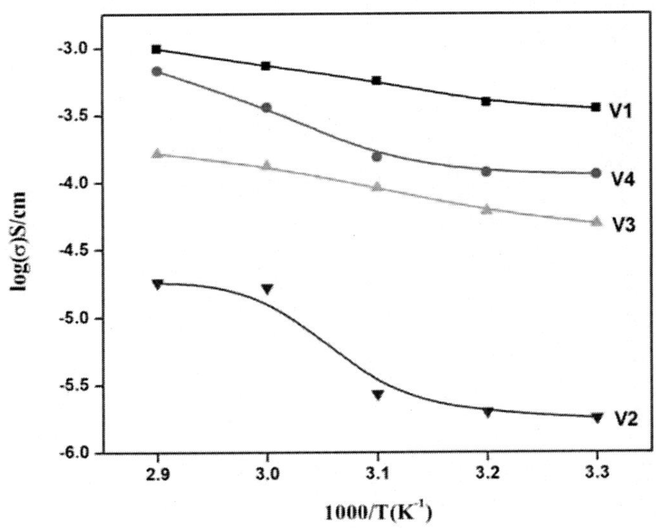

Fig. 4: Plot of Log σ versus 1000/T

CONCLUSIONS

PVCAC/$LiClO_4$ polymer electrolytes with different plasticizers (EC, PC, GBL and DMC) were prepared using solvent casting technique. Ionic conductivity of the polymer electrolyte has been measured in the temperature range 303-343K. The maximum conductivity value has been obtained for EC-based complex because of its higher dielectric constant compared to other plasticizers used in the present study. The maximum ionic conductivity of (3.5148×10^{-6} Scm^{-1}) at 303K has been observed for PVCAC /$LiClO_4$/EC complex system. The complex formation has been confirmed by XRD and FTIR spectroscopic studies.

REFERENCES

[1] S. Rajendran, Ravi shanker Babu, K. Renuka devi, *Ionics* (2009) 15:61-66

[2] S. Rajendran V. Shanthi Bama, M. Ramesh prabhu, *Ionics* (2010)16:283-287

[3] T. Nagatomo, C.Ichikawa, O.Omoto, *J. Electrochem. Soc*, 134 (1987) 305.

[4] D. Dolphin, A. Wick, *Tabulation of Infrared Spectral Data*, Wiley, New York, 1977.

[5] R.G.J. Miller, H.A. Willis (Eds.), Infrared Structural Correlation Tables and Data Cards, Heyden & Son Ltd., Spectrum House, London, 1969.

[6] S. Rajendran, O.Mahendran, R.Kannan, *J. Phys.Chem.Solids* 63 (2002) 303.

[7] D. Ostrovshii, A.Brodin, M.Torell, G.B.Appetecchi, B.Scrosati, *J. Chem.Phys.* 109 (1998) 7618.

SAMPLE PREPARATION OF CHEMICALLY TREATED COIR FIBER

ALVEERA KHAN AND SHIRISH JOSHI

Department of Physics Govt. M.V.M. College, Bhopal
E-mail: khan.alveeera@gmail.com

ABSTRACT

Natural renewable materials have been great attention in recent years due to their low cost, ecofriendliness, light weight and world wide environmental awareness. In the present work chemically treated coir fiber samples were made using annealing method. Raw coir fiber was chemically treated with aluminum nitrate salt and ammonium chloride in order to change the morphology of coir fibre.Xrd studies of the sample revealed gamma alumina as the major phase. The use of coir fiber revealed the change in the morphology of samples when characterized by SEM.the importance of using annealing method for preparation of alumina is also manifested in a shorter reaction time.

INTRODUCTION

Most of the developing countries are very rich in agricultural and natural fiber. Except a few exceptions, a large part of agricultural waste is being used as a fuel. India alone produces more than 400 million tones of agricultural waste annually. It has got a very large percentage of the total world production of rice husk, jute, stalk, and baggage and coconut fiber [21]. All these natural fibers have excellent physical and mechanical properties and can be utilized more effectively in the development of composite materials for various applications.

In the present work coir fiber is used as reinforcing material since it is abundant in nature and has a minimal effect on the environment due to their biodegradable properties [1,2,3] coir

V. Rajendran, P. Manivasakan and K.E. Geckeler (eds.)
Advanced Nano Materials for Industrial Applications, pp. 185-190 (2012).

fiber is cheapest lignocelluloses fiber. It is the hardest natural fiber among all fibers. one of the major drawback of coir fiber is its hydrophilic nature, responsible for moisture absorption and poor interfacial bonding between matrix and fibre.To overcome this problem coir fibre was chemically treated with aluminum salt as it is easily available at low cost.

This paper gives an overall view of coir fiber when embedded with inorganic salts. Coir fiber has been selected for this study for four reasons. first coir fiber have an attractive combination of strength, stiffness and elongation [4], second, coir fiber have the highest lignin content (35%) of any natural fiber[5,6] making them quite resistant to burning, odor producing decay due to fungal attack and degradation induced by other micro organisms.third,coir fiber are generally very durable in wet environments, which are well demonstrated by their widespread use in the production of large diameter ropes that are used to tie-up ocean-going vessels.fourth,there are 50 million coconut that grow each year and vast majority of them are burned as trash, damaging the environment and depriving the poor coconut farmers of some additional income to $500/year on average that they currently make[7] V.G. Geethamma et al. **[2]** studied on the short coconut fiber natural rubber composites and the effect of fiber loading, orientation and chemical modification on the overall properties of the composites. They treated the fiber with alkali like 5% solution of sodium hydroxide and sodium carbonate for 48 hrs and washed to remove the excess alkali content. From another study from the same authors **[8],** dynamical mechanical behavior of short coir fiber natural rubber composites was revealed. Short fiber reinforced rubber composites can be used in vibration dampers, tires etc. so the study on dynamic mechanical properties are of great interest. The study carried out on various treated and untreated coir fibers in which, NaOH treatment, Resorcinol Formaldehyde Treatment, Bleaching etc were done. In the treatments provided, the one with bleaching exhibited good dynamic mechanical properties. Chanakan Asasutjarit et al **[9]** studied on the treated coir fiber green composites. The treatment done to coir fibers were washing in boiling water and then washing in cold water. In the first treatment coir fibers were thoroughly washed in excess of water till the water pH reached 7. by this the water procedure removes a part of traneous components, such as inorganic compounds, tannins, gums, sugars and coloring matter present in coir. The hotwater procedure removes, in addition, starches. From the morphology obtained it was found that in the surface of treated fibers there was the formation of small pits which increased the total surface area which eventually increases the interaction between binder and the filler. From the two treatments boiling and then washing in water gave much better properties than the first one.

Recently, there has been an increasing interest in the synthesis of g Al_2O_3 owing to its important applications in catalysis, miniature power supplies, space crafts etc. [10,11,12]several reports of physical and chemical processes for the synthesis of alumina particles in literature. In many chemical routes, boehmite and bayerite have been used as the starting materials for the preparation of alumina. Boehmite is prepared by various methods such as hydrothermal treatment of gibbsite [13] hydrolysis of alkoxides [14] aluminum nitrate solutions [15], etc.

Aluminum metal has also been used as a precursor for the synthesis of aluminum hydroxides and these hydroxides are used to prepare alumina particles [16, 17] the use of microwave radiation has been widely accepted in many chemical reactions for different applications in the last few years (18, 19). The main aim of this study is to make composites of biodegradable natural fibre.

MATERIALS AND METHODS.

Aluminum nitrate and ammonium chloride salts are used for the chemical treatment of fibre, the coir fibre used are collected from rural area of India. It contains 42.44% of cellulose, 45.4% legnin, 0.25% of cellulose, 3% pectin and other constituents.

Treatment of Coir Fibre

Coir fibre was washed with tap water and then dried at room temperature for 24 hours and then kept in a sealed container. to have aluminum nitrate and ammonium chloride in the ratio 10:4 in 500 ml of distilled water. A 100 gm of fibre was submerged into the solution and then pour 100 drops of ammonia to it and left it for one hour and then fire the sample in muffle furnace at 1000 degree and kept constant for 15 min at the same temperature.

NATURE AND STRUCTURE OF THE SAMPLE ATFER FIRING

The material formed is very porous in nature It is white in colour Giving foamy and coral structure X-ray diffraction studies for the present investigation were performed using an

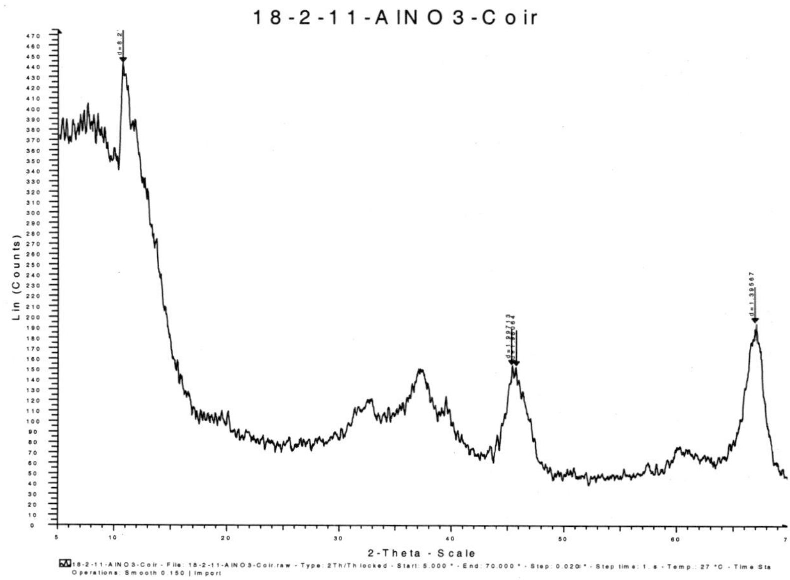

AXS Bruker D5005 diffractometer (Germany) with a vertical goniometer. X-ray generator was operated at 40 kV and 30 mA. CuKα (1 = 1·54056 Å) radiation was used with Ni filter. Electron microscopy study was carried out using a Hitachi H600 (Japan) electron microscope to characterize the alumina particles.

Xrd Pattern of Alumina(Al_2O_3) Particles when Embedded with Coir Fibre when Fired at 1000 Degree and Kept Constant at the Same Temperature for Fifteen Minutes

It is observed from the xrd pattern of the nanoparticles of alumina obtained by annealing method contains gamma alumina (Al_2O_3) (ICCD PDF card no 29-0063).along with gamma alumina some peaks of boehmite phase (ICDD PDF No 21-1307) along with bayerite peaks (ICDD PDF No 20-0011).xrd analysis shows that the powders contain both boehmite and bayetite phases with beyerite as the major phase. the sharp peaks of bayerite indicate the presence of highly crystalline and coarse particles a.k sivadasan et.al in his work found the gamma alumina when these particles were calcined in air to produce gamma alumina [20].

Scanning Electron Microscopy of the Sample

The morphology and average grain size of the powder alumina were examined by means of scan electron microscope

The morphology composed of net like structure

RESULT AND DISCUSSIONS

Alumina parcels have been successfully prepared through annealing method. The alumina particles were characterized using xrd diffraction. The importance of using annealing method for the processing of alumina particles is manifested in a shorter reaction time.

REFERENCES

[1] Brahmakumar M, Pavithran C, Pilai, R.M. Coconut fibre reinforced polyethylene composites; Effect of natural waxy surface layer of the fibre on fibre/matrix interfacial bonding and strength of composites, *Compos.Sci.Technol.* 2005,65,563-569.

[2] V.G. Geethamma et al., *Dynamic mechanical behavior of short coir fibre reinforced Natural Rubber composites: Composites: Part A* 36 1499 (2005).

[3] Yousif, B.F.; Ei-Tayeb N.S.M. Adhesive wear performance of TOPRPand UROPRP composites. Tribol. Lett. 2008, 32, 199-288.

[4] K M Rao and K M Rao 'Extraction and tensile properties of natural fibres'vakka,date and bamboo, composite structures 77, 288-295(2007).

[5] K. Sindhu and J Kuruvilla, degradation studies of coir fibre/polyester and glass fibre/ polyester composites under different conditions journal of reinforced plastics and composites, prepublication copy (2007)

[6] Polyolefin composites 2008, edited by domacius nwabunma and Thein kyu (newyork wiellyinterscience; A John Wielly and Sons Inc, Publications.

[7] R N Aracon, Jr Asian and Pacific Coconut Community

[8] V.G. Geethamma et al., *Composites: Part A* 36 1499 (2005).

[9] Chanakan Asasutjarit et al., *Composites: Part B* (2009).

[10] Xu Z, Xiao F S, Purnell S K, Alexeev O, Kawi S, Deutsch S E and Gates B C 1994, *Nature* 372 346.

[11] Drost M.K., Call C.J., Cuta J.M. and Wegeng R.S. 1997 J. Microscale Thermophys. Eng. 1, 321

[12] Prins R. 2001 Adv. Catal. 46 399

[13] Inoue M., Kazuhisa K., Hirokazu T., Hiroyuki N. and Tomoyuki I, 1989, *Clays and Clay Miner.* 37 71.

[14] Yoldas B.E. 1975 *Am. Ceram. Soc. Bull.* 54, 289.

[15] Mani T.V., Varma H.K., Warrier K.G.K. and Damodaran A.D. 1991 *J. Am. Ceram. Soc.* 74, 1807.

[16] Kannan T.S., Panda P.K. and Jaleel V.A. 1997 *J. Mater. Sci.* Lett. 16, 830.

[17] Mo C and Yuan Z 1993 Nanostruct. Mater. 2 47

[18] Michael D., Mangos P. and Baghurst D.R. 1991 *Chem. Soc.* Rev. 20, 1.

[19] Komarneni S., Pidugu R., Li Q.H. and Roy R. 1995 *J. Mater. Res.* 10, 1687.

[20] A.K. Sivadasan, (I Packia Selvam and Sankara Narayanan Potty.

[21] Amit Rai & C.N. Jha, Natural Fibre Composites And Its Potential As Building Materials, ED, BMTPC.

STUDY OF SURFACE POSITION CHANGES OF MATERIALS USING LASER SPECKLE PATTERN INTERFEROMETRY TECHNIQUE

R. Balamurugan and S. Muruganand

Department of Electronics & Instrumentation, Bharathiyar University, Coimbatore
E-mail: kctbala@gmail.com

ABSTRACT

A simple technique of speckle photography has been applied to measure small changes/deformation of the surface of laser scattering materials. Low-cost commercial charge coupled device photo camera provides the images of Laser speckle pattern with the beam splitter arrangement. A speckle pattern has been taken with the system at rest and then second image captured after the deformation was made in the surface of the material. By simple subtraction of the digital pictures a fringe pattern obtained called as specklegram; it gives information about modification of the position of surface of the material.

INTRODUCTION

Quick measurement of surface shape and deformation of mechanical parts of various materials is the ardent need of industries. Experimental methods in solid mechanics highly depend on surface displacement measurements. The conventional instruments cannot be used for those of soft materials and complex shape. Scanning method of mechanical probes takes much time hence it is not suited for quick and in-process measurement. Interferometric techniques are free from these issues have been mainly applied to optically smooth surfaces of materials. Speckle metrology, a simple and widely used non-destructive evolution tool in metrology [1] is used in surface deformation analysis under mechanical or thermal loading conditions and

V. Rajendran, P. Manivasakan and K.E. Geckeler (eds.)
Advanced Nano Materials for Industrial Applications, pp. 191-196 (2012).

determination of in plane translation [2].Speckle interferometry is used to measure the deformation of micro electromechanical systems [3, 4]. Other applications range from medical studies on bone dynamics to quality inspection of various products [5] and the measurement of the refractive index of a liquid in a cell was reported [6]. The most important ones are based on liquid penetrant, ultrasound, magnetic particle, eddy current, acoustic emission, radiology, active thermography and optical methods [7]. When an optically rough surface is illuminated with a coherent beam, a high contrast granular structure, known as speckle pattern is formed in the space is known as 'objective speckle pattern'. It can also be observed at the image plane of a lens and it is then referred as 'subjective speckle pattern' as shown in Fig. 1. The scattering regions are statistically independent and uniformly distributed between $-\pi$ and π. The speckles in the pattern undergo both positional and intensity changes when the object is deformed. The randomly coded pattern that carries the information about the object deformation provided to develop a wide range of methods, which can be classified into three broad categories: speckle photography, speckle interferometry and speckle shear interferometry.

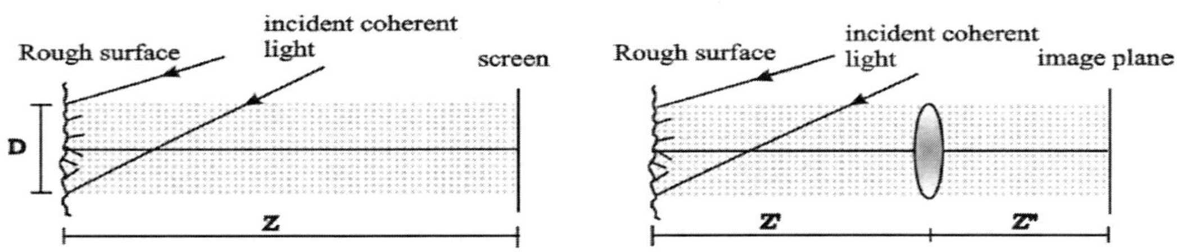

Fig. 1: Basic for Speckle Image Formation (objective and subjective pattern)

Speckle photography includes all those techniques where positional changes of the speckles are monitored, whereas speckle interferometry includes methods that are based on the measurement of phase changes and hence intensity changes. If instead of phase change, we measure its gradient, the technique falls into the category of speckle shear interferometry. All these techniques can be performed using digital/electronic detection using a CCD and imaging processing system. Illumination of a rough surface with coherent light produces a random intensity distribution in front of the surface, called speckle pattern [8]. Because the speckle pattern follows the movement of the scattering surface the speckle can be used for displacement/ deformation measurement [9]. Beam division and combination can be analysis on the basis of either by amplitude (Michelson, Fizeau, Mach-Zehnder and Jamin) or by wave front division (Young and Fresnel- biprism) methods.

SPECKLE INTERFEROMETRY

The optical setup for speckle interferometry is based on the Michelson interferometer [10]. The pattern that results by imaging a rough surface with a lens is itself a speckle field. The minimum size ρ_s of the image speckles is related to the optical system f-number F and the magnification M, and is given by:

$$\rho_s = 1.2(1 + M)\lambda F \qquad \text{... (1)}$$

where λ is the wavelength of the laser, and ρ_s is the radius of the Airy disc that is formed for the given optical imaging configuration.

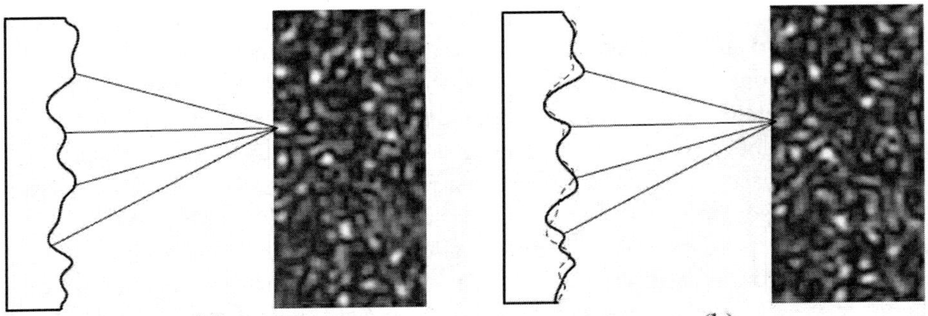

(a) (b)

Fig. 2: (a) Basic Speckle Pattern Formation (b) Changes in the Microstructure of the Surface (the original surface is indicated by the dashed line)

The resultant intensity of each point of the object before deformation is given by:

$$I_{before} = I_{obj} + I_{ref} + 2\sqrt{I_{obj}}\sqrt{I_{ref}}\cos(\varphi_0) \qquad \text{... (2)}$$

where I_{obj} and I_{ref} are the local intensities of the object and reference beams respectively and φ_0 is the unknown, and random, initial phase distribution of the speckle pattern at that point. When the object moves, the phase distribution undergoes a change and the intensity at the object becomes:

$$I_{after} = I_{obj} + I_{ref} + 2\sqrt{I_{obj}}\sqrt{I_{ref}}\cos(\varphi_0 + \varphi_{obj}) \qquad \text{... (3)}$$

where $\Delta\varphi_{obj}$ is the phase change induced by object deformation, which is directly related to the out-of-plane displacement of the object according to the sensitivity vector theory:

$$u_z = \Delta\varphi_{obj}/\,2\pi\cdot \lambda/(1 + \cos\theta i) \qquad \text{... (4)}$$

where λ is the wavelength used and θi is the incidence angle. From equation (3) it is possible to determine the displacement, once $\Delta\varphi_{obj}$ is known. A simple approach to measure $\Delta\varphi_{obj}$ is to record I_{before} and I_{after} and to calculate the difference is:

$$\Delta I = I_{after} - I_{before} = 2\sqrt{I_{obj}}\sqrt{I_{ref}}\,[\cos(\varphi_0 + \Delta\varphi_{obj}) - \cos\varphi_0] \qquad \text{...(5)}$$

which can be rewritten as

$$\Delta I = 4\sqrt{I_{obj}}\sqrt{I_{ref}}\sin\{\varphi_0 + \Delta\varphi_{obj}/\,2\}\sin\Delta\varphi_{obj}/2 \qquad \text{... (6)}$$

The intensity change ΔI depends on the random initial phase distribution φ_0 as well as on the deformation induced phase change $\Delta\varphi obj$, therefore a statistical consideration is required to find a unique relationship between intensity change ΔI and phase change $\Delta\varphi obj$ during deformation.

EXPERIMENTAL SETUP

It basically uses a setup analogous to a Michelson interferometer in which both mirrors are replaced by scatter surfaces. The resultant speckle pattern in the image plane is formed by the interference of the two speckle fields issuing independently from the two scatter surfaces. One of the surfaces is subjected to deformation and the other serves as the source of reference speckles [11]. The interested object is a rectangular aluminium plate with a size of 220 mm and 165 mm. The light source is a 5-mW laser diode at the wavelength 680 nm. The beam splitter divides the laser beam into two parts, each of which illuminates a different surface. The beam splitter also serves to combine the light diffused by the two surfaces. An image forming optical system, for example a converging lens is used to collect the diffused light and project the image onto a screen. For these conditions there are two overlapping images of the diffusing surfaces on the screen, each being a speckle field. Because the scattering angle is typically wide, such surfaces do not need to be exactly aligned. We used a polished glass plate with partial metal coating on one face as a beam splitter.

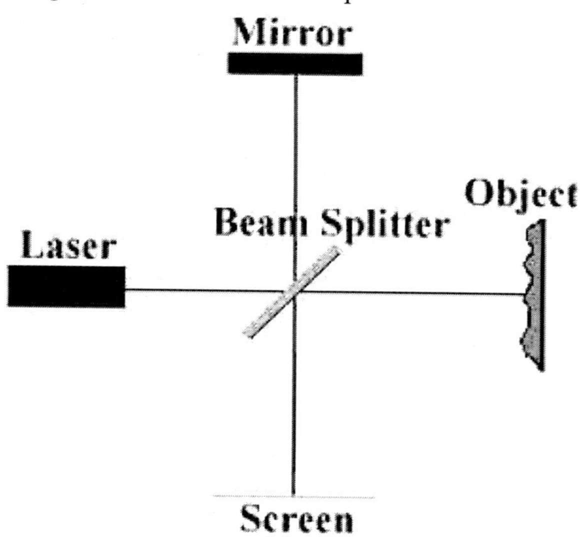

Fig. 3: Basic Model Interferometer

RESULT AND ANALYSIS

Laser speckle interferograms of deformation on the aluminium plate are obtained before and after deformation. The subtraction cancels out the speckle grains that are unchanged leaving a dark area in their place. Overall a noisy bright fringe pattern that appears shows the area displaced by odd multiples of $\lambda/4$. Contigious fringes differ by integer, it means by a step displacement of $\lambda/2$. Now the fringe decoding provides the information of deformation.

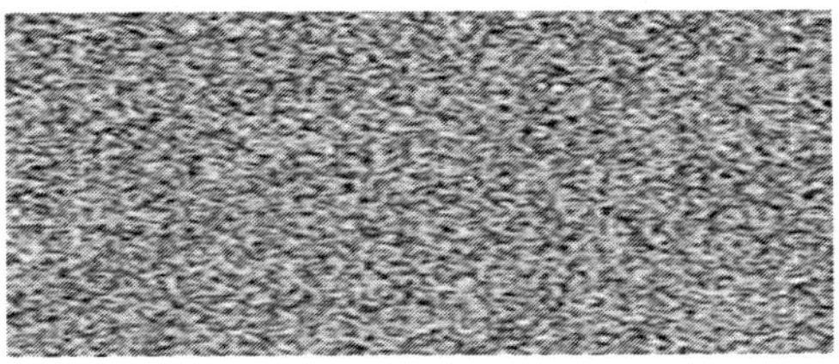

Fig. 4: (a) Before Deformation

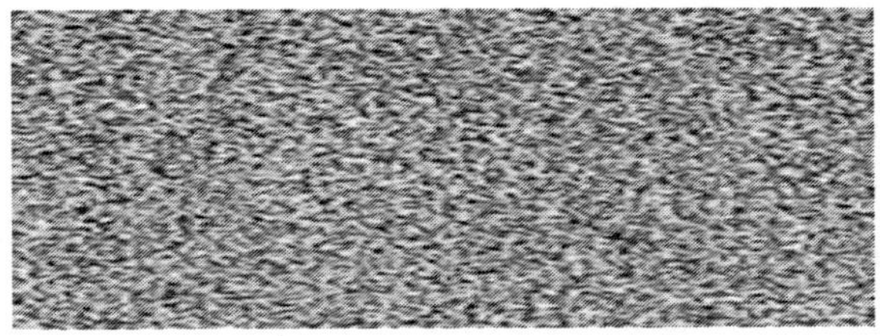

Fig. 4: (b) After Deformation

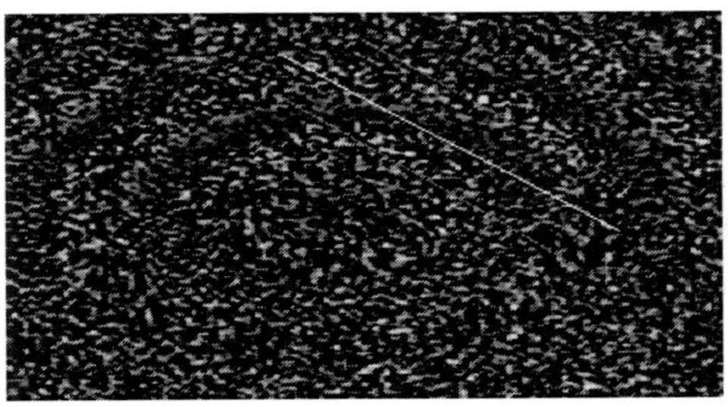

Fig.4: (c) Fringe Pattern – Subtracted

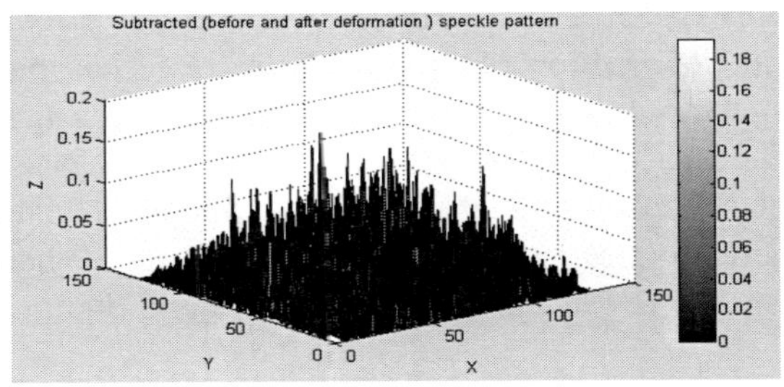

Fig. 5: Correlation using Matlab

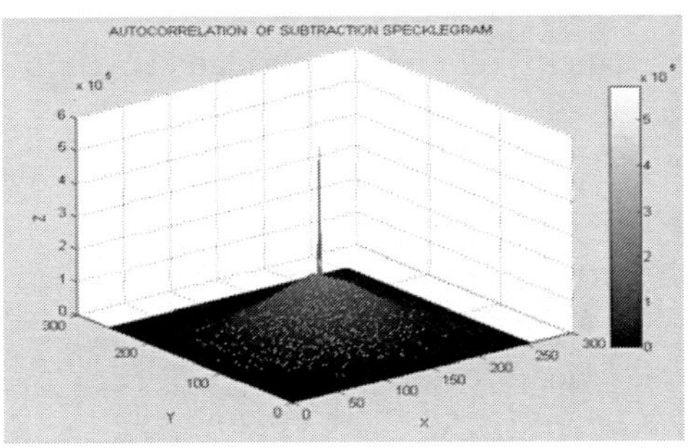

Fig. 6: Peak Indicates Deformation

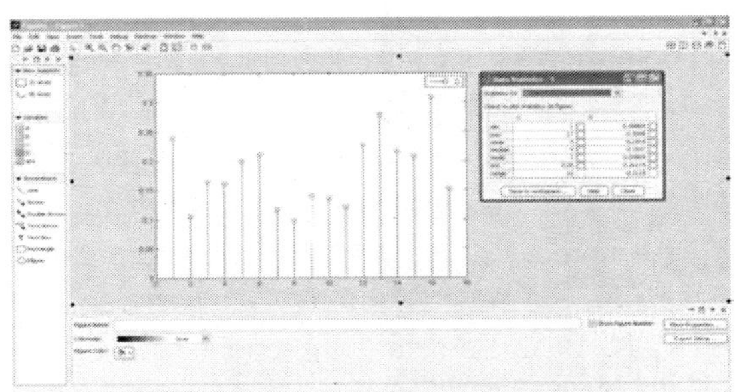

Fig. 7: Deformation Measurement

To image the surfaces and detect the speckles, we used a CCD photo camera, adjusting its operation. The major requirement is that most of the speckle grains are resolved, that is, their size exceeds that of the camera pixels so we must adjust the speckles [12]. To find the deformation by the fringe analysis method, two red lines are drawn to be tangent to the center of two adjacent fringes. The yellow line is drawn to pass through the center of one of the fringes where it intersects the edge of the test piece. The distance between the two red lines is the width of one fringe which we call "X" (here measured to be 5.02 in arbitrary units) and the distance from the center of the lower fringe to the yellow line we have called "Y" (here measured as 1.24). The flatness of the test piece is then: Y/X = 1.24/5.02 = 0.247 fringes. Hence this is approximately equal to 6ìm displacement/deformation. Strain gauges and extensometers available to determine the small deformation, but they are not suitable in all environment with accurate measurements.

ACKNOWLEDGEMENT

The authors would like to thank management of Kumaraguru College of technology for the continuous support to this work.

CONCLUSION

In this paper, a simple double exposure electronic speckle photography technique to detect small deformation of an aluminium plate has been done. Speckle displacement is detected from the position of correlation peak. Its results agree well those obtained from the conventional methods. The components for this work are typically inexpensive and the image processing of the laser speckle pattern is simple. Digital correlation analysis of speckle patterns arising from coherent optical systems is used effectively for non contact and non destructive measurements by means of low cost photonic device and digital technology.

REFERENCES

[1] J.W.Goodman, "Some fundamental properties of speckle", *J opt.Soc.Am*, 1145-1149 (1966).

[2] Jones R Wykes C, "Holography and speckle interferometry", 2nd ed. Cambridge University Press, New York, USA, 122-142, (1989).

[3] R. S. Sirohi "Speckle Metrology", Marcel Dekker, New York (1993).

[4] R.K. Erf "Speckle Metrology", Academic Press, New York (1978).

[5] P.Aswendt, C. D. Schmidt, D. Zielke, and S. Schubert "ESPI solution for non contacting MEMS-on-wafer testing", *Opt. Lasers Eng.* 40, 501–515 (2003).

[6] P. K. Buah-Bassuah, F. Francini and G. Molesini, "Measurement of refractive index by double exposure speckle pattern recording", *Am. J.Phys.*57, 366– 370, (1989).

[7] P Shull, "Non-destructive Evaluation", Marcel Decker, New York (2002).

[8] J.W. Goodman, "Statistical Optics", John Wiley and sons, New York (1985).

[9] A.R.Ganesan, Sharma D.K., and Kothiyal M. P, "Electronic speckle shearing phase shifting pattern interferometer", *Appl. Opt.*, 27, 4731 (1988).

[10] M. Franc, "Laser Speckle and Applications in Optics", Academic Press, New York (1979).

[11] J. C. Dainty, "Laser Speckle and Related Phenomena", Springer- Verlag, Berlin, 2nd ed. (1984).

[12] R.S.Sirohi, Tay, C.J., Shang, H.M., Boo, W.P. "Non-destructive assessment of thinning of plates using digital shearography", *Opt. Eng.* 38 (9), 1582–1585 (1999).

SILVER NANOPARTICLE SIZE AND SPACING CONTROL FOR LOW-REFLECTIVITY STRUCTURE DEVELOPMENT

NAZME MOUSHUMY[1], KAMAL ALAMEH[1,2], V. RAJENDRAN[3] AND YONG TAK LEE[2]

[1]*Electron Science Research Institute, Edith Cowan University, 270 Joondalup Dr, Joondalup, WA 6027, Australia*
[2]*Department of Nanobio Materials and Electronics, Gwangju Institute of Science and Technology, World-Class University (WCU), Gwangju Institute of Science & Technology (GIST), Gwangju 500-712, South Korea*
[3]*Centre for Nano Science and Technology, K S R College of Technology, K S R Kalvi Nagar, Tiruchengode, Namakkal (Dt.), Tamil Nadu, India*
E-mail: k.alameh@ecu.edu.au, nmoushum@our.ecu.edu.au

ABSTRACT

In this paper, we investigate the impact of silver thin film thickness and annealing temperature for the fabrication of silver nano-particles of controlled size and spacing distributions. We also use these measured distributions to predict the performance of subwavelength grating structures developed using dry and isotropic etching of semiconductor substrates. Silver (Ag) thin films of different thicknesses are deposited on Si and GaAs semiconductor substrates and annealed at different temperatures. Experimental results demonstrate that by annealing the Ag thin films with different temperature profiles it is feasible to develop Ag nanoparticles of an average diameter ranging from 50 nm to 400 nm on silicon substrates and 100 nm to 500 nm on GaAs substrates.

In addition, different subwavelength structures developed by etching of the Ag nanoparticle deposited Si and GaAs substrates are simulated using a Finite-Difference Time Domain (FDTD) software package], and the simulation results show that substantial reduction in light reflection can be achieved by optimizing the heights of the subwavelength structures through the control of the etching time.

V. Rajendran, P. Manivasakan and K.E. Geckeler (eds.)
Advanced Nano Materials for Industrial Applications, pp. 197-202 (2012).

INTRODUCTION

Subwavelength grating (SWG) structures have gained an enormous interest over the last decades in the field of the photovoltaic due to several interesting advantages. For example, integrating a SWG onto a solar cell device provides almost a lossless reflecting surface that enhances the solar cell's efficiency. The SWG often takes a one- and/or two-dimensional periodic configuration; If the pitch (or period) of a single grating structure is less than the wavelength of the incident light, it behaves like a homogeneous medium with an effective refractive index [1]. Thus, SWG structures give gradual changes in the refractive index assuring an excellent antireflective medium along with a light trapping phenomena in comparison to the planar thin films [2-4]. A nanorod structure acts as a single layer antireflective (AR) coating, while the triangular (conical) and parabolic shaped grating structures are more advantageous since they behave like a multilayer broadband AR coating [2].

In this paper, we report on the silver nanoparticle size and spacing control for low reflectivity structure development. Silver nanoparticles are fabricated by thermal annealing and predict the performance of conical shaped SWG structure realized through the etching of Si or GaAs substrates whereon the developed silver nanoparticles are deposited. Several silver (Ag) thin films of thickness 10nm, 8nm and 5nm are deposited on GaAs and Si substrates and annealed at different temperatures leading to the development of randomly distributed silver nanoparticles on the surface of each sample. Different nanoparticle diameter and spacing are achieved by controlling annealing temperature and film thickness that gives realization of SWG structures having different grating width and spacing. FDTD simulation of that SWG structures confirm the low reflection in comparison with the flat substrates.

ANTIREFLECTIVE COATING AND SWG STRUCTURE

The etching of Si or GaAs substrates on which Ag nanoparticles are embedded, results in the formation of SWGs which act as antireflective coatings that enhance the transmission through the substrates. Minimum reflection can be obtained through multilayer thin-film coating [1], however, in addition to their complex fabrication process [1-3], optical coatings have practical drawbacks, such as adhesion and material mismatching, making the thin-film mechanically unstable and susceptible to thermal fluctuations.

The use of SWG structures has been inspired by the moth's eyes. The surface of a moth's eye is covered with nano-structured film that absorbs most of the incident light with a minimal reflection. The nano-structured film consists of a hexagonal pattern bump which is 200nm high acting as an AR coating, because the bumps are smaller than the wavelength of the visible light. The refractive index between the air and the surface changes gradually in which case the reflection of the light is also decreased. This model is being applied in solar cells to increase the cell conversion efficiency by allowing large amount of electromagnetic waves to reach the embedded charge carriers zone [4].

In the rectangular-shaped grating structure shown in Fig. 1, the refractive index changes very sharply from air (n = 1) to the grating zone or structure (approximately 2.5 at TM mode) [1]. The reflection loss can be calculated easily by Fresnel's equation:

$$R = \left(\frac{n_2 - n_1}{n_2 + n_1} \right)^2 \quad \text{... (1)}$$

where n_1 is the refractive index of first medium and n_2 is the refractive index of second medium.

From Fig. 4, the total reflection of the grating structure can be calculated using the following equation:

$$R_{total} = R_1 + R_2 \quad \text{... (2)}$$

where R_1 is the reflection at the interface of air and grating structure and R_2 is the reflection at the interface of the grating structure and substrate.

If the grating structure has a shorter period than the wavelength of the incident light, then it acts as a homogeneous medium with an effective refractive index [3]. According to Fig. 1, the reflection is approximately 20.2% for a rectangular-shaped grating structure, however, for the conical shaped SWG structure shown in Fig. 1, the effective refractive index changes gradually resulting in reflection below 8% [3]. For a parabolic-shaped SWG structure, the reflection can be as low as 5% over the entire solar spectrum [1].

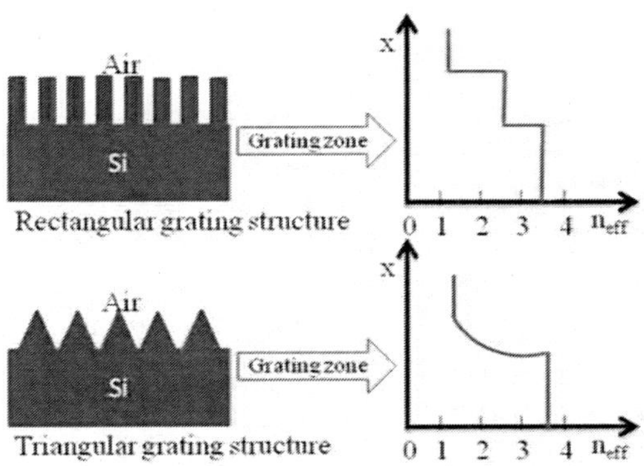

Fig. 1: Grating Geometry and Plot of Subwavelength Grating (SWG) Height Versus Refractive Index (n) in Silicon (Si)

SWG FABRICATION STEPS

There are several conventional methods for the fabrication of conical SWG structures, such as photolithography, colloidal formation, nanoimprint and nanoparticles. Unfortunately, perfect conical shapes cannot be obtained by any of these methods. The formation of uniformly distributed nanoparticles and the use of dry etching in conjunction with additional isotropic etching processes can produce SWG with almost conical shapes. Figure 2 illustrates the fabrication steps for creating the SWG structure, which were reported by Song et al. [2].

At first the nano-particles can be formed using the annealing process, then CF4 or O2 gas is flown through them using dry plasma etching process. To shape the formed nanorods in a perfect conical shape an additional isotropic etching process can be used as reported in [2, 4]. In this paper, the fabricated Ag nano-particles are proposed to be used as a mask to realize the conical shapes in two-dimensional grid-like distribution to realize the SWG structure.

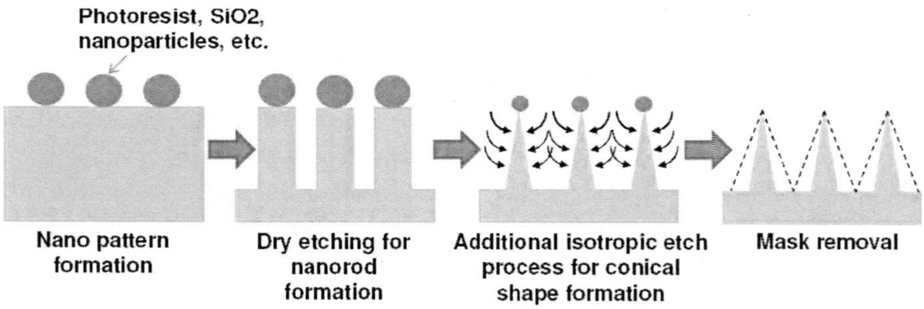

Fig. 2: Steps for the Formation of SWG Structure

EXPERIMENTAL RESULTS

Silver (Ag) Nanoparticles on GaAs and Silicon were fabricated through sputtering and annealing method subsequently. Annealing was performed in an oven placed in atmospheric environment. After setting the temperature, time and ramp, the oven was allowed to start to reach the set temperature and then cool down to the atmospheric temperature.

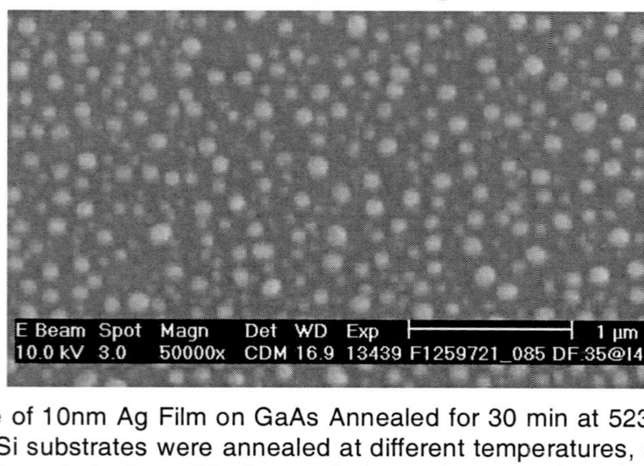

Fig. 3: SEM Image of 10nm Ag Film on GaAs Annealed for 30 min at 523K 10nm, 8nm, 5nm Ag on GaAs and Si substrates were annealed at different temperatures, 523K, 573K, 623K, 673K and 723K. Characterization of that annealed samples were done by using SEM method and ImageJ software. Figure 3 shows SEM image of Ag nanoparticles on GaAs substrates.

The average diameter and average distance between nearest nanoparticles were measured by ImageJ software. With these data we calculated the reflection of a predicted SWG structure having grating width equivalant to nanoparticle diameter and spacing between gratings equivalent to spacing between nanoparticles by FDTD method (softeware that devolped by Optiwave Inc.) [5], which numerically solves the Maxwell's equaitons within a certain medium, the light propagation, scattering, reflection and polarization were simulated for the conical shaped SWG structure shown in Fig. 4.

Figure 4 also schematically illustrates the electric field density distribution and the procedure for simulating the conical shaped SWG structure. The incident light hits directly on top of the SWG grating structure (or nano-structure). A major portion of incident electromagnetic wave is absorbed by the grating zone due to the gradual change of refractive index, while the rest of the wave is reflected and transmitted.

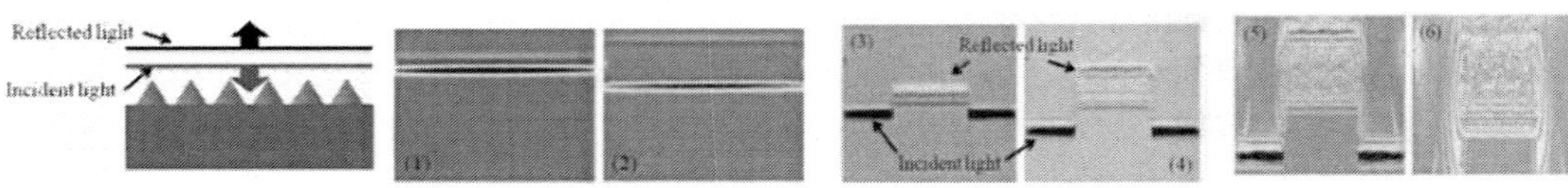

Fig. 4: Schematic Diagram of Conical Shaped Subwavelength Grating (SWG) Structure and the Simulation Steps of the Propagation of the Electromagnetic Waves Across the Simulated Structure

Figure 5 shows the average spacing between the silver nanopartciles versus the annealing temperature for 10 nm thick Ag thin film deposited onto Silicon and GaAs substrates. It is obvious that for the silicon substrate, the average spacing of 200 nm is maintained over the temperature range of 520-575 K, whereas for a GaAs films the spacing increases with increasing the temperature (this is, however, useful for some applications requirind some degrees of spacing control).

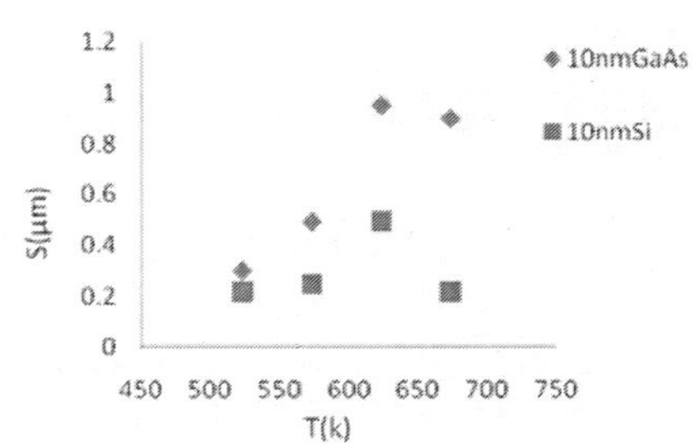

Fig. 5: Average Spacing S(μm) Between the Ag Nanoparticles Versus Annealing Temperature for a 10 nm Ag thin film Deposited onto Si and GaAs Substrates

Figure 6 shows the reflection spectrum for Ag films of different thicknesses deposited onto Si substrates, and annealed at different temperatures. It is seen that a 10nm thick Ag film on Si annealed at 523K yields maximum reflection as low as 12% while the 8nm thick Ag film annealed at 573K results in maximum reflection of 13% and the 5nm thick Ag film annealed at 623K produces maximum reflection below 7% over a wide range of wavelength.

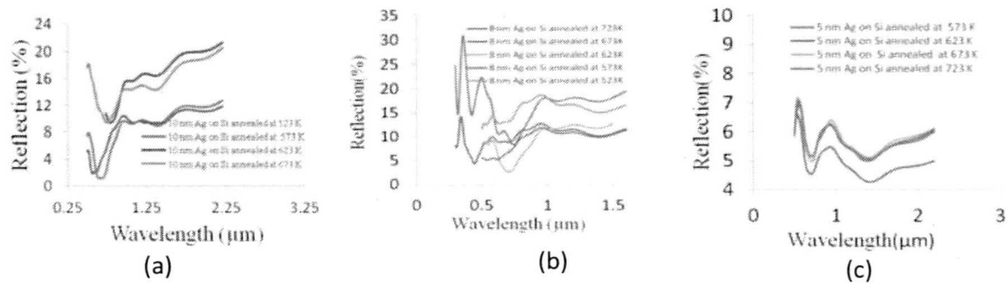

Fig. 6: Reflection Spectra against Wavelength for Various SWG Structures of 10nm, 8nm and 5nm Ag on Si Substrates

CONCLUSIONS

Ag thin films with different thicknesses have been deposited on semiconductor substrates and annealed at different temperatures to fabricate Ag nano-particles that can be used to develop cost-effective AR coatings for semiconductor devices. We have found that desired diameter and spacing between nanoparticles can be achieved by controling annealing temperatute and Ag film thickness that can be ultimately used to fabricate SWG structures having desired grating width and spacing between gratings which ensures the minimum reflection.

ACKNOWLEDGEMENTS

This research is supported by the Faculty of Computing, Health and Science, Edith Cowan University, and the World-Class University Program funded by the Ministry of Education, Science, and Technology through the National Research Foundation of Korea (R31-10026).

REFRENCES

[1] Song. Y.M. and Lee. Y.T. "Simulation of antireflective subwavelength grating structure for optical device applications," In proceedings of the *9th International Conference on Numerical Simulations of Optoelectronic Devices 2009* (NUSOD 09), pp. 103-104, 2009.

[2] Song. Y.M., Jang. S.J., Yu. J.S., and Lee. Y.T., "Bioinspired parabola subwavelength structures for improved broadband antireflection," *Small*, vol. 6, pp. 984-987, 2010.

[3] Schmid. J.H., Cheben. P., Janz. S., Lapointe. J., E.Post, Delage. A., Densmore. A., Lamontagne. B., Waldron. P., Xu. D.-X., "Subwavelength grating structures in planar waveguide facets for modified reflectivity," In proceedings of the Photonics North 2007 conference, Ottawa, Ontario, Canada, Proc. SPIE, vol 6796, 67963E, 2007.

[4] Lee. Y.T. "*High-efficiency GaInP/LCM-GaInP/GaAs triple-junction solar cells with antireflective subwavelength structures,*" Gwangu Institute of Science and Technology (GIST), Republic of Korea, 2009.

[5] "Finite Difference Time Domain Photonics Simulation Software," OptiFDTD Technical background and Tutorials. Version 8.

SYNTHESIS AND THERMALLY STIMULATED LUMINESCENCE STUDIES OF -IRRADIATED $CdSiO_3$: Ni^{2+} NANOCRYSTALLINE PHOSPHORS

D.V. Sunitha, C. Manjunatha[1], H. Nagabhushana, S.C. Sharma, B.M. Nagabhushana[1] and R.P.S. Chakradhar[2]

Centre for Nano Research (CNR), Tumkur University, Tumkur
[1]Department of Chemistry, M.S. Ramaiah Institute of Technology, Bangalore
[2]National Aerospace Laboratories (CSIR), Bangalore
E-mail: sunitha_30386@rediffmail.com; bhushanvlc@gmail.com

ABSTRACT

Nanocrystalline Ni^{2+} (1-7mol %) doped $CdSiO_3$ phosphors have been synthesized for the first time by a low temperature solution combustion technique using oxylyldihydrizide (ODH) as a fuel. The powders were well characterized by powder X-ray diffraction (PXRD. The crystallite size estimated by Debye Scherer's method and is found to be in the range 40-80 nm. Thermoluminescence properties of $CdSiO_3$ doped with Ni^{2+} of different concentration (1-7 mol%) have been studied by irradiating the samples with 6kGy γ-dose at a heating rate of 5 °C s^{-1}. Two TL glow peaks at 160 °C and 252 °C was recorded for all irradiated samples. The maximum TL intensity was observed for 7 mol% Ni^{2+} doped sample. Further, $CdSiO_3$:Ni^{2+} (7 mol%) was exposed to ã-irradiation in the dose range 1-6 kGy. Up to 2 kGy a single glow peak at 160°C was recorded. Upon increase the γ-doseto 2 kGy an additional TL glow peak at 252°C was observed. The glow peak intensity at 160°C and 252°C increases with increase in γ-dose. Effect of heating rates (5-15°Cs^{-1}) was also studied by irradiating the samples for 3kGy. The TL intensity decreases with the increase of heating rate and glow peak was shifted towards higher temperature side. Kinetic analysis of the experimental TL glow curves has been estimated using Chen's glow peak shape method. The activation energy (E) and frequency factor was found to be ~1.09eV and 1.69x10^{09} s^{-1} respectively.

V. Rajendran, P. Manivasakan and K.E. Geckeler (eds.)
Advanced Nano Materials for Industrial Applications, pp. 203-208 (2012).

INTRODUCTION

Materials with nanosize finds wide applications due to its enhanced physical, chemical, structural and optical properties. Further, materials doped with rare earth/transition metal ions finds potential applications in the field of luminescence such as cathode ray tubes, plasma display panels, TLD materials, etc [1]. Thermoluminescence (TL) is a simple technique for obtaining information on the distribution of artificially created or naturally occurring point defects. The light emission during TL process depends on the type of ionizing radiation (X-rays, γ-rays, ions, etc.) used [2]. Silicate materials are found to be the best host materials than oxides, sulphides and aluminates, due to their excellent stability, high resistance to acid, alkali and oxygen and easy preparation, etc. [3].

Nanomaterials can be synthesised by different techniques. Among them, combustion synthesis is very popular and advantageous due to wide choice of fuels, produce homogeneous product in a short time without the use of expensive high temperature furnaces. This synthesis technique makes use of the heat energy liberated by the redox exothermic reaction at a relatively low igniting temperature between metal nitrates and fuels [4]. Here, we report the synthesis the $CdSiO_3:Ni^{2+}$ nanophosphor by solution combustion technique and thermoluminescence (TL) properties of the ã-irradiated samples.

EXPERIMENTAL AND INSTRUMENTS USED

Cadmium nitrate ($Cd(NO_3)_2.4H_2O$), Nickel nitrate ($Ni(NO_3)_2.4H_2O$) , fumed silica (SiO_2 , 99.9%, surface area 200 m^2/g) and fuel oxylyldihydrizide (ODH: $C_2H_6N_4O_2$) as starting materials used for combustion method. The detailed synthesis was given elsewhere [5]. The obtained product was then annealed at 800 °C for 2h and used for characterization. The phase purity and crystallinity of the samples were examined by using powder X-ray diffraction (PXRD), which was recorded on a Philips X'pert PRO PAnalytical X-ray diffractometer with graphite monochromatized Cu Kα radiation source (λ = 1.5418 Å) operated at 40 kV and 30 mA. The XRD patterns were collected in the range of $10^\circ \leq 2\theta \leq 70^\circ$. The Thermoluminescence (TL) measurements were measured on a Nucleonix TL Reader with a heating rate of $5^\circ Cs^{-1}$ in the temperature range from 40 – 300°C. Before measurement, the powder sample was exposed to ã-radiation in the range 1-5 kGy and the measurements were carried out at room temperature.

RESULTS AND DISCUSSION

The XRD pattern of $CdSiO_3:Ni^{2+}$ (7 mol%) sample is shown in the Fig. 1 and the diffraction patterns was well matched with the standard JCPDS card no. 35-0810 [6]. The average crystallite size of the $CdSiO_3:Ni^{2+}$ was estimated from the line broadening of the XRD peaks using Scherer's method [7]

$$D = \frac{k\lambda}{\beta \cos\theta} \quad \text{... (1)}$$

where d is the average diameter of the nanoparticles, λ is the wavelength of CuKα (1.54 Å) radiation, β (in radians) is full width at half maxima (FWHM) and θ is the Bragg angle. And, the average crystallite size was found to be 40-80 nm.

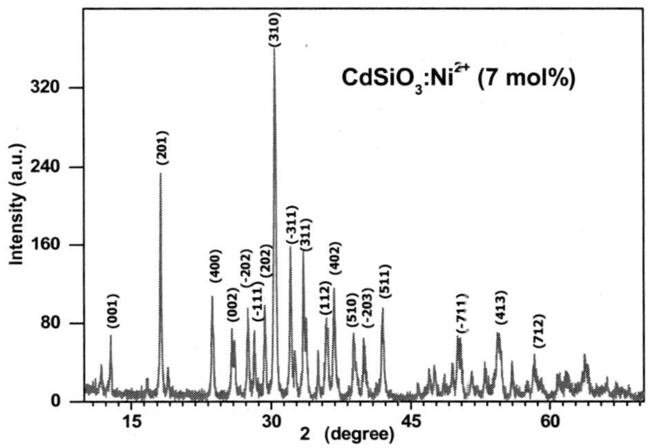

Fig. 1: PXRD Pattern of $CdSiO_3:Ni^{2+}$ (7 mol%) Phosphor Calcined at 800 °C for 2hr

THERMOLUMINESCENCE

Figure 2 shows the TL glow curves were measured for $CdSiO_3:Ni^{2+}$ (1-7 mol%) nanophosphor exposed to 6 kGy γ-dose at a warming rate of 5°Cs^{-1}. As can be seen from the inset of Fig. 2 the TL intensity increases with increase of Ni^{2+} concentration. Two TL glow peaks at 160 and 252°C was recorded for all concentrations of Ni^{2+} doped $CdSiO_3$. Further, the 252°C glow peak is well resolved for 7 mol% Ni^{2+}.

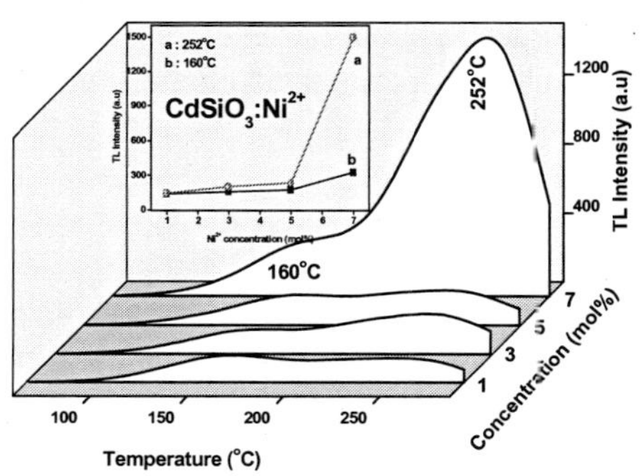

Fig. 2: TL Glow Curves of $CdSiO_3:Ni^{2+}$ (1-7 mol%) Nanophosphor Irradiated with 6 kGy γ-dose (Inset: Variation of TL Intensity with Concentration)

Effect of different γ-dose (1-6 kGy) on $CdSiO_3:Ni^{2+}$ (7 mol%) was studied. TL glow peaks at 160 °C (Shouldered) and 252 °C (well resolved) was recorded at a warming rate of 5°C s^{-1} (Fig. 3). The 252°C peak was observed only after 2 kGy and was more prominent then 160°C glow peak which increased sublinearly with the γ-dose (Inset Fig. 3). And, also the peak at 160°C was also increasing sublinearly with γ-dose. However, the sublinear, response was attributed to the to the changes in the population of the luminescent and trapping centers [8] due to the use of ionizing radiation (γ-rays). Further, we have also tried to study the effect of different warming rates on the TL glow curves 3kGy γ-irradiated $CdSiO_3:Ni^{2+}$ (7 mol%) nanophosphor in the range 5-15 °Cs^{-1}. With the increase in the warming rate the the the two glow peaks were reduced to one peak with shift towards higher temperature side with decrease in TL intensity (Fig.4), which may be due to the well known phenomena of thermal quenching [9].

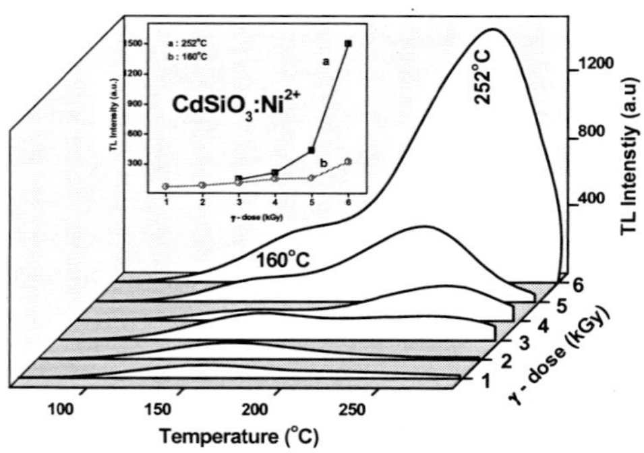

Fig. 3. TL Glow Curves of Ni^{2+} Doped (7 mol %) in $CdSiO_3$ Nanocrystalline Phosphor γ- irradiated for 1-6kGy (Inset: Variation of TL intensity (I_m) as a function of γ- dose (1-6kGy)).

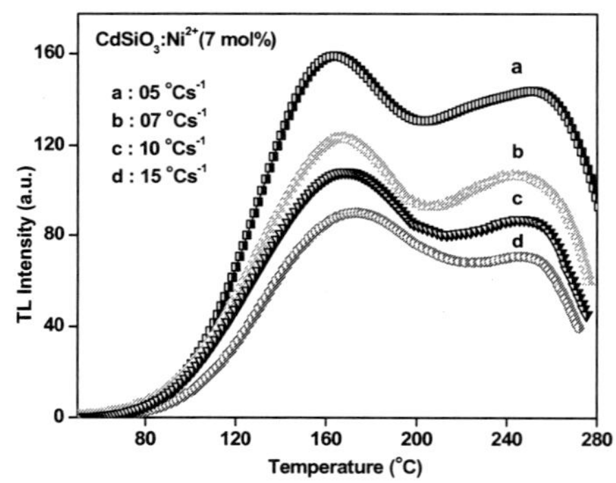

Fig. 4: The Effect of Warming Rate on TL Response of $CdSiO_3:Ni^{2+}$ (7 mol%).

The determination of kinetic parameters has been an active area of research and various techniques have been developed to derive these parameters from the glow curve. However, we apply peak shape method to the whole experimental glow curve and evaluate the parameters which define as

$$\delta = T_2 - T_m \quad \omega = T_2 - T_1 \quad \tau = T_m - T_1 \quad \mu_g = \frac{\delta}{\omega} \qquad \text{... (2)}$$

where T_m is the maximum temperature and T_1 and T_2 are low and high half-intensity temperatures of the glow curve at the ascending and descending side of the peak, μ_g gives the order of kinetics whose value is 1 for $0.42 > \mu_g > 0.48$ and 2 for $0.48 > \mu_g > 0.52$. Therfore, Chen's peak shape method [10] was used to calculate the trapping parameters by glow curve deconvolution using ORIGIN 8.1 software. The trap depth and frequency factor are calculated by using the following equations.

$$E_\alpha = C_\alpha\left(\frac{KT_m^{\ 2}}{\alpha}\right) - b_\alpha(2KT_m) \qquad \text{... (3)}$$

where α = τ, δ, ω

$$C_\tau = 1.51 + 3.0(\mu_g - 0.42)\ C_\delta = 0.976 + 7.3(\mu_g - 0.42)$$

$$C_\omega = 2.52 + 10.2(\mu_g - 0.42)\ b_\tau = 1.58 + 4.2(\mu_g - 0.42)\ b_\delta = 0\ b_\omega = 1$$

Once E and b are known, s can be evaluated by using Kittis relation [11] given by

$$\frac{\beta E}{kT_m^{\ 2}} = s\exp\left\{\frac{-E}{kT_m^{\ 2}}\right\}[1 + (b-1)\Delta_m] \qquad \text{... (4)}$$

Where $\Delta = \frac{2kT}{E}; \Delta_m = \frac{2kT_m}{E}$ and β is the linear heating rate and k is Boltzmann constant (8.6×10^{-5} eV K^{-1}). The trapping parameters (E, b, s) for $CdSiO_3:Ni^{2+}$ with different warming rate are estimated and shown in Table 1.

Table 2: Estimated trap Parameters of 3 kGy γ-irradiated $CdSiO_3:Ni^{2+}$ (7 mol%) for Different Warming Rate

Heating Rate($^oCs^{-1}$)	*Peak*	*T_m(^oC)*	*b (μ_g)*	*Activation Energy(eV)*				*Frequency Factor(s^{-1})*
				E_τ	E_δ	E_ω	E_{ave}	
05	1	160	2(0.50)	0.606	0.665	0.638	0.636	2.5E+08
	2	233	2(0.50)	1.050	1.078	1.07	1.066	5.4E+11
	3	270	2(0.48)	1.996	1.957	1.991	1.981	5.5E+19
07	1	157	2(0.49)	0.532	0.602	0.569	0.568	4.0E+08
	2	229	2(0.49)	0.907	0.961	0.938	0.935	2.8E+10
	3	262	2(0.48)	2.448	2.367	2.426	2.414	1.6E+24
10	1	162	2(0.50)	0.524	0.592	0.560	0.559	2.6E+08
	2	239	2(0.50)	1.292	1.303	1.305	1.300	1.0E+14
	3	264	2(0.48)	1.055	1.922	1.015	1.997	5.0E+19
15	1	161	2(0.48)	0.478	0.556	0.518	0.518	8.4E+08
	2	242	2(0.48)	1.153	1.194	1.18	1.175	4.6E+12
	3	266	2(0.50)	3.466	3.248	3.380	3.365	1.6E+23

CONCLUSION

The TL properties of $CdSiO_3$ doped with Ni^{2+} (1-7 mol%) nanophosphors has been investigated for 6kGy γ-irradiation. Two glow peaks were observed at 160°C and 252°C. With the increase in concentration the TL intensity increases and maximum intensity was observed for 7 mol% Ni^{2+} concentration. And also, the effect of dose in the range 1-6 kGy was carried ot for which the Tl intensity increase sublinearly with the dose. Further, the effect of warming rate on the 3kGy ã-irradiated $CdSiO_3:Ni^{2+}$ was observed where the TL intensity decreases and the glow peaks shifted towards higher temperature side with increase in warming rate.

ACKNOWLEDGEMENTS

One of the author DVS thank IUAC, New Delhi for the award of research fellowship under UFR-47314 scheme.

REFERENCES

[1] D.D. Jia, J. Zhu, B.Q. *Wu, J. Lumin.* 93 (2001) 107-114.

[2] S. Xiaolin, H.Guangyan, D. Xinyong, X.Dong, Z.Guilan, T.Guoqing, C.Wenju, *J. Phys and Chem.* Solids 62 (2001) 807.

[3] C. Barthou, J. Benoit, P. Benalloul, *J. Electrochem.* Soc.141 (1994) 524-528.

[4] K .C. Patil, S.T. Aruna, S. Ekambaram, Curr. Opin. *Solid State Mater. Sci.* 2 (1997) 158-165

[5] C. Manjunatha, D.V. Sunitha, H. Nagabhushana, S.C. Sharma, S. Ashoka,J.L. Rao, B.M. Nagabhushana, R.P.S. Chakradhar, *Matt. Res. Bull.* (2012) In press.

[6] B.F. Lei, Y.L. Liu, J. Liu, Z.R. Ye, C.S. Shi, *J. Solid State Chem.* 177 (2004) 1333-1337.

[7] P. Klug, L.E. Alexander, in: X-ray Diffraction Procedure, Wiley, New York, 1954

[8] J L Lawless, R Chen, D Lo and V Pagonis, *J. Phys.: Condens. Matter 17*, 737 (2005).

[9] Akselrod, M.S.; Larsen, N.; Whitley, V.; Mckeever, S.W.S. *J. Appl. Phys*. 1998, 84, 2636–2640.

[10] R.Chen, Y.Krish, Analysis of Thermally Stimulated Processes, Pergamon Press, New York (1981).

[11] C. Furetta, Handbook of Thermoluminescence, World Scientific, Singapore (2003).

PZT BASED NANOSTRUCTURED FUNCTIONAL MATERIALS FOR DEFENCE APPLICATIONS

B. Praveen Kumar, H.H. Kumar, M. Balasubramanian[1] and B.S. Murty[1]

PZT Centre, Armament Research & Development Establishment, Pune
[1]Indian Institute of Technology, IIT Madras
E-mail: praveen0406@gmail.com

ABSTRACT

The ferroelectric $Pb(Zr_xTi_{1-x})O_3$, PZT ceramics are well known piezoelectric materials for electro-mechanical transducers. Rare earth doped PZT compositions near morphotrophic phase boundary show significant piezoelectric properties. In the present studies, Ta doped PZT ceramics were prepared by kinetic energy processing and solid-state oxide route. X-ray diffraction studies are done to determine the presence of perovskite PZT in sintered PZT ceramics. Microstructural studies were carried out by using scanning electron microscope (SEM) and atomic force microscope (AFM). Detailed dielectric and piezoelectric studies were carried out for the sintered samples. PZT ceramics prepared by kinetic energy processing route followed by sintering has higher dielectric constant, compared to conventionally prepared PZT. Piezoelectric studies were also carried out to investigate the effect of grain size on piezoelectric properties. These nanocrystalline PZT ceramics are suitable candidates for underwater transducers, seismic sensors and energy harvesting applications.

INTRODUCTION

Recently there has been a great interest in the field of ferroelectric materials for their ability to exhibit piezoelectric behavior. Lead zirconate titanate (PZT) based ceramics are the vital ferroelectric materials because of their excellent dielectric and piezoelectric properties [1]. PZT based ceramics are widely used in ultrasonics, underwater transducers, health monitoring

---

V. Rajendran, P. Manivasakan and K.E. Geckeler (eds.)
Advanced Nano Materials for Industrial Applications, pp. 209-214 (2012).

and energy harvesting applications [2]. These ceramics are conventionally prepared by solid-state reaction method, where using oxides as the starting materials. The solid-state reaction method demands a calcination process at higher temperature to synthesize the designed PZT compound. It makes the process complicated and time consuming. PZT ceramics are also been synthesized by chemical processing routes like sol-gel, co-precipitation, polymer precursor and hydrothermal process [3-5]. However most of these chemical routes are not distinctively different from the solid-state route, in terms of processing steps like calcination. It also demands high purity, sensitive and costly chemicals for processing which results in costly manufacturing process.

Very recently this kinetic energy processing technique (high energy ball milling technique), has been successfully used to synthesize nanocrystalline ferroelectrics and other ceramics. The dielectric properties of nanocrystalline piezoelectric materials were found superior to the microcrystalline materials. This kinetic energy processing technique [6] is superior to other procession routes due to its (i) low cost oxide starting materials (ii) no calcination is required, because phase formation occurs during milling (iii) low sintering temperature, due to their atomic or molecular state homogeneity of nanocrystalline powders (iv)no PbO loss, due to the phase formation occurs at low temperatures.

The electro-mechanical properties of PZT are highly depend on the microstructure, processing methods, nature and concentration of dopants [7]. PZT modified by different dopants have been comprehensively studied by different researchers. The effect of various dopants like Sm, Ba, Fe, Nb, La, and Bi etc [8-12] on PZT ceramics are studied by different researchers and interesting results are reported. Properties like Curie temperature, dielectric constant and dielectric loss are normally affected by dopants and grain size. In donor doping (Ta doping), to maintain electro neutrality the addition of Ta^{5+} ion is usually compensated by the formation of Pb^{2+} vacancies. these vacancies enhances the domain orientation of PZT ceramics and increases the dielectric and piezoelectric properties.

The aim of this work is to study the effect of Ta doping on PZT ceramics processed by both solid-state and kinetic energy processing route. In addition the comparison studies of structural, dielectric and piezoelectric properties of PZT ceramics will also be studied in detail.

EXPERIMENTAL

The chemicals used in the present study were of high purity grade (at least 99% pure). The starting materials used in the experiment were lead oxide (PbO), zirconium dioxide (ZrO_2), titanium dioxide (TiO_2) and tantalum oxide (Ta_2O_5). The raw materials were mixed in stoichiometric proportion $Pb_{1-(x/2)}(Zr_{0.53}Ti_{0.47})_{1-x}Ta_xO_3$ to obtain PZT based piezoelectric ceramics. These compositions were prepared by solid-state reaction route. The raw materials were ball milled for 24 h and then calcined at 1050°C for 2 h. The calcined powders were compacted in to disks and followed by sintering at 1200-1300°C. the sintered samples were electroded by silver paste and poled in silicon oil bath by applying a DC field of 2-3 kV/mm.

The crystal structure of sintered PZT specimens were confirmed by X-ray diffraction (XRD) studies. Based on the obtained structural and electrical properties, optimized composition was selectively chosen and they were prepared by kinetic energy processing route. The kinetic energy dry ball milling was carried out in Fritsch Pulverisette P5 planetary ball mill at room temperature for 20 h. The powders prepared by kinetic energy processing route were compacted and sintered, without calcination. The difference in the processing route leads to large variation in grain size. The properties obtained from both the processing routes are compared with respect to grain size and dopant. Scanning electron micrographs of the sintered specimens were recorded using the JEOL, JSM 5800, scanning electron microscope. The microstructure of sintered nanocrystalline PZT samples were analyzed by using tapping mode of atomic force microscope (AFM). Agilent, Picoplus, AFM. Dielectric and piezoelectric properties of Ag electroded specimens were measured using Hioki 3532 LCR HiTester.

RESULTS AND DISCUSSION

Bright field transmission electron microscopy (TEM) image of Ta doped nanocrystalline PZT powder prepared by kinetic energy processing route is shown in Fig. 1. The TEM image clearly depicts the nano particles of PZT powder with particle size of around 20-50 nm. The particle size of Ta doped PZT powder prepared by solid-state route is studied by using laser diffractometer based particle size analyzer. The measured average particle size is of around 1 μm.

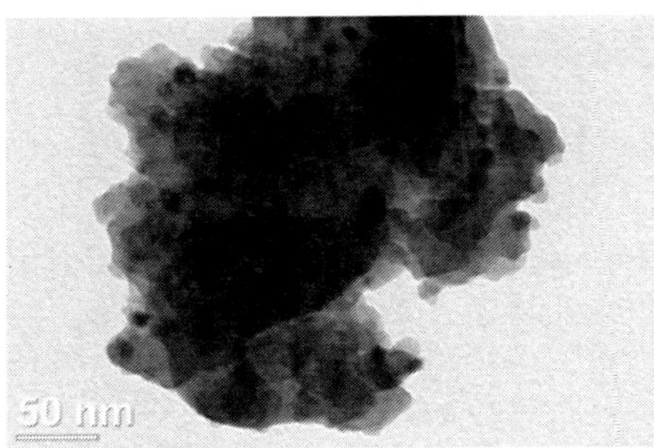

Fig. 1: Bright Field TEM Image of Ta Doped Nanocrystalline PZT Powder

XRD patterns of microcrystalline Ta doped PZT prepared by solid-state route and sintered at 1250°C are shown in Fig. 2. The formation of perovskite phase is clearly indicated. The XRD studies clearly show no other unwanted phase is present in the material. The basic structure of PZT is not disturbed by addition of Ta dopant. The splitting of peaks corresponding to (200) and (002) planes clearly indicated the existence of intensified tetragonal structure in Ta 1% doped PZT. It is observed by addition of Ta, the phase changes from intensified tetragonal structure to combined existence of rhombohedral and tetrahedral structure. The addition of Ta concentration results in large structural deformation and that result in crystal structure transforming from tetragonal to combine existence of rhombohedral and tetragonal [13].

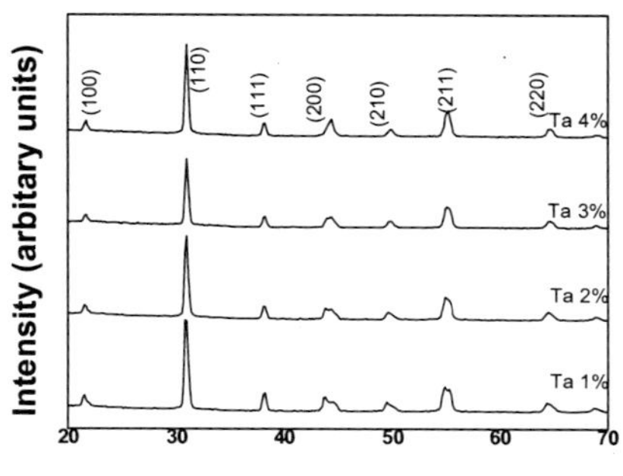

Fig. 2: XRD Patterns of Solid-State Route Prepared Ta Doped Sintered PZT

Fig. 3: (a) SEM of Microcrystalline. and (b) AFM of Nanocrystalline Ta (1 mol%) Doped Sintered PZT

SEM and AFM images of microcrystalline and nanocrystalline Ta (1 mol%) doped PZT is shown in Fig. 3 (a) and (b), respectively. SEM image of solid-state route prepared and sintered PZT reveals the grain size of PZT is around 2-3 μm. The grains are coarser and homogeneous in nature. AFM image of kinetic energy route processed and sintered PZT shows the average grain size is of around 200-300 nm. It clearly reflects that the grain size of kinetic energy processed PZT is very less compared to solid-state route prepared PZT and the same effect is observed in dielectric and piezoelectric properties.

The effect of Ta dopant on dielectric constant and dielectric loss for solid-state and kinetic energy route prepared PZT sintered at 1250°C is shown in Fig. 4. Dielectric constant decreases with increase in Ta content in PZT. The phases present in the PZT material and the dopant concentration highly influences the domain wall motion. It has been discussed earlier in XRD patterns, the increase in the amount of Ta decreases the amount of tetragonal phase present in the PZT material. When Ta^{5+} ions of higher valency are substituted for Zr^{4+}/Ti^{4+} ions, vacancies are formed on the A-site in PZT. The incorporation of Ta and a vacancy may break the long range structural interaction of the atoms. It is observed in kinetic energy processing route prepared PZT doped with Ta (1 mol.%) has higher dielectric constant compared to solid-state samples doped with Ta (1 mol.%). The increase in dielectric constant with respect to reduced grain size is due to increase in the number of defect and electric dipoles. The dipole-dipole interaction depends on location of dipoles, and when grain size of confined system is reduced, the fraction of dipoles at the interfaces increases significantly [4].

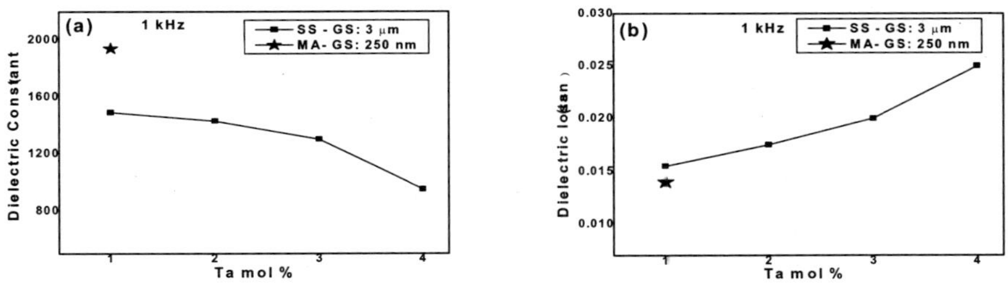

Fig. 4: Variation of (a) Dielectric Constant and (b) Dielectric Loss in Ta Doped PZT

Dielectric loss increases with addition of Ta concentration. In general, the partial addition of dopant increases the conductivity of the material (high concentration of charge carriers) which in turn increases the dielectric loss of the material. Further increase in dielectric loss is due to the decrease in dielectric constant. In general, the material with higher dielectric constant will have lower dielectric loss at particular frequency. It is also observed that nanocrystalline grain size PZT samples has lower dielectric loss.

It is seen from Table 1, that kinetic energy process route prepared nanocrystalline Ta doped PZT has lower d_{33} and K_p values compared to solid-state rote prepared PZT. The decrease in piezoelectric properties is due to confinement in grain size. With decrease in grain size, the coupling effect between grain boundaries and the domain wall will inhibit the domain orientation and severely constrain the domain wall motion. this decrease in domain wall motion leads to decrease in piezoelectric properties.

Table 1: Piezoelectric Properties of Ta Doped (1 mol.%) PZT

Process	*Grain Size*	*Piezo Coefficient (d_{33}), pC/N*	*Coupling Coeff. (K_p)*
Solid-state	3 μm	325	0.54
Kinetic energy	250 nm	270	0.48

CONCLUSION

Ta doped PZT ceramics are successfully prepared by solid-state and kinetic energy processing routes. The optimized composition of Ta doped PZT is chosen based on dielectric and piezoelectric properties. Structural, dielectric and piezoelectric properties of Ta doped PZT are compared and studied in detail. Improvement in dielectric properties and decrement in piezoelectric properties are observed in nanocrystalline PZT. The prepared PZT sensor samples are tested for structural health monitoring applications.

ACKNOWLEDGEMENTS

Authors are highly grateful to Shri. Anil M Datar, Outstanding Scientist & Director, ARDE, Pune for his constant support and encouragement.

REFERENCES

[1] Raman, K. and M. Lopez, Microstructure, dielectric and electromechanical properties of PLSZFT nanoceramics for piezoelectric applications. *J. Mater. Sci: Mater. Electron.*, 19, 1140, 2008.

[2] Wang, X. , Piezoelectric nanogenerators-Harvesting ambient mechanical energy at the nanometer scale. *Nano energy*, 1, 13-24, 2012.

[3] Yang, J.K., W.S. Kim and H.H. Park, Effect of grain size of $Pb(Zr_{0.4}Ti_{0.6})O_3$ sol-gel derived thin films on the ferroelectric properties. *Appl. Surf.* Sci., 169-170, 544, 2001.

[4] Jiang, A.Q., G.H. Li and L.D. Zhang, Dielectric study in nanocrystalline $Bi_4Ti_3O_{12}$ prepared by chemical coprecipitation. *J. Appl. Phys.*, 83, 4878, 1998.

[5] Mandoki, N.T., C. Courtois, P. Champagne and A. Leriche, Hydrothermal synthesis of doped PZT powders: sintering and ceramic properties. *Mater. Lett.*, 58, 2489, 2004.

[6] Schaffer, G.B. and P.G. McCormick, On the kinetics of mechanical alloying. *Metall. Trans.* A, 23, 1285, 1992.

[7] Raman, K. and M. Lopez, Effect of strontium doping on dielectric, ferroelectric and piezoelectric properties of PLZTN ceramics. *Mater. Sci. Eng. B*, 145, 41, 2007.

[8] Sangawar, S.R., B. Praveenkumar, H.H. Kumar and D.K. Kharat, Effect of Fe and Fe-Ba substitution on the piezoelectric and dielectric properties of lead zirconate titanate ceramics. *Mater. Sci. Eng. B*, 176, 242, 2010.

[9] Pandey, S.K., O.P. Thakur, D.K. Bhattacharya, C. Prakash and R. Chatterjee, Structural and electrical properties of Sm^{3+} substituted PZT ceramics. *J. Alloys Compd.*, 468, 356, 2009.

[10] Pdungsap, L., N. Udomkan, S. Boonyuen and P. Winotai, Optimized conditions of La dopant in PZT ceramics. *Sensors Actuators A*, 122, 250, 2005.

[11] Rai, R. and S. Sharma, Structural and dielectric properties of (La, Bi) modified PZT ceramics. *Solid State Commun.*, 129, 305, 2004.

[12] Chu S.Y., T.Y. Chen and I.T. Tsai, Effects of sintering temperature on the dielectrics and piezoelectric properties of Nb doped PZT ceramics and their applications. *Integr. Ferroelectr.*, 58, 1293, 2003.

[13] Choi, G.P., J.H. Ahn, W.J. Lee, T.H. Sung and H.G. Kim, Phase formations and electrical properties of doped $PZT/PbTiO_3$ films deposited by reactive sputtering using multi targets. *Mater. Sci. Eng. B*, 41, 16, 1996.

ELECTRICAL AND MAGNETIC PROPERTIES OF Cu-SUBSTITUTED $Ni_{0.5-x}Zn_{0.3}Mg_{0.2}Fe_2O_4$ NANOCRYSTALLINE FERRITE

M.R. Bhandare, A.M. Sheikh[1] and Arif D. Sheikh[2]

Department of Physics, S. M. Joshi College, Pune, Maharashtra
[1]Department of Electronics, The New College, Kolhapur, Maharashtra
[2]Physical and Materials Chemistry Div, NCL ,Pune, Maharashtra
E-mail: milindbhandare@gmail.com

ABSTRACT

Ferrites having chemical formula, $Ni_{0.5-x}Cu_xZn_{0.3}Mg_{0.2}Fe_2O_4$ with x = 0.0, 0.1, 0.2, 0.3, 0.4 and 0.5 have been prepared by auto-combustion route. Selected area electron diffraction (SAD) patterns confirm the spinel structure of the ferrite. Transmission electron microscopy (TEM) images show a well defined nano-crystallite state with an average particle size of around 28 to 39 nm. The electric properties have been studied by means of DC resistivity. The increase in (σ_{AC}) is observed with increase in Cu-content, indicates dispersion with frequency at room temperature. The magnetic hysteresis were done using vibrating sample magnetometer (VSM) which show that saturation magnetization (M_s), remanent magnetization (M_r) and coercivity (H_c) decrease with increasing Cu-content exhibiting superparamagnetism characteristic at room temperature. The frequency dependence initial permeability was studied in the range of 100Hz to 5MHz. This shows high frequency stability after dispersion occurred. The graph of magnetic loss tan (δ_μ/μ_i) supports dispersion. Variation of loss tan (δ_μ/μ_i) against frequency exhibit resonating behavior.

INTRODUCTION

Spinel nanocrystalline ferrites are one of the best magnetic semiconductors which have a great importance in telecommunication, microwave and electronics engineering. They are used in

---

V. Rajendran, P. Manivasakan and K.E. Geckeler (eds.)
Advanced Nano Materials for Industrial Applications, pp. 215-222 (2012).

variety of applications ranging from simple function devices, like small permanent magnets, to sophisticated device used for the electronic industry Some interesting applications of these materials are found in computer peripherals, telecommunication equipments, permanent magnets, electronic and microwave devices, magnetic media used in computers, recording devices and magnetic cards. The electrical and magnetic properties are the most important properties of ferrites, which depend on the processing conditions, sintering temperature, chemical composition and the amount of the additives [1]. High electrical resistance and lower saturation magnetization (M_s) are the characteristics properties of soft ferrites which are not found in traditional soft ferromagnetic materials like silicon, steels or perm alloys [2].

In recent years, nanocrystalline ferrite materials are becoming a subject of intense research because of their unique properties, both from the fundamental and the application point of view [3]. Soft spinel ferrite ($M\ Fe_2O_4$, M = Co, Ni, Zn, Mn, etc.) nanoparticles have been intensively investigated due to their remarkable magnetic and electrical properties and wide practical applications in ferrro fluids technology magnetic drug delivery, magnetic high-density information storage etc. [4, 5, 6].

EXPERIMENTAL

We have successfully synthesized Cu-substituted Ni-Mg-Zn ferrite using auto-combustion technique. The details about the synthesis parameters and structural analysis are reported in our previous paper [7]. TEM images and selected area electron diffraction (SAD) patterns were carried out by (Model: PHILIPS CM200). The DC resistivity measurements were carried out using two probe methods. AC conductivity (σ_{AC}) was carried out in the temperature range 30°C-600°C at different frequencies viz. 1kHz, 10kHz, 100kHz and 1MHz, using LCR meter bridge (Model: HIOKY3532–50LCR). The hysteresis and magnetization measurements were performed using a vibrating sample magnetometer (VSM; 9600-1LDJ, USA) For permeability measurements were carried out at room temperature using LCR meter-bridge. Initial permeability was calculated using relation [8].

$$\mu_i = \frac{2\Pi L}{N^2 h \mu_o \ln \frac{R}{r}} \quad \ldots (1)$$

where, L is the inductance in Henry. R and r is outer and inner radius of the toroid in meter respectively. **N** is number of turns of Inductor. h is the height of the toroid in meter. μ_0 is the magnetic permeability of vacuum ($4\pi \times 10^{-7}$ Henry per meter).

All the measurements were carried out at room temperature.

RESULTS AND DISCUSSION

TEM Study

Transmission electron microscopy (TEM) micrographs are shown in Figs.1 (a) and (b) of the nano-sized Ni-Cu ferrite with Cu-content of x = 0.0 and x = 0.5. These macrographs are almost spherical in shape and found to be loosely agglomerated. The agglomeration can be attributed to the magnetic dipole interaction arising between ferrite particles [9]. The estimated average particle size for x = 0.0 is found to be around 28 nm and that for x = 0.5 is found to be 39 nm. This reveals that as Cu-content increases the particle size also increases. The present results suggest that, the substitution of Cu helps for the phase formation of spinel cubic lattice. It means that the temperature required for the phase formation of the ferrite with Cu doping is low as compared with undoped ferrite. One function of copper is to decrease the sintering temperature [10]. In the present case, the sintering temperature and sintering time of all the ferrite samples is same (i.e. 800°C for 4h) even though, the Cu doped samples shows higher sized grains as compared to the remaining samples. These results suggest that, addition of Cu enhances the grain growth process, which may due to the soft nature of Cu as compared to Ni. Similar results were also observed by A. Ghasemi et al. [11] in case of Cu substituted nickel ferrite prepared by a reverse micelle process.

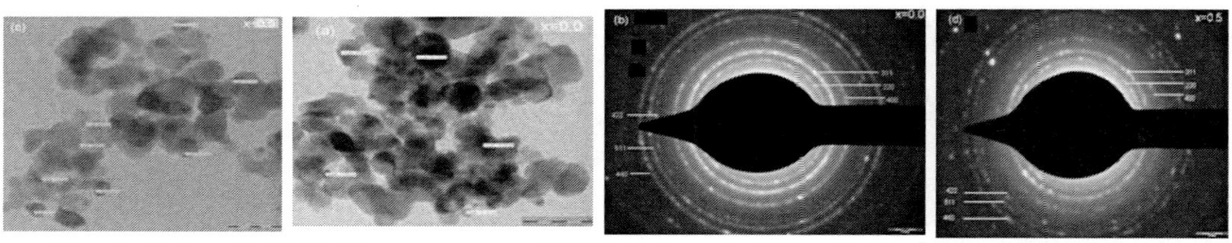

Fig. 1: (a-b-c-d) TEM Micrographs and SAD Patterns of $Ni_{0.5-x}Cu_xZn_{0.3}Mg_{0.2}Fe_2O_4$ Ferrite Nanopowders for *x* = 0.0 and x = 0.5

Fig. 1(c) and (d) shows corresponding selected-area diffraction (SAD). The diffraction patterns are composed of series of concentric circles marked with miller indices, which show ferrite nanoparticles, which are highly polycrystalline with spinel structure. The measured average particle size is found to be in good agreement with the sizes calculated by Scherer formula.

Electrical Properties

AC conductivity (σ_{AC}) shows two different conduction regions I-region (583K to 873K) and II-region (303K to 583K) as depicted in Fig. (2) (a) and (b). It can be seen in I-region which is at a higher temperature (873K) the magnitude of AC conductivity seems to be frequency-independent and increases linearly with increasing temperature while little dispersion is observed at a relatively lower temperature (873K). In region (II) at lower temperature (303K) the magnitude of AC conductivity increases with increasing frequency. Generally AC conductivity is sum of two components [13,14]. First component is due to DC electrical conductivity related to the drift mobility and independent on frequency. Second component is due to dielectric relaxation caused by bound charge carriers. It obeys a power law. [12, 14]. A.M.Abo et al. [15] have also observed similar behavior in rare-earth substituted Li-Co ferrites. These composition shows semiconducting trend as commonly seen in most of the ferrites [16]. The increase in AC conductivity with increase in temperature may be attributed to the increase in the drift mobility of the thermally activated charge carriers according to the conduction mechanism, i.e. increase in temperature thermally activated electron exchange between $Fe^{2+} \rightarrow Fe^{3+}$ ions on octahedral sites.

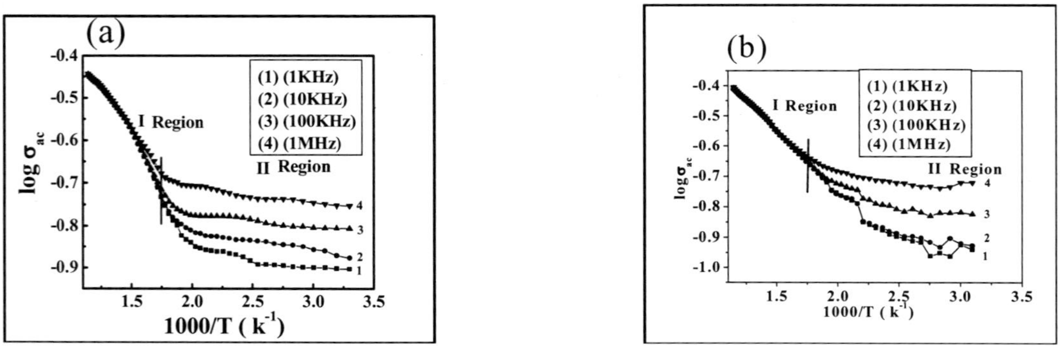

Fig. 2: (a-b) Variation of AC Conductivity (σ_{AC}) of $Ni_{0.5-x}\,Cu_x\,Zn_{0.3}\,Mg_{0.2}\,Fe_2O_4$ Ferrites with Temperature for Fixed Frequencies 1,10,100 KHz and 1 MHz for (a) x = 0 and (b) x = 0.5

M-H Hysteresis Loop

The room temperature magnetization measurements of $Ni_{0.5-x}\,Cu_x\,Zn_{0.3}\,Mg_{0.2}\,Fe_2\,O_4$ nanoparticles with different Cu-contents (x = 0.0,0.1,0.3, 0.4, and 0.5) were performed on a vibrating sample magnetometer (VSM). M-H curve is typical for a soft magnetic material. It is evident from the figure that the hysteresis loop for theses nanocrystals gets saturated with strong magnetic field of 10k Oe for all samples indicating their ferromagnetic behavior. The saturation magnetization decreases with increasing copper concentration. This decrease is almost slow in the region from x = 0.0 to 0.2, after which a rapid change in the saturation is observed from x = 0.3 up to a copper content of 0.5. A sharp decrease to 31.42 emu/g was obtained for the tetragonal phase copper ferrite. The remanent magnetization almost decreases with increasing Cu-content. the present saturation magnetization data are found to be in quite good agreement with those reported for Ni-Cu ferrites prepared using other different methods [17] The decrease in saturation magnetization values are explained by assuming that the A-B interaction favors anti-parallel coupling of A-site and B-site groups and is strong enough to overcome any tendency for A-A or B-B couplings to disrupt the parallel alignment of all spins within the A-site or B site groups separately. The magnetic order in the ferromagnetic spinels is due to a super-exchange interaction occurring between the metal ions in the tetrahedral A-sites and octahedral B-sites [18]. The replacement of Ni^{2+} ions (2.18 B.M.) by Cu^{2+} ions (1.0 B.M.), which have preferential octahedral site occupancy, results in the reduction of the super-exchange interaction between the A and B sites. In other words, as the copper concentration increases, the magnetization of the B-site decreases and the net magnetization decrease.

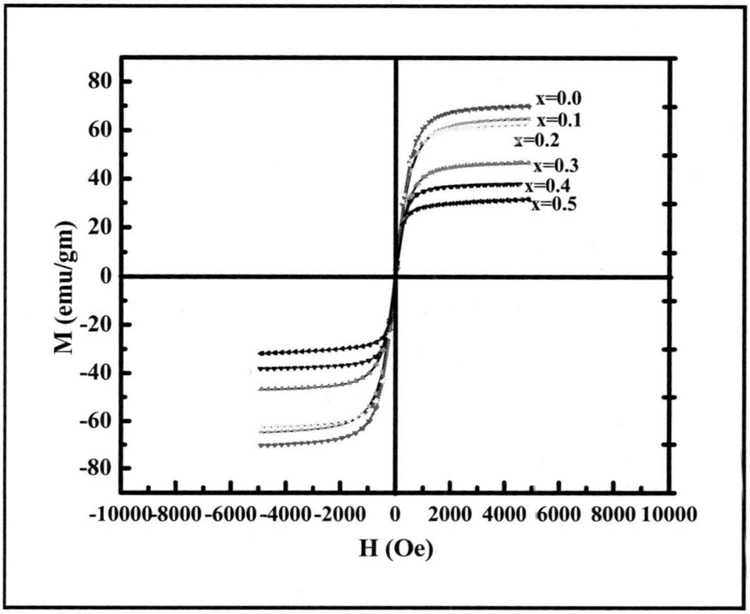

Fig. 3: M-H Curve of $Ni_{0.5-x}Cu_xZn_{0.3}Mg_{0.2}Fe_2O_4$ Nanoparticles with Different Cu-contents (x = 0.0, 0.1, 0.3, 0.4, and 0.5) Using a Vibrating Sample Magnetometer (VSM) at Room Temperature

Initial Permeability

Fig. 4. It is seen from fig that all the samples show a flat profile in the frequency range 100Hz to 1MHz for x = 0.0, x = 0.1, x = 0.3 and x = 0.5. This indicates high frequency stability. Then after the dispersion occur. In general it is observed that the initial permeability remains almost constant up to certain frequency after which it increases to a maximum value and then decreases rapidly to low value. This phenomenon is known as dispersion of initial permeability, which is attributed to either domain wall displacement or domain rotation or both of these. It is also observed for the sample (x = 0.5) that at low frequency range of 100kHz-1MHz, domain wall motion plays a predominant role in the magnetizing process and loss mechanism. For the sample x = 0.5 wall relaxation occur in the 3 MHz while for the sample x = 0.0 and x = 0.1 relaxation occur at 1MHz. The rapid decrease in initial permeability at high frequencies indicates of the onset ferromagnetic resonance [19]. When the frequency of the applied magnetic field equal the Larmor precession frequency of the electron spin, at this position resonance occurs and the energy is transferred from the field to the system in orienting magnetic dipoles. Similar behaviour observed in the case of Cu substituted Mg-Zn ferrite [20].

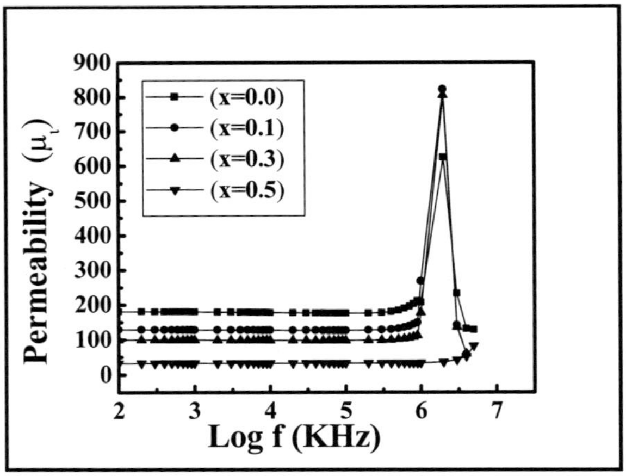

Fig. 4: Variation of Initial Permeability with Frequency of $Ni_{0.5-x}$ Cu_x $Zn_{0.3}$ $Mg_{0.2}$ Fe_2O_4 Ferrites for x = 0.0, x = 0.1 x = 0.3 and x = 0.5

The variation of relative loss factor (RLF) with frequency is show in Fig. 5. RLF is the ratio of the magnetic loss tangent (δ_μ) to initial permeability (μ_i). For high value of (μ_i) and low value of tan (δ_μ), the RLF is low. The RLF should be as low as possible for some high frequency application. In the present case RLF is low and it is still decreasing with increasing in frequency. It also shows that as Cu-content increases RLF decreases. The fairly constant initial permeability (δ_μ) value over a large frequency range shows the compositional stability and quality of the ferrites prepared by the auto-combustion method. This is a desirable characteristic for various applications such as broadband pulse transformer and wide band read-write heads for video recording [19].

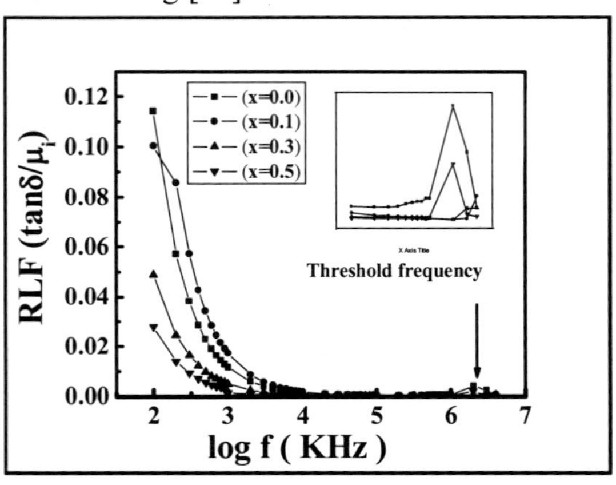

Fig. 5: Relative Loss Factor (RLF) with Frequency for x = 0.0, 0.1, 0.3 and 0.5 of $Ni_{0.5-x}$ Cu_x $Zn_{0.3}$ $Mg_{0.2}$ Fe_2O_4 Ferrites

CONCLUSIONS

The auto-combustion method is found to be convenient for the synthesis of nanocrystalline Ni–Cu ferrites. The formation of single-phase cubic spinel ferrites was confirmed by SAD diffraction patterns. TEM micrographs confirm that the particles are in nanometer range. DC electrical resistivity has been explained according to site's preferences. AC conductivity (σ_{AC}) explained with help of two regions one regions. Decreases in M_s with increasing Cu-content and approaches the superparamagnetic behavior. The fairly constant initial permeability (μ_i) value anlow RLF value shows the compositional stability and quality of the ferrites prepared by the auto-combustion method.

ACKNOWLEDGMENT

Authors are thankful to U.G.C., WRO Pune for the financial support and gratefully acknowledge to Dr. V. L. Mathe, Department of Physics, University of Pune, for providing the dielectric measurement facilities and Principal N.L. Jadhav. S.M. Joshi College, Hadapsa, Pune for his encouragement.

REFERENCES

[1] M. Guyot, *J. Magn. Magn. Mater*. 18,925-926, 1980.

[2] S. Yan, J. Yin and E. Zhou, *J. Alloys Compd*. 450,417–420, 2008.

[3] P. Dutta, M.S. Seehra, S.Thota, J. Kumar and *J. Phys. Condens. Matter* 20,1-8 2008.

[4] C. Holm and J.J. Weis, *Curr. Opin. Colloid. Interfac. Sci.* 10,133-140, 2005.

[5] P. Tartaj, *Curr. Nanosci*. 2,43-55, 2006.

[6] M.H. Kryder, E.C. Gage, T.W. Mcdaniel, W.A. Challener, R.E. Rottmayer, G.P. Ju, Y.T. Hsia and M.F. Erden, *Pro. IEEE*. 96, 1810-1835, 2008.

[7] M.R. Bhandare, H.V. Jamadar, A.T. Pathan, B.K. Chougule, A.M. Shaikh, *J. Alloys. Compd*. 509,L113–L118, 2011.

[8] A.C. Razzitte,S. E. Jacobo W.G. Fano, *J. Appl. Phys*. 87,6232-6234, 2000.

[9] L. Li, C. Xiang, X. Liang, B. Hao, *Syth. Met*. 160, 28-34, 2010.

[10] J.J. Shrotri, S.D. Kulkarni, C.E. Deshpande, A. Mitra, S.R. Sainkar, P.S. Anil Kumar, S.K. Date, *Mater. Chem. Phys*. 59,1-5, 1999.

[11] A. Ghasemi, E. Ghasemi E. Paimozd, *J. Magn.Magn.Mater*. 323, 1541-1545, 2011.

[12] S.A. Mazen, H.M. Zaki S. F. .Mansour, *J. Pure. Appl. Phy*. 3 40-48. 2007

[13] A.M. Abo ElAta M. A. Ahmed, *J. Magn. Magn. Mater*, 208, 27-36, 2000.

[14] H. Kojimo in E.P. Wohlfarth (Ed.), Ferromagnetic Materials, vol. 3, North-Holland Pub. Co, Amsterdam, 1982, p. 305.

[15] A.M. Abo ElAta, M. K. El. Nimr, S. M. Attia, D. ElKony A.H. Al-Hammadi, *J. Magn. Magn*. Mater, 297, 33-43, 2006.

[16] M. A. Ahamed, M.A. El Hiti, M. K El Nimr, M.A. Amar, *J. Magn. Magn. Mater*. 152, 391-395, 1996.

[17] S. M. Hoquea, Md.A. Choudhury M.F. Islam, *J. Magn. Magn. Mater*. 251 292-295, 2002.

[18] Y. Kseoglu, A. Baykal, F. Gzüak H. Kavas *Polyhedron* 28, 2887, 2009.

[19] T. Jahanbin, M. Hashim, K.A. Matori, S.B. Waje, *J. Alloys Compd*. 503 111-117, 2010.

[20] M. Manjural Haque, M. Heq M.A. Hakim, *Mater. Chem*. Phys. 112 580-586, 2008.

MEMS AND NANOTECHNOLOGY IN RADAR APPLICATIONS

M.K. Anand Kumar

SSO-I, Office of the Dy. Director General, DGAQA, Ministry of Defence
C/o. HAL (Nasik Division), Ojhar, Maharashtra
E-mail: ocrink.dgaqa@nic.in; jayand_ocri@rediffmail.com

ABSTRACT

This papers basically highlights the advent of MEMS & Nanotechnology in Radar applications & their advantages over other existing technologies.

High frequencies switches & sensors used in radar applications demand low insertion loss, high Q factor, high isolation, reliability, etc. Also inductive components are variably used in high frequency electronic devices, which suffer from high eddy current losses, core laminations, increased size & weight due to laminated core, etc.

This paper talks about (1) The use of MEMS switches & sensors having both electronic & mechanical devices on a monolithic micro chip to produce superior performance over other types of switches.(2) The evolvement of metal/ceramic nano composites for the next generation high frequency magnetic applications facilitating a better core material which offers high resistivity, resulting in reducing eddy current losses for the high frequency inductive components Further reduction in heat generation (as eddy current loss is reduced) reduces the cooling requirement of the components

INTRODUCTION

Nanotechnology is the act of manipulating matter at the atomic scale. It is the rapidly emerging technology of building machines and mechanisms at the next level of precision and miniaturization, and it is the scaling down of existing technologies to the nanoscale.

V. Rajendran, P. Manivasakan and K.E. Geckeler (eds.)
Advanced Nano Materials for Industrial Applications, pp. 223-230 (2012).

Nanotechnology means materials and devices in which every atom is assigned a place and having every atom in the right place will be essential for the functioning of the device.

MEMS Techniques: Micro electromechanical systems are integration of mechanical elements sensors actuators and electronics on a common silicon substrate through micro fabrication technology. MEMS are reactive devices, which translate an event into usable data.

There are various types of RF MEMS components, such as RF MEMS resonators and self-sustained oscillators with small form factor and low phase noise, RF MEMS tunable inductors, and **RF MEMS switches**, switched capacitors and varactors.

MEMS APPLICATION IN RADAR

The proposed RF MEMS switches finds a variety of usefulness in RF circuits and systems, including wireless Communication devices, antenna switching, T/R (transmitter/receiver) switching, band selection adjustable gain amplifiers radar systems for military applications, phase shifters Phased Array Antennas, Measurement Equipments, Impedance Matching Circuits, etc.

Like solid-state switches, RF MEMS switches can be divided into two categories depending on circuit configurations series type and a shunt type, as illustrated in Fig. 1 It is not always the case but in most of the cases series-type switches incorporate metal-to-metal contacts and shunt-type switches utilize capacitance changes. Both of them offer low insertion Loss across a very wide frequency range. The metal contact series switches, however offer superior isolation in the low frequency range while the capacitive contact switches offer superior performance in the high frequency range.

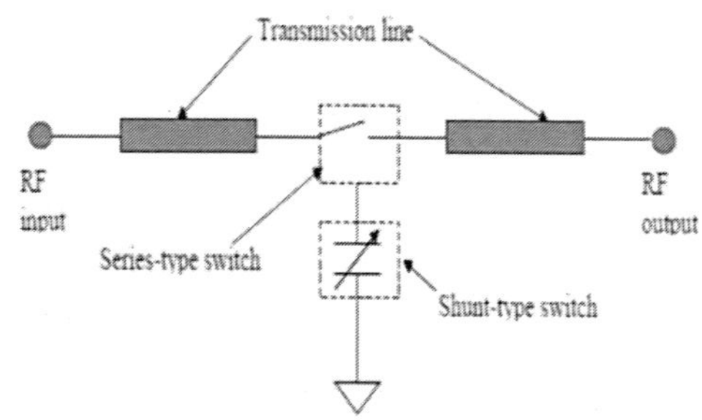

Fig. 1: A schenatic Drawing of a Series and a Shunt-type Swich

Metal Contact Switch

A SPST series metal contact switch consists of a relatively thin metal cantilever beam suspended over an underlying actuation electrode (see Figure 2). Applying a bias between the actuator electrode and the suspended beam creates an electrostatic force. This force deflects the cantilever beam downward until the dimples on the underside of the free end of the cantilever make contact with the signal line on the right side to close the signal path. The dimples serve two purposes: 1) the dimples prevent the beam from shorting to the actuation electrode, and 2) the dimples are used to reduce the apparent contact area between the cantilever beam and the contact pads to minimize stiction. The mechanical structure of the beam provides the restoring force necessary to return the switch to the separated (up) position once the applied bias is removed. Since the tip of the switch only moves a few microns, contact can be made or broken in a matter of microseconds.

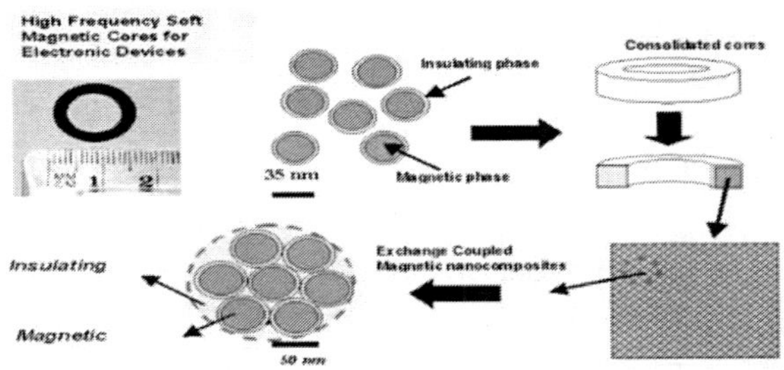

Fig. 2

Capacitive Switch

A shunt capacitive MEMS switch is also shown in Figure 1. This switch consists of a thin metal membrane "bridge" suspended over the centerline conductor of a CPW and fixed at both ends to the ground conductors of the CPW line. A thin dielectric on the signal line, is used to DC isolate the bridge from the CPW centerline conductor. When the switch is in the up-state, the switch is viewed as a small shunt capacitance to ground. Once the switch is pulled down and makes contact with the underlying dielectric, the shunt capacitance increases by a factor of 20-100, creating an RF short .

For the shunt capacitive switch shown in Figure 3, the Underlying electrode consists of evaporated gold with a thickness of 0.5 mm. Directly above the electrode lies the dielectric which consists of a 0.3mm thick layer of sputtered SiN. Finally, the 0.6mm thick evaporated gold beam "bridge" is anchored at both ends and suspended approximately 2-3 mm above the dielectric layer. For both the metal contact and shunt capacitive switches, the beam height and beam thickness are establish such that switch actuation occurs for bias voltages less than 50 V.). As can be seen, the isolation is very poor at low frequencies, but above 10-15 GHz is very good.

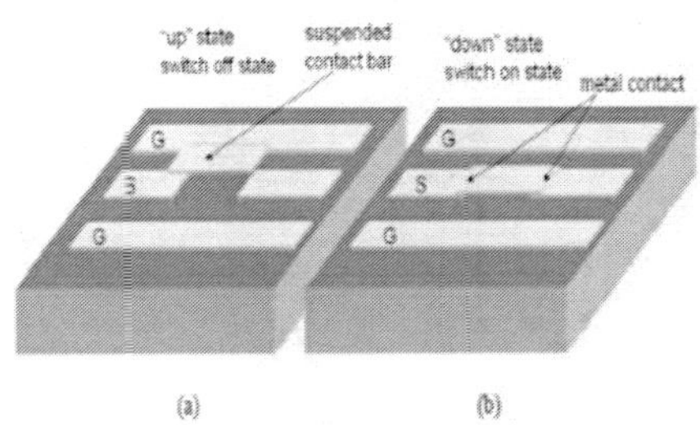

Fig. 3: A metal-to-metal-contact MESM swich (a) at an off state ('up' or blocking state), and (b) at an on state ('down or pass-through state')

Types of RF MEMS Switches

Applications

RF MEMS switches have numerous potential applications covering frequencies from DC through millimeter waves. This wide range of frequencies is due to the high performance over a very wide bandwidth. Two areas where RF MEMS switches offer significant potential benefits are large lightweight electronically scanned arrays (ESAs) and tunable filters. Typically, large lightweight ESAs, are in systems with very tight power and weight requirements. In such cases, the ability of RF MEMS switches to provide two critical advantages over competing technologies: 1) prime power savings due to reduced RF loss and reduced drive power, and 2) weight reduction due to reduced thermal dissipation, smaller transmitters, and smaller power supplies. For tunable filters, RF MEMS switches offer very high linearity and low loss.

Switch Type	Properties[1]					
	Insertion Loss	Isolation	Power Consumption	DC Voltage	Speed	Bandwidth
PIN/SCHOTTKY	~.15 dB	45 dB	1-5 mW per device	1-10 V	1-5 ns	Narrow/ Wide
GaAs FETs	1-2 dB	~20 dB	1-5 mW per device	1-10 V	2-10 ns	Narrow/ Wide
HBT/PIN	0.82 dB	25 dB	1-5 mW per device	1-10 V	1-5 ns	Narrow/ Wide
Best FET	0.5 dB	70 dB	5 mW	3.5 V	2 ns	Narrow/ Wide
MEMS – Sergio Shunt	0.06 dB @ 20 GHz	30 dB	~1μW	12-14 V	> 30 μs	Wide (1-40 GHz)
MEMS – Scott/Jeremy(Shunt)	0.3 dB @ 30 GHz	50-60 dB	~1μW	~20 V	> 30 μs	Moderate (10-40 GHz)

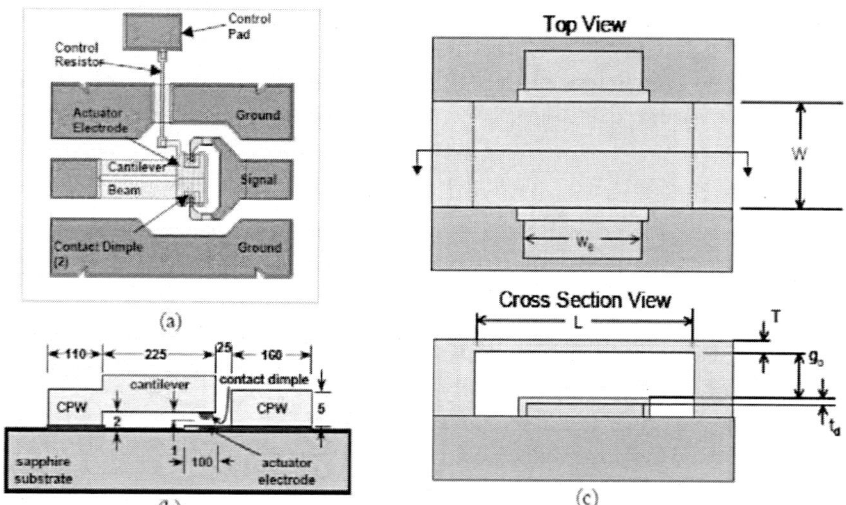

Fig. 4: RF MEM Switchas (a) Metal Contact Layout, (b) Metal Contact Schematic Cross Section and Capacitive Switch Layout and Schematic Cross Section. Dimanslonas are in μm, not to Scale.

Phased Array Antennas: Recent developments in MEMS provide the potential for significant reduction in the cost and mass of ESAs for Space Based Radar (SBR) and communications applications. MEMS phase shifters have a significant efficiency, cost and weight advantage for large array sizes that are consistent with SBR applications. The development of multi-mission SBR concepts is strongly dependent upon large space-deployable ESA antennas. Due to satellite prime power limitations and the relatively long ranges that are involved, large antennas are required to provide detection and tracking of small targets in the presence of clutter and jamming. These large antennas, however, must satisfy strict mass and volume constraints imposed by the launch vehicle.

Tunable Filters

In addition to ESA applications, tunable filters are rapidly becoming a key microwave circuit device for Radar applications. Numerous universities are actively performing research in the use of MEMS varactors for various tunable filter designs. Some key initial results from this research include that maintaining a tunable resonator Q values above 200 requires a varactor series resistance of less than 0.4 ohms. Under this requirement, tunable low loss (1-2 dB) and narrow (3-5%) filter using MEMS varactors and planar resonators will be difficult to design. Numerous universities have operational tunable filters at 100 MHz to 30 GHz with bandwidths of 3-12% with corresponding insertion losses of –3 to –6 dB. In the design of tunable filters, it is imperative that careful attention is placed on the voltage and current handling capabilities of the MEMS components in low-loss tunable filters.

A summary of published millimeter-wave MEMS tunable filters performance characteristics include: frequency range of 15-32 GHztypically have a 1-dB bandwidth ranging from 3-8% with an average loss of -3.3 to -4.9 dB for continuous switching.

There are a number of other RF MEMS switches reported with some variations, but by and large, they are similar to those two examples just described. These RF MEMS switches with out-of-plane actuators have several drawbacks. Actuation voltage is usually high (typical, 15-80 V) because most of them use electrostatic actuators. Instead, electromagnetic actuators can be used, but they have an issue of high standing power consumption. They encounter failure at high RF power (> 1W). Capacitive-shunt type switches have narrowband isolation characteristics.

NANOTECHNOLOGY IN INDUCTIVE COMPONENTS

Inductive components are extensively used in high frequency (> 1MHz) electronic devices from radar, satellite, telecommunication home radio stereos. Metallic alloys and ferrites are the only choice as core materials though it offered very low resistivity (~10-6 W-cm). This gives high eddy current losses. The only way to reduce eddy current loss is use of laminated core which increases size & weight of the components. Metal/ceramic nanocomposites have been developed for the next generation of high frequency magnetic applications. Nanocomposite processing has provided a new approach for fabricating soft magnetic materials. In a magnetic/ ceramic nanocomposite, the resistivity can be drastically increased (without changing other properties), leading to significantly reduced eddy current loss.

Nanocomposite Composition

Chemical synthesis of Ni-Fe/SiO_2, Co/SiO_2, Fe-Co/SiO_2, Fe/nickel-ferrite, Ni-Zn-ferrite/SiO_2, Fe-Ni/ polymer and Co/polymer magnetic nanocomposites.

The Processes

Consolidate these magnetic nanocomposite powders into exchange coupled (>90% theoretical density) bulk components via vacuum hot press consolidation. Consolidate these magnetic nanocomposite powders into exchange coupled (>90% theoretical density) bulk components via tape casting process. The synthesis and processing of metal/insulator magnetic nanocomposite is a radical departure from the conventional metallic and ferrite materials. A schematic of magnetic nanocomposite technology is illustrated in Fig. 4, which includes powder synthesis (insulated coated magnetic nanoparticles), consolidation of the powder into exchange coupled cores. A completely new phenomenon has been observed when reducing the particle size and the separation between neighboring particles into a nanometer scale in composite matters. For example, it has been found that a Co- or Fe-based nanocomposite can possess a permeability much higher than that obtainable from the bulk Co or Fe metal. This large enhancement in permeability is due to the exchange coupling effect.

The exchange interaction which leads to magnetic ordering within a grain also extends out to neighboring environments within a characteristic distance. For a traditional powder material of large particle sizes, exchange coupling effect is negligibly small in determining magnetic properties. However, when the particle size plus the separation between particles is reduced intergrain exchange coupling plays a dominant role and the material will possess a variety of properties different from the bulk size material. when the particle-particle separation is significantly less the intergrain exchange interaction makes all the neighboring particles coupled. This coupling averages out the magnetic anisotropy of individual nanoparticles. As a consequence, the permeability of an exchange-coupled nanocomposite can be even much higher than the permeability of its bulk counterpart.

Powder synthesis: chemical synthesis route for metal/insulator nanocomposite magnetic materials provides a unique opportunity to modify the magnetic as well as electrical properties of the complex material in a rather large scale. For example, a nanocomposite consisting of a magnetic phase and an insulating phase such that the magnetic particles are embedded in the insulating matrix, the material now has essentially no overall electric conductivity. In addition, since the magnetic particle is in a nanometer size, the eddy currentproduced within the particle is also negligibly small. Therefore, conductivity of the magnetic constituent is no longer a factor in the material selection consideration and metallic materials can be used as magnetic phase.

THE ADVANTAGES & APPLICATION

MEMS

1. **Insertion loss**: The decrease in transmitted signal power resulting from the insertion of a device, at an on-state is smaller over most of frequency range because of good ohmic contacts or low capacitive coupling. Both a p-n junction and a semiconductor channel have non-negligible resistance that causes significant insertion loss.
2. MEMS switches have very high linearity with third-order intercept point of 66-80 dBm since they are essentially passive devices. This implies improved distortion characteristics
3. RF MEMS switches have significantly higher signal isolation at an off-state than their solid-state counterparts. Finally, MEMS switches can be fabricated on various kinds of substrates.

Nanotechnology

The advantage of magnetic nanocomposite include, (1) reduction in total core power losses, (2) the high flux capabilities at elevated temperatures that the nanocomposite cores are expected to support, thereby enabling attractive to reduce the cost and size of the current magnetic components effect in nanomagnetic materials is of 10 times larger in magnitude than in macroscopic materials.

Radar Applications

This carbon based nanocomposites are great evaluation in the field of Radar application since Inductive components are extensively used in high frequency (> 1 MHz) electronic devices of radar systems (like High Voltage Transformer, primary and secondary power supply systems and other inductive coils used in various filter circuits, etc. This gives significant reduction in eddy current loss, reduction in weight of the core, reduction in size of the core, reduction in heat generation as eddy current loss is reduced, reduction in cooling requirements hence reduction in overall weight, size & cost of components/units. Low weight & size are the crucial required parameter in airborne radar.

CONCLUSION

This year nanotechnology survey gives a momentary overview of civil and defence-oriented technology developments. The main findings are:

- exploitation of nanotechnology is already part of society, with high investments worldwide
- many nano products already exist on the market (additives in cosmetics, paints, polymers, catalysts, filters).

- some products are expected to come soon (within 5 years), especially wireless tags and nano sensor systems.
- others are expected to come later, in a timeframe of 5-15 years (high-strength nanomaterials, smart/adaptive materials).
- more and more nanotechnology is being converged with other technologies, especially biotechnology, information technology and cognitive science.
- the defence organization can gain a lot of benefit from these new technologies and should actively participate in these developments.

REFERENCES

[1] Nanobook By Frank Simonis & Steven Schilthuizen.

[2] Nanotechnology by Hari Singh Nalwa.

[3] Modern Radar Systems Hamish Meikle.

APPLICATION OF NANOTECHNOLOGY IN AVIATION INDUSTRY

MUKESH VERMA

Off. of the Dy. Director General(N), DGAQA, Ministry of Defence
C/o. HAL (Nasik Division), Ojhar, Nasik, Maharashtra
E-mail: ocrink.dgaqa@nic.in; mukeshverma10@yahoo.co.in

ABSTRACT

Nanotechnology is the rapidly emerging technology used in Electronics, Materials, Chemical, Life Science, Aviation, building machines and mechanisms at the next level of precision and miniaturization, and it is the scaling down of existing technologies to the nanoscale.

This paper presents trends in, (1) Nano materials for Aerospace Industry. (2) Polymer Nano-composites for Aerospace industry. (3) Nano material for Avionics

This paper identifies some of the technical challenges and the key research efforts in the field of nano materials for aerospace applications. Specifically, it focuses on carbon nanotube-reinforced polymers and materials produced by severe plastic deformation (SPD). Selected European projects and world conferences related to aerospace are included. The state of the art of polymer nano composite research is also reviewed.

INTRODUCTION

Nanotechnology is the science and engineering of creating materials, functional structures, and devices on a nanometer scale. In scientific terminology "nano" means 10 to the power –9. One nanometer is one billionth of a meter.

The deliberate manipulation of matter by certain chemical and/or physical processes (referred to as "bottom-up" production) to create materials with specific properties is nanotechnology that are not displayed in their larger forms.

V. Rajendran, P. Manivasakan and K.E. Geckeler (eds.)
Advanced Nano Materials for Industrial Applications, pp. 231-238 (2012).

Why are we so fascinated with downsizing materials to a nanoscale?

According to the **Hall-Petch relationship**, the yield strength is a linear function of the inverse of the square root of grain diameter (d-1/2) which implies a 10 times higher grain boundary strengthening when the grain size is reduced by 2 orders of magnitude.

For example, 6-nm copper grains show five times more hardness than conventional copper.

Molecular Building Blocks of Nanotechnology

Nanotechnology research today uses three types of molecular building blocks: proteins, polymers and carbon nanotubes.

- Proteins: They are building block of DNA. These are mainly one-dimensional structure can be accurately grown in repetitive patterns according to rules designed into to their "seeds"—similar to crystal growth patterns. Rearranging proteins into atomically precise "designer" molecule could serve as a nano assembler's building block.
- Polymers: Brownian motion- the random mixing of molecules in a liquid or gas can exhibit auto assembly. Small molecules with the edges pretreated so that they can bond together with the correct parts, auto-assemble into atomically precise molecular building block.
- Carbon nanotubes: Most promising of all building blocks for nanotechnology. They are unique nanostructures with remarkable electronic and mechanical properties. Details is covered latter.

MANUFACTURING TECHNIQUES

1. *Arc Discharge:* Nanotubes can be produced by flowing a precursor gas through a plasma discharge at very high temperature. This arc discharge technology produces high-quality nanotubes, but they are often accompanied by a large volume (up to 50%) of contaminants.
2. *Laser Ablation:* Laser ablation produces nanotubes by directing a high-energy laser at a carbon target. This technology produces high quality nanotubes with fewer contaminants than arc discharge, but the production rate remains low. Furthermore, this production technology is capital-intensive and is likely limited to research quantities.
3. *Chemical Vapor Deposition:* Chemical vapor deposition produces nanotubes by heating a precursor gas and flowing the gas over a reactive metal surface. This technology can produce nanotubes in good yield—occasionally over 90%—with few contaminants. However, the tubes produced have a large number of defects—places where atoms and bonds are missing. This can be an advantage or disadvantage, depending on the application, since defects can also be viewed as reactive sites on the molecule.

MATERIAL CHALLENGES FOR AEROSPACE INDUSTRY

In the aerospace industry, there is a great need for new materials which exhibit improved mechanical properties. Materials possessing high strength at a reduced mass and size make lighter aircraft with lower fuel consumption. The development of new materials with tailored properties is a primary goal of today's materials science and engineering. However, the possibility of obtaining improved mechanical properties by the conventional methods of cold working, solution hardening, precipitation hardening, etc., has been almost exhausted. The current trend is to integrate intelligence and multi-functionality into the varied components of aerospace systems and vehicles.

It would be important to introduce the working conditions that the airborne equipments are subjected during service applications.

Operating Conditions and Stresses

The typical application of an aircraft involves variations in stress and temperature during flight. The modern day aircraft experience temperature variations due to the effect of aerodynamic heating at supersonic speeds. The skin temperature drops initially as the aircraft climbs due to exposure to decreasing ambient air temperatures. As the speed increases the temperature begins to increase, the skin temperature is in the range of 120°C to 150°C in the speed. The exposure to extremes in ambient temperature ranging from about 60°C to 50°C, moisture from extreme humidity to desert conditions.

Operating Environment and Corrosion

Corrosion is most concerned factor which damage aircraft fuselages. Aircraft operating in marine environment deserve special mention in this context.

High Reliability Requirement

The nature of task to be accomplished by an aircraft system leaves no scope for laxity in the quality related issues and reliability of the component. Malfunctioning of even a small component can lead to loss of hundred of crores and human life.

Service Life

Average life of an aircraft system is 25 to 40 years. It is desirable that the component sees through the expected life of system.

NANOMATERIALS IN AEROSPACE

Polymer Nanocomposites

The reinforcement of polymers (thermoplastics, thermosets, elastomers) using fillers, whether inorganic or organic, is common in the production of modern plastics. Polymer composites are strong, yet remarkably lightweight and so they are leading the field in aerospace applications. Nanocomposites based on commodity plastics, such as polypropylene, polyethylene and PVC, will dominate the market.

Polymer nanocomposites - PNCs represent an alternative to conventional-filled polymers or polymer blends. In contrast to the conventional systems, where the reinforcement is on the order of microns, PNCs take advantage from unique effects of the addition of nanometre-sized inorganic materials to a polymer matrix. These effects however, are driven not only by the small size but unusual shapes and aspect ratios (L/h > 300) of the additives and include extraordinarily high interfacial areas or highly aligned phases of the additive. The constituent inorganic additives can be applied in a form of particles, tubes and wires, two-dimensional platelets and porous materials. Their application brings improvements in mechanical strength and aging resistance, reduction of wear and flammability, barrier to diffusion, optical transparency, and unprecedented morphologies such as interpenetrating networks.

In general, polymer nanocomposites fall into three categories, depending on the form of nanoparticles being used: layered silicate or nanofibres/carbon nanotube-polymer nanocomposites and high-performance PNCs resins.

Severe Plastic Deformation

Grain refinement down to the nanometre scale thus offers good prospects for a new generation of high strength materials. To produce such high strength, nanocrystalline materials, the development of new processing methods is necessary. Nanomaterials can be produced following bottom-up methods (such as inert gas condensation, consolidation of nanopowders, electro deposition or crystallization from an amorphous state), but it is only possible to produce small items of such materials, usually with a diameter and length not greater than a few millimetres. Therefore, it is probable that only a topdown approach can offer good "technological" prospects. Such a concept consists in the transformation of metals or alloys possessing a conventional grain size into bulk materials with a submicron- or nanoscale structure by the application of severe plastic deformation (SPD).

The advantages of the SPD methods are:

(*i*) a 100% dense nanostructured material is obtained

(*ii*) conventional materials are used as precursors,

(*iii*) there are no toxicological issues involving the use of nanopowders.

Thus, research in this field has attracted the attention of numerous scientific groups throughout the world including representative European institutions.

Nanostructured Metals

Nanostructured metals have nanosized grains, which gives them greater strength and hardness. Heralded as alternatives to toxic materials like chromium for coatings and for structural applications, their use can be hampered by their increased brittleness and complex processing requirements.

Nanostructured metals can provide very hard coatings that are resistant to corrosion, useful for applications including aerospace components, such as landing gear and construction equipment such as drill bits and bulldozer blades. Low volume, high margin applications for the aerospace and defence industries, and high-end sporting goods are largely driving the development of nanostructured materials.

Carbon Nanotubes (CNTs)

CNT is a tubular form of carbon with Diameter as small as 1nm and Length from few nm to microns. In configuration, CNT is equivalent to a two dimensional graphene sheet rolled into a tube.

Electronic properties can be tailored through application of external magnetic field or application of deformation.

There are two basic types of nanotubes.

Single walled Nanotubes (SWNT): have one shell of carbon atoms in hexagonal arrangement. An ideal nanotube can be thought of as a hexagonal network of carbon atoms that has been rolled up to bake a seamless cylinder. Just a nanometer across, the cylinder can be tens of microns long, and each end is "capped" with half of fullerene molecule.

Multiwalled Nanotubes (MWNT): MWNTs consists of multiple concentrically nested carbon tubes, similar to the rings of a tree trunk.

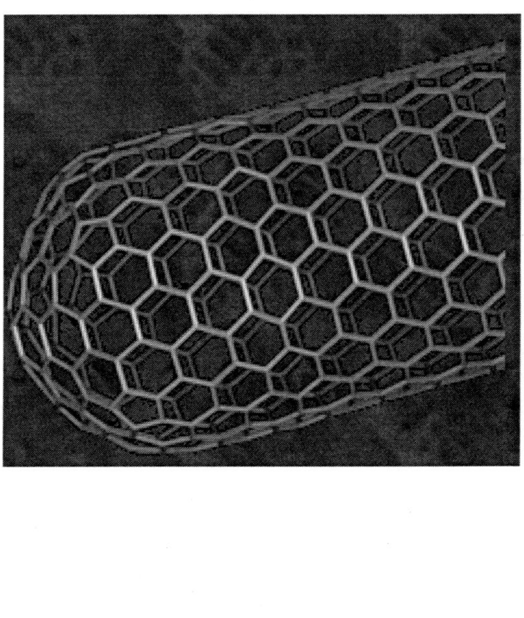

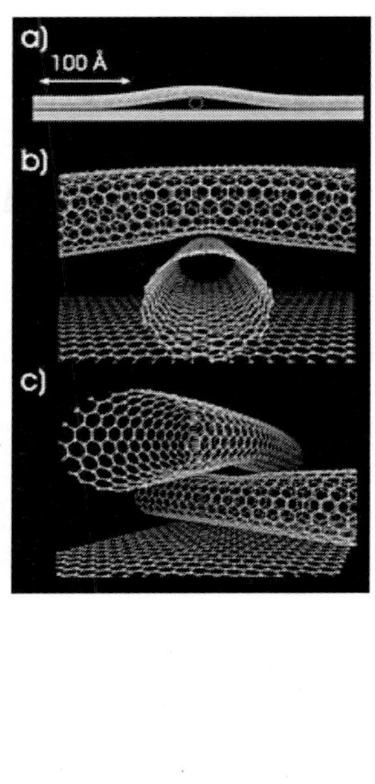

CNT exhibits extraordinary mechanical properties

- Excellent Metallic conductivity.
- Superconductivity with T_c's of up to 33K.
- Ferromagnetism.
- Non-linear optical activity.
- Polymerization to form a variety of 1-, 2- and 3D polymer structures.
- The chart compares the tensile.
- Strength of SWNT's to some.
- Common high-strength materials.
- These one-dimensional fibers exhibit electrical conductivity as high as copper, thermal conductivity as high as diamond,
- Strength 100 times greater than steel at one sixth the weight, and high strain to failure ratio.
- Current length limits are about one millimeter.

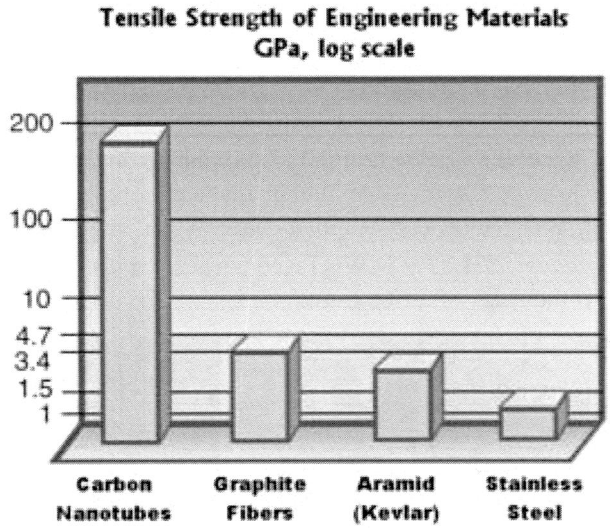

APPLICATION IN AEROSPACE INDUSTRY

Engines

Improvements in aircraft engine efficiency can be reached by materials which allow higher operating temperatures, lower engine weights, higher pressures and increased rotor operating stresses. The application of high temperature nanoscale materials to aircraft engines may lead to an increase of the thrust-to-weight ratio of up to 50 percent and fuel savings of 25 percent for conventional engines.

- Nanomaterials are being applied as coatings on aircraft engine blades. Research is ongoing to manipulate the properties of the coatings down to the molecular level making them adhere more firmly to the surface of the metal blade and allowing the engines to run hotter.
- Nano-phase ceramics are being tested for use as thermal barrier coatings (TBC). Improved TBC protective coatings have wide application in aircraft engines, aero-structures, turbine engines, and chemical processing. The coating system consists of an outer layer that is chemically resistant, deposited on an underlying strain-resistant layer that can deform without cracking. Both layers are made of perovskite oxide ceramic layers. If successful, higher fuel efficiency can be reached due to longer lasting TBCs that do not peel off. (Navy, 2006).
- The enhanced creep, fatigue and sulphidation resistance of grain boundary engineered components is expected to significantly increase the time between engine overhaul/ refurbishment.
- Because of the high surface area, nanoparticles can act as very efficient catalysts, even for liquid and solid aerospace engine fuels. Fuels used at present can be improved by the addition of nano-sized energetic particles, which allow a higher combustion temperature, faster energy release rates, a shortened ignition delay, shortened burn times resulting in more complete combustion, a greater flexibility in designing new energetic fuel/propellants, replacing inert or low-energy gellants, and a rapid energy release.
- Aluminium nanoparticles are used with liquid jet and rocket fuel to increase the propulsion energy.
- Iron oxide nanoparticles can act as a catalyst for solid propellants.
- Nano-sized energetic metals and boron particles possess desirable combustion properties such as a high combustion temperature and fast energy release rates.

Sensors

Sensors are needed in aircraft for measurements of velocity, acceleration, position, temperature, and flow properties.

- The conductivity of wires with diameter of a few nanometres is very sensitive to small changes in electrochemical potential. Because of this, they can be used as very sensitive sensors for different gases. Nanosensors can be used for the early detection of fires.
- Gyroscopes are used to track an airplane's position. Microscopic structures are now being built into chips that perform the same function at far less weight and space.

Fuels

Fuels used at present can be improved by the addition of nano-sized energetic particles, which allow a higher combustion temperature, faster energy release rates, a shortened ignition delay, shortened burn times resulting in more complete combustion.

Aluminium nanoparticles are used with liquid jet and rocket fuel to increase the propulsion energy.

Nano-sized energetic metals and boron particles possess desirable combustion properties such as a high combustion temperature and fast energy release rates.

- According to information from a wiring company, a Boeing 747 has approximately 225 km of wire weighing approximately 1600 kg.
- Theoretically, replacing that wire with conductive polymer will bring the wiring weight alone down to well below approximately 454 kg, which will positively impact the range and fuel efficiency of the aircraft.

CONCLUSION

To conclude the main Advantages are:

- Weight Reduction.
- Increases impact resistance
- Increased safety.
- Reduced emissions.
- Reduced noise.
- Increased capacity.
- Increased range.
- Enhanced payload.
- Higher speed.
- Lower operating and maintenance costs.
- Better overall management of the aircraft and its use.
- High strength-to-weight ratio.
- EMI/EMC Capability.

POLYMER-CHAIN-DISTRIBUTION REORGANIZATION FOR IMPROVING THE POWER CONVERSION EFFICIENCY OF POLYMER SOLAR CELLS

Bao-Fu Ding and Kamal Alameh

Electron Science Research Institute, Edith Cowan University, 270 Joondalup Dr, Joondalup, WA 6027, Australia
E-mail: b.ding@ecu.edu.au; k.alameh@ecu.edu.au

ABSTRACT

We investigate the influence of the post solvent evaporation time delay on the performance of polymer solar cell (PSC) devices employing a bulk heterojunction photoactive polymer layer of regioregular poly(3-hexylthiophene) as electron donor and polymer [6,6]-thienylC61 butyric acid methyl ester as an electron acceptor. Characterization of the fabricated solar cell devices clearly demonstrates balanced transport of electrons and holes in the blend and confirms increased surface roughness and crystallinity of the films when post solvent evaporation time delay is optimised. An optimum device performance is obtained with 72 hours of post solvent evaporation time delay, achieving a power conversion efficiency of 4.1%, which is 0.9% higher than similar devices made without enough time for polymer-chain-distribution reorganization.

INTRODUCTION

The power conversion efficiencies of thin-film polymer solar cells (PSCs) have significantly increased over the last decade, mainly through the combination of new organic materials and enhanced device structures. While the power conversion efficiency (PCE) of these thin-film

V. Rajendran, P. Manivasakan and K.E. Geckeler (eds.)
Advanced Nano Materials for Industrial Applications, pp. 239-246 (2012).

organic solar cells is still lower than that attained by their inorganic counterparts, the attractive features of low cost, simple fabrication and wide choice of materials have been the driving force for the wide acceptance and commercialization of PSCs for the Green Energy market [1,2].

The improvement of the PCE of PSCs has recently become a challenging research area [3-5]. Among the many approaches used for improving the PCE of PSCs, the widely adopted one is based on solvent annealing after blending and spin-coating the polymer layers [1,6].

With the use of a sufficient self-organization time after annealing, the mobilities of electron and hole can be made more balanced, resulting in negligible carrier accumulation, and thus, high PSC performance.

In this paper, we propose a new approach based on introducing an additional (Step III) annealing time for polymer-chain-distribution reorganization (PCDR), which enables a higher PCE to be attained, in comparison with conventional two-step solvent annealing processes.

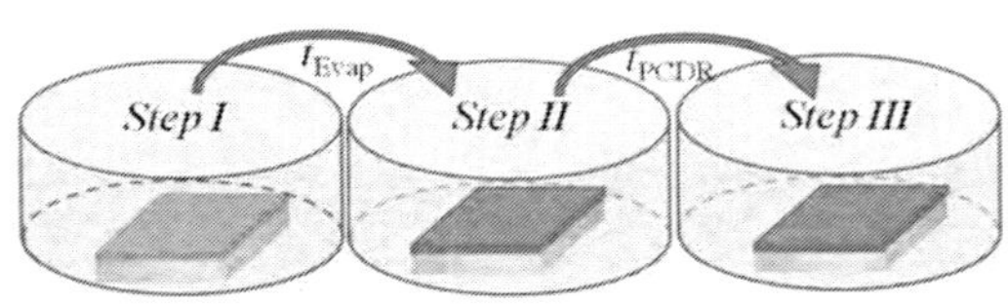

Fig. 1: Processing Steps for the Development of Polymer Solar Cells
(i) Fresh Spin Coating of Blended Polymers, (ii) Phase Separation, After Solvent Evaporation Time, (iii) Proposed Additional Annealing time for PCDR

EXPERIMENT

Figure 1 shows the proposed processing steps for the development of polymer solar cells, namely (i) fresh spin coating of blended polymers, (ii) phase separation, achieved by evaporating solvent at room temperature for around 20 mins, and (iii) additional annealing time for tens of hours in order to reorganize the polymer chain distribution, and hence increase exciton dissociation, leading to a higher PCE.

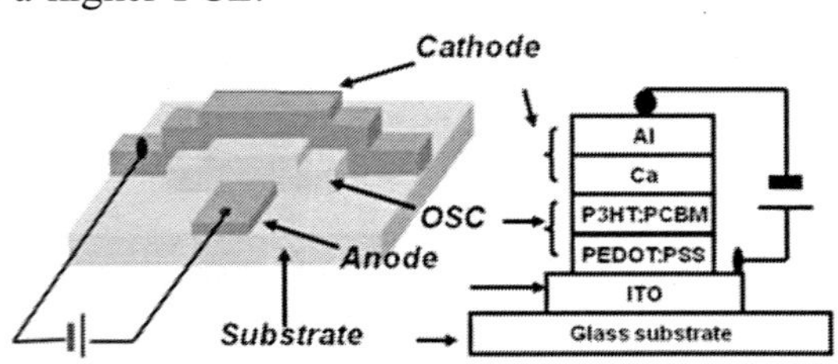

Fig. 2: Schematic View and Architecture of the PSC

The developed PSC structure is shown in Fig. 2. It comprised the following layers: (i) ITO, (ii) Poly(3,4-ethylenedioxythiophene) poly(styrenesulfonate) (PEDOT: PSS) (42 nm),

(iii) Poly(3-hexylthiophene) (P3HT): phenyl C61 butyric acid methyl ester (PCBM) (205 nm), (iv) Ca(20nm) and (v) Al (90nm). The ITO-coated glass substrate, which had a sheet resistance of 22 ohm, were cleaned as described in Ref. [7]. A thin layer (~30 nm) of PEDOT/PSS (Baytron P VP A1 4083) was spin-coated at a spinning speed of 3500 rpm for 40 seconds to make the ITO surface smoother and increase its work function. The ITO/PEDOT/PSS substrate was backed at 130 °C for 15 mins, and then transferred to a nitrogen-filled glove box (O_2<1 p.p.m. and H_2O<1 p.p.m.) for further processing. P3HT was first dissolved in 1,2-dichlorobenzene (DCB) to make 18 mg/ml solution, which was then blended with PCBM in 50 wt% and stirred for ~18 h at 35°C. Subsequently, the three processing steps shown in Fig. 1 were carried out, namely:

I. 60 μ the solution was spin coated on the PEDOT:PSS-coated substrate with a spinning speed of 700 rpm for 30 s to form the active layer of the PSC. The thickness of this active layer was around 190 nm as measured with a profiler.

II. The sample was put into a special Petri dish of 2.5 cm diameter and 1.5 cm height for solvent evaporation [1]. The evaporation time, t_{Evap}, is the time needed to change color of the top film from red to black.

III. The sample remained in the same Petri dish for PCDR for a time t_{PCDR}.

After the above three steps, the sample was thermal annealed, using a hot plate, at 120°C for 15 minutes. Finally, Ca and Al layers were thermally evaporated at 0.8Å/s and 1.4Å/s, respectively, in a vacuum chamber at a pressure of 6×10^{-6} Pa to form a composite cathode of Ca(20nm)/Al(90nm). The top Al layer crossed the bottom ITO to form active device area of approximately 1.5 mm². Five similar PCDR devices were fabricated and processed with different polymer-chain-distribution reorganization times, namely, t_{PCDR} = 0 h, 24 h, 48 h, 72 h and 240 h, and their performances were experimentally evaluated.

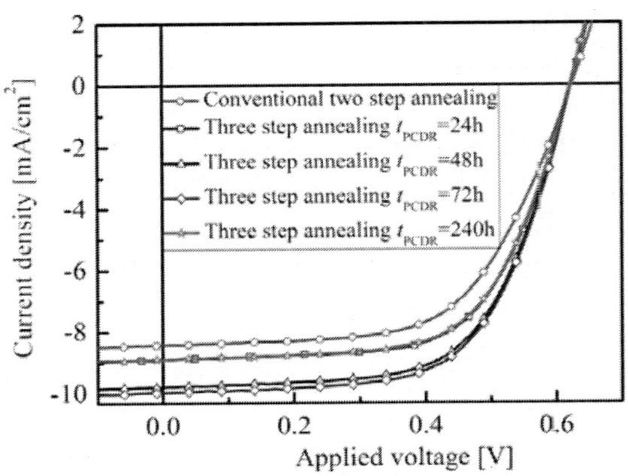

Fig. 3: Current Density Versus Applied Voltage for Different t_{PCDR}. Illumination = 1 sun. t_{PCDR} = 0 Corresponds to a Conventional two-step Process

Table 1: Summary of Device Performance for Various PV Devices in the Work

t_{PCDR} (h)	J_{sc} (mA/cm^2)	FF (%)	V_{oc} (V)	PCE (%)
0	8.4	60.0	0.63	3.2
24	8.9	64.3	0.62	3.5
48	9.8	63.6	0.62	3.9
72	10.2	63.6	0.62	4.1
240	8.9	63.5	0.62	3.6

RESULTS AND DISCUSSION

The current density versus voltage characteristics of the various PSCs fabricated with different t_{PCDR} are shown in Fig. 3 and summarized in Table 1. It is obvious that PSC that was post processed with t_{PCDR} = 72 h displays the best performance. Compared to a conventional PSC fabricated with a two-step process, the short circuit current (J_{sc}) and PCE of the PSC post processed with t_{PCDR} = 72 h increase from 8.4 mA/cm^2, 3.3% to 10.2 mA/cm^2, 4.1% at t_{PCDR} = 72 h. Noticeably, the fill factor (FF) increases from 60% (t_{PCDR} = 0 h) to 64.3% (t_{PCDR} = 24 h), then stays almost the same as t_{PCDR} increases.

For the improvement of FF, charge balance plays a key role. When the charge transport for electrons and holes in the device is unbalanced, the accumulation of majority carriers in the bulk of the polymer layer results in the space-charge-limited current. In this case, the photocurrent is governed by a square-root dependence on bias [8-10], and hence, FF will have a small value. Therefore, a balanced charge transport is the prerequisite to achieving a high FF. To measure the mobilities of electrons (μ_e) and holes (μ_h) for the PSCs, the space-charge-limited-current (SCLC) method was employed, which is comprehensively described in Refs [8,11]. An electron-dominant device of ITO/Ca/P3HT: PCBM/Ca/Al and a hole dominant device of ITO/PEDOT:PSS/P3HT:PCBM/Au were fabricated to measure μ_e and μ_h, respectively. For the electron-dominant device, the low work functions of Ca (2.8 eV) on both sides of the polymer P3HT:PCBM active layer make them efficient for injecting electrons and blocking holes into the polymer layer [12,13]. Therefore, the majority carries in the organic bulk are electrons. Oppositely, the high work functions of PEDOT:PSS (5.2eV) and Au (5.3 eV) favor hole injection from electrodes to the polymer layer [1,14], making holes the majority carriers for the latter device. By using the SCLC expression given in [8,11] *J-V* curves were generated for the above-described devices and fitted with experimentally measured curves, leading the evaluation if the electron and hole moblities μ_e and μ_h, which were 9.6×10^{-4} cm^2/Vs, 4.5×10^{-4} cm^2/Vs, respectively, for the two-step PSC device (t_{PCDR}=0 h) and 6.5×10^{-4} cm^2/Vs and 6.3×10^{-4} cm^2/Vs for the three-step optimized PSC device (t_{PCDR}=72 h). Here, the charge-balance ratio is defined as $\mu_e\mu_h$. For the two-step film, the ratio is around 2.1, which is similar to the value reported in [1]. On the other hand, for the 3-step PSC device, the ratio is significantly reduced to 1.03. Such a close-to-unity charge-balance ratio further improves the FF and results in a higher PCE. Generally, the charge-balance ratio depends on the self-organization ability of the polymer chain. According to [1], the degree of self-organization is determined by t_{Evap}, which is the time needed for the wet films to solidify. After solidification, t_{PCDR} allows for the reorganization of polymer chain distribution as well as self-organization which can be monitored by measuring the surface roughness of the active layer [15].

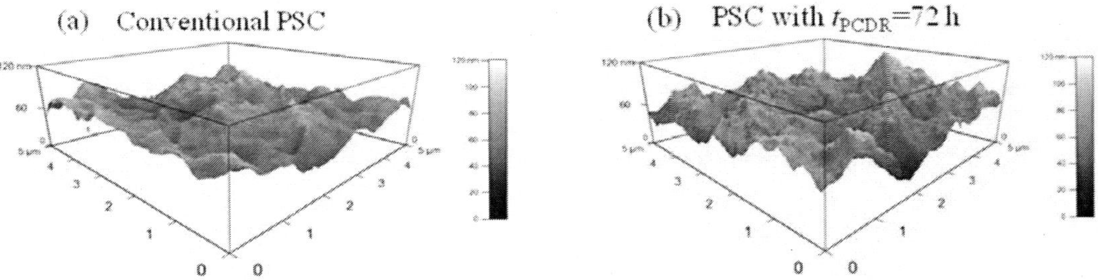

Fig. 4: Effects of t_{PCDR} on the Morphology of the Polymer Layer; AFM Height Images of the P3HT/PCBM Composite Films Showing a 5 μm → Á5 μm Surface Area. (a) Conventional PSC with t_{PCDR} =0 h, (b) PSC with t_{PCDR} =72 h

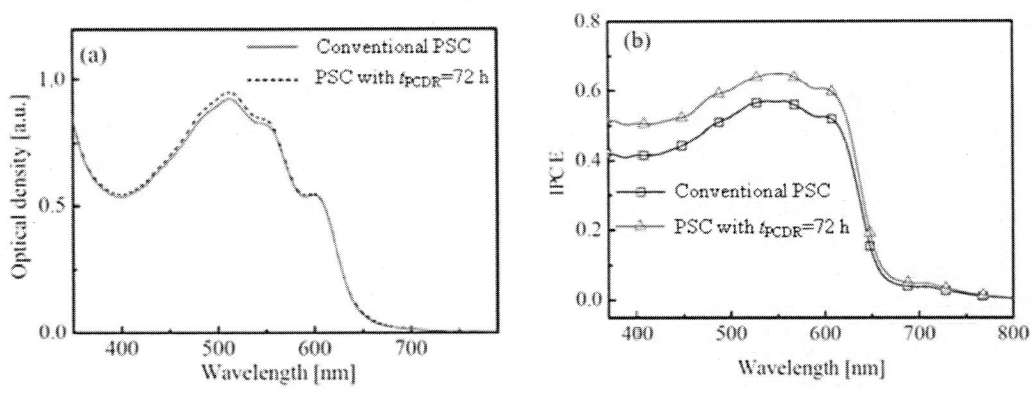

Fig. 5: Effects of t_{PCDR} on the (a) Absorbance of the P3HT/PCBM Films and (b) IPCE of the Conventional and Optimized PSC Devices

Figures 4(a) and 4(b) show atomic force microscopy (AFM) images of the conventional two-step PSC device and the optimized three-step PSC device, respectively, revealing three features that contribute to increasing the PSC performance: 1) For the two-step film, the surface is relative smooth, with an r.m.s. roughness of 9.9 nm (Fig. 4(a)). This result is in agreement with that reported by Li Gang et al. [1]. For the three-step film, a rougher surface is observed with roughness of 16.9 nm. This agrees with the earlier discussion that the extra time t_{PCDR} allows self-organization of the active layer leading to a higher surface roughness. 2) The rougher morphology likely increases the contact area between the active layer and the cathode (Ca/Al composite), which enhances electron extraction to the cathode. 3) The peak-to-valley polymer height for the three-step PSC device is around 120 nm, corresponding to ~60% of the mean thickness, while the height is only 80 nm for the conventional two-step PSC device. Such large peak-to-valley height effectively reduces the charge-transport distance and thus increases J_{sc}.

To examine whether the introduction of step III influences the composition of the polymer layer, the optical absorption spectra, which depend on the composition of the bulk polymer, were measured for both the conventional two-step PSC device and the optimized three-step PSC device. As shown in Fig. 5(a), almost no change is observed in the absorption spectra, which implies that the thickness and composition of the film are not altered during step III. On the other hand, the internal photon-to-charge efficiency (IPCE) shown in Fig. 5(b) obviously increases after step III. Generally, the enhancement of absorption by the active layer is crucial to increase the IPCE. Figure 5(a) demonstrates that the composition of the active layer is insensitive to step III.

Thus the enhancement in absorption is due to the surface roughness of the active layer as confirmed by Figs. 4(a) and 4(b). To measure the absorption spectra, new two-step and three-step PSC devices were fabricated with the Ca/Al composite metal layer removed so that the devices were transparent. However, IPCE measurements were performed on PSC devices with highly reflective Ca/Al composite metal layers.

As seen in Figs. 4(a) and 4(b), the rougher surface of the active layer can scatter the incident light at the interface between the polymer layer and the cathode back into the active layer leading to enhanced absorption and increase in IPCE.

CONCLUSIONS

We have proposed and investigated the influence of the post solvent evaporation time delay on the performance of polymer solar cell (PSC) devices fabricated using the composite polymer P3HT:PCBM as the active layer. Experimental results have demonstrated a charge-balance ratio close to unity through the use of three processing steps, namely spin coating, solvent evaporation and polymer-chain-distribution reorganization time delaying. An optimum device performance has been obtained with 72 hours of polymer-chain-distribution reorganization time delay, displaying a power conversion efficiency of 4.1%, which is 0.9% higher than that of conventional two-step PSC devices.

ACKNOWLEDGEMENTS

This research is supported by Edith Cowan University, the Department of Industry, Innovation, Science, Research and Tertiary Education, Australia.

REFERENCES

[1] Li, G., Shrotriya, V., Huang, J.S., Yao, Y. and Moriarty, T. et al., High-efficiency solution processable polymer photovoltaic cells by self-organization of polymer blends. *Nat. Mater.* 4, 864-868, 2005.

[2] Yu, G., Gao, J., Hummelen, J.C., Wudl, F. and Heeger, A.J., Polymer Photovoltaic Cells: Enhanced Efficiencies via a Network of Internal Donor-Acceptor Heterojunctions. *Science* 270, 1789-1791, 1995.

[3] Woo, C.H., Beaujuge, P.M., Holcombe, T.W., Lee, O.P. and Freichet, J.M.J., Incorporation of Furan into Low Band-Gap Polymers for Efficient Solar Cells. *J. Am. Chem. Soc.* 132, 15547-15549, 2010.

[4] Kekuda, D., Lin, H.-S., Chyi Wu, M., Huang, J.-S., Ho, K.-C. et al., The effect of solvent induced crystallinity of polymer layer on poly(3-hexylthiophene)/C70 bilayer solar cells. *Sol. Energ. Mat. Sol.* C 95, 419-422, 2011.

[5] Price, S.C., Stuart, A.C., Yang, L., Zhou, H. and You, W., Fluorine Substituted Conjugated Polymer of Medium Band Gap Yields 7% Efficiency in Polymer"Fullerene Solar Cells. *J. Am. Chem. Soc.* 133, 4625-4631, 2011.

[6] Li, G., Yao, Y., Yang, H., Shrotriya, V., Yang, G. et al., "Solvent annealing" effect in polymer solar cells based on poly(3-hexylthiophene) and methanofullerenes. *Adv. Funct. Mater.* 17, 1636-1644, 2007.

[7] Ding, B.F., Zhan, Y.Q., Sun, Z.Y., Ding, X.M., Hou, X.Y. et al., Electroluminescence and magnetoresistance of the organic light-emitting diode with a $La_{0.7}Sr_{0.3}MnO_3$ anode. *Appl. Phys. Lett.* 93, 183307, 2008.

[8] Davids, P.S., Campbell, I.H. and Smith, D.L., Device model for single carrier organic diodes. *J. Appl. Phys. 82*, 6319-6325, 1997.

[9] Mihailetchi, V.D., Xie, H.X., de Boer, B., Koster, L.J.A. and Blom, P.W.M., Charge transport and photocurrent generation in poly (3-hexylthiophene): Methanofullerene bulk-heterojunction solar cells. *Adv. Funct. Mater.* 16, 699-708, 2006.

[10] Reid, O.G., Munechika, K. and Ginger, D.S., Space Charge Limited Current Measurements on Conjugated Polymer Films using Conductive Atomic Force Microscopy. *Nano Lett.* 8, 1602-1609, 2008.

[11] Mihailetchi, V.D., van Duren, J.K.J., Blom, P.W.M., Hummelen, J.C., Janssen, R.A.J. et al., Electron Transport in a Methanofullerene. *Adv. Funct. Mater.* 13, 43-46, 2003.

[12] Jiang, C.Y., Sun, X.W., Zhao, D.W., Kyaw, A.K.K. and Li, Y.N., Low work function metal modified ITO as cathode for inverted polymer solar cells. *Sol. Energ. Mat. Sol.* C 94, 1618-1621, 2010.

[13] Zhao, D.W., Liu, P., Sun, X.W., Tan, S.T. and Ke, L. et al., An inverted organic solar cell with an ultrathin Ca electron-transporting layer and MoO[sub 3] hole-transporting layer. *Appl. Phys. Lett.* 95, 153304-153303, 2009.

[14] Sun, Z., Ding, B., Wu, B., You, Y., Ding, X. et al., LiF Layer at the Interface of Au Cathode in Organic Light-Emitting Devices: A Nonchemical Induced Carrier Injection Enhancement. *J. Phys. Chem. C* 116, 2543-2547, 2011.

[15] Lei, B., Yao, Y., Kumar, A., Yang, Y. and Ozolins, V., Quantifying the relation between the morphology and performance of polymer solar cells using Monte Carlo simulations. *J. Appl. Phys.* 104, 024504-024506, 2008.

PIEZOELECTRIC -POLYMORPH STABILIZATION BY INTEGRATION OF IONIC LIQUID MODIFIED MULTIWALLED CARBON NANOTUBE IN POLY(VINYLIDENE FLUORIDE)

AMIT MANDAL AND ARUN K. NANDI

Polymer Science Unit, Indian Association for the Cultivation of Science, Jadavpur, Kolkata
E-mail: psuakn@iacs.res.in

ABSTRACT

Homogenous dispersion is most efficient condition to get full potential of carbon nanotubes (CNTs) for fabrication of polymer nanocomposites. b polymorph of Poly(vinylidene fluoride) (PVDF) has drawn significant attention towards its stabilization and its technological utilization due to its piezo and pyro-electric properties. Here we report the ionic liquid (IL) functionalized multiwalled carbon nanotube (MWNT-IL) reinforced composites materials with PVDF. The TEM micrographs showed the average coating thickness of MWNTs by IL is about 8-10 nm and the covalent linkage of IL is confirmed from FTIR spectra. The FE-SEM, TEM micrographs indicates homogeneous dispersion of MWNTs by IL in the PVDF matrix. FTIR and XRD results indicate the most important β-crystal formation of PVDF. It is demonstrated that the specific interaction of MWNT-IL with the >CF_2 groups in PVDF chains could facilitate the orientation of PVDF chains on the MWNTs surface with all zigzag conformations to produce the piezo-electric β polymorph.

INTRODUCTION

After the discovery of carbon nanotubes (CNTs),[1] it is attracted extensive attention among researchers as advanced reinforcing fillers for their one-dimensional structure, higher aspect

V. Rajendran, P. Manivasakan and K.E. Geckeler (eds.)
Advanced Nano Materials for Industrial Applications, pp. 247-254 (2012).

ratio, superb mechanical and excellent conducting properties.[2-4] The outstanding properties of CNTs due to their strong covalent bonds and unique atomic structure have provided in demonstrate to their wide application in many fields. It afford great opportunities for the developing high-performance multifunctional materials.[5-7] Large quantity of work has done but the effective processing of nanotubes and polymers to produce new ultra-strong composite materials is still a great challenge.[8,9] The large aspect ratio, good dispersion, alignment of the nanofillers and interfacial stress transfer in polymer matrix are main requirements for the reinforcement properties in polymer nanocomposites.[9-11] The dispersion quality and interfacial interaction is often poorest in fabrication procedure by physical blending.[3,10] The functionalization of CNTs is most significant approaches to overcome these problems due to the strong covalent bonding between the grafted polymer and CNTs. The process include radical, living anionic polymerization and direct attachment of the functional polymer.[12-14] Usually the functional groups such as -COOH or -OH are generated on the sidewalls of the CNTs by oxidation of the CNTs.[15,16]

Poly(vinylidene fluoride) (PVDF) and its copolymer have been extensively studied because of their largest piezo and pyro-electric coefficient. It has five different crystalline polymorphs (α, β, γ, δ and ε) having unique properties.[17] Amongst them a polymorph is the most common having $TGT\overline{G}$ chain conformation and g phase having $T_3GT_3\overline{G}$ conformation is produced at high temperature and pressure.[18] The most important β polymorph has all trans conformation, where all the fluorine atoms located on the same side of the polymer chain and shows piezo and pyro-electric properties. So, the β-phase of PVDF has most important towards its stabilization and its technological utilization. The β-PVDF obtained by melt crystallization, mechanical deformation and composites preparation with clay, metal nanoparticles and functionalized CNTs as nanofillers.[11,19-22]

In PVDF nanocomposites, the β-polymorph of PVDF is produced when made from solution casting method but this β polymorph is not always stabilized. Priya [19] and Giannelis et al. [20] have reported conversion of β polymorph in PVDF/clay composites. The PVDF/ester functionalized MWNTs composite a maximum of 50% β polymorph PVDF was achieved.[22]

Ionic liquids (IL) have strong interactions with PVDF which afford new way to fabricate CNTs in PVDF matrix. However, few studies have succeeded in dispersing IL-modified CNTs in polymer matrices.[23] Here, we have modified the multi-walled carbon nanotubes (MWNTs) with a ionic liquid, (3-aminoethyl)imidazolium bromide, which act as a compatibilizer for the homogeneous dispersion of MWNTs in PVDF matrix. The MWNT-IL was then incorporated into PVDF matrix by the solution method to prepare the PVDF/MWNT-IL nanocomposite films. Here MWNT-IL are homogeneously dispersed in the PVDF matrix and the PVDF nanocomposites materials with appropriate content of MWNT-IL show the formation of β-crystal form.

EXPERIMENTAL SECTION

Materials and Synthesis of MWNT-IL

PVDF ($\overline{Mw}$ = 180,000, PDI = 2.57), 1-methylimidazole, 2-bromopropylamine hydrobromide, Multiwalled carbon nanotubes (MWNTs) (purity 95%,), Dicyclohexylcarbodiimide (DCC), *N,N*-(Dimethylamino)pyridine (DMAP) are used as received.

Carboxyl-functionalized multiwalled carbon nanotubes (MWNT-COOH) were prepared by oxidation of pristine MWNTs using HNO_3 and H_2SO_4 acid mixture. The mixture was diluted with deionized water and filtered through a 0.22 μm polycarbonate membrane to produce MWNT-COOH. Amine terminated Ionic liquid (IL-NH_2) was prepared by 1-methylimidazole with 2-bromoethylamine hydrobromide in methanol. (3-aminoethyl)imidazolium bromide functionalized MWNTs (MWNT-IL) was prepared by coupling reaction between MWNT-COOH and IL-NH_2 in presence of DCC and DMAP in DMF solvent.

Preparation of PVDF/MWNT-IL Composites

1 gm of PVDF was dissolved in *N,N* dimethyformamide (DMF) solution and heating the mixture at 70 ^{0}C for 2 hr. The required amount of MWNT-IL was dispersed in DMF solution of PVDF by sonication. They were homogenized and the films were cast by evaporation of the solvent on flat dishes at 70 ^{0}Cin controlled air stream to produce solvent cast samples. The designations of the composite samples are PMLx(x = weight percentage of MWNT-IL in PVDF).

RESULTS AND DISCUSSION

MWNTs Modification with Ionic Liquid and Characterization

The functionalization and covalent bonding of MWNTs with amine terminated imidazoluim salts (IL-NH_2) are demonstrated by FTIR spectroscopy (Fig. 1a).

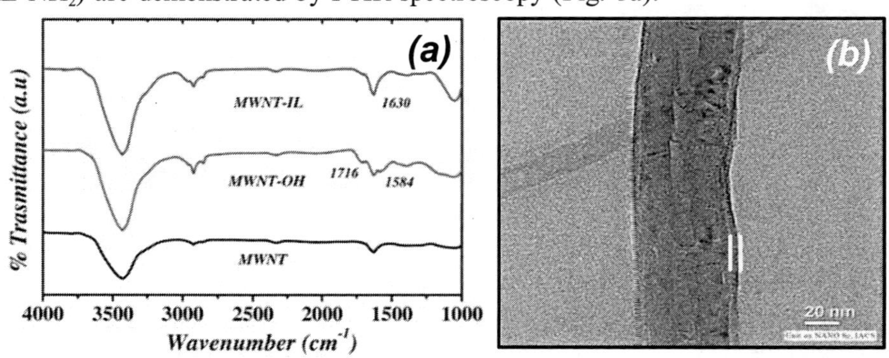

Fig. 1: (a) FTIR spectra of MWNTs, MWNT-COOH and MWNT-IL respectively (b) HR-TEM images of IL functionalized MWNT (MWNT-IL)

In MWNT-COOH the > C = O bands appeared at 1716 and 1584 cm^{-1} indicating the presence of carboxylic acid groups. After amidation reaction with MWNT-IL, the >C=O band of the carboxylic acid group at 1716 cm^{-1} shifts to 1630 cm^{-1}, indicating the formation of the amide linkage between MWNT-COOH and IL-NH_2. The morphology of the MWNT-IL observed in HR-TEM microscope is shown in Fig. 1b. It indicates that the MWNT-IL are thicker than the pristine MWNTs and well coated with ionic liquid (IL). It also indicates that the average thickness of IL present on the MWNTs surface is about 8-10 nm which suggests a significant co-valent functionalization of the ionic liquid on the MWNTs surface.

Characterization of PVDF/MWNT-IL (PML) Composites

Morphological Studies

From the FE-SEM micrographs (Fig. 2(a-b)) of PVDF/MWNT-IL nanocomposites shown that the MWNTs are found to homogeneously dispersed within the PVDF matrix.

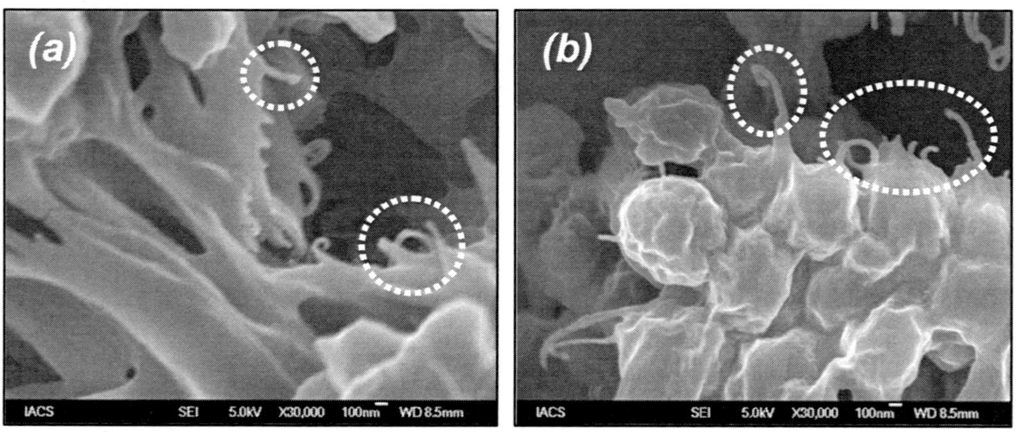

Fig. 2: FE-SEM images of PML Nanocomposites: (a) PML0.1 and (b) PMI0.5 Samples Respectively

Most of MWNTs are found to be embedded in the matrix with the highly exfoliated state and exhibits three-dimensional network structure in the composites. The IL functionalized MWNTs have improved the compatibility and widely dispersed in the polymer matrix. It also indicates that ionic liquid improve the dispersion of MWNTs in PVDF matrix. In the corresponding TEM micrographs (Fig. 3(a-b)) the MWNTs are found well dispersed within the PVDF matrix. The density of MWNTs gradually increases with increase of MWNT-IL concentration and the polymer layers surrounding the MWNTs are larger than the MWNT-IL (Fig. 1b).

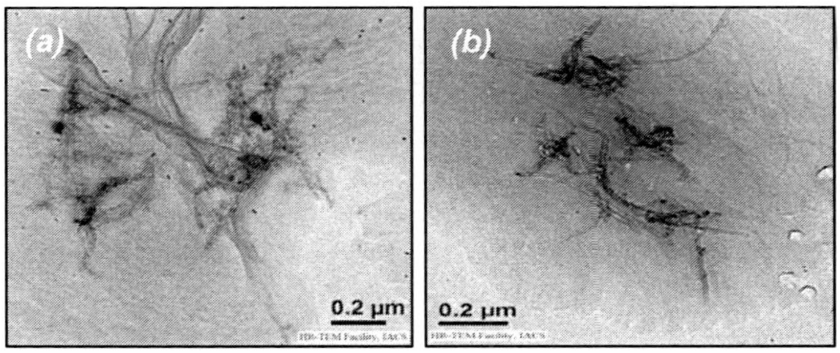

Fig. 3: TEM images of PML Nanocomposites (a) PML0.1, and (b) PMI0.5 Samples Respectively

Probably, this is due to the specific interaction between the MWNT-IL and the PVDF chain,[22,23] where IL act as a compatibilizers for the dispersion of MWNTs in PVDF matrix.

Structural Aspect

FTIR spectra of PVDF and nanocomposite films (PML) with different concentration of MWNT-IL are presented in Fig. 4. The peaks at 532, 615, 796, and 976 cm^{-1} are the α-phase and peaks at 445, 510, 840 and 1275 cm^{-1} for the α-phase of PVDF.[20-22,25,26] PVDF are showing mainly the α-crystals. In these PML samples, the intensity of the α-peaks decreases and the intensity of α-peaks increases gradually, i.e conversion of α-phase to α-phase on addition of MWNT-IL. In PML1 sample the complete conversion of â-phase. Here, the dipolar interaction take place between -CONH groups of MWNT-IL with the >CF_2 groups of PVDF. Due to the presences of specific interaction, the -CF–CH –CF bending vibration of PVDF at 876 cm^{-1} shift to higher energy (882 cm^{-1}) in the composites with increasing MWNT-IL concentration.[25,26]

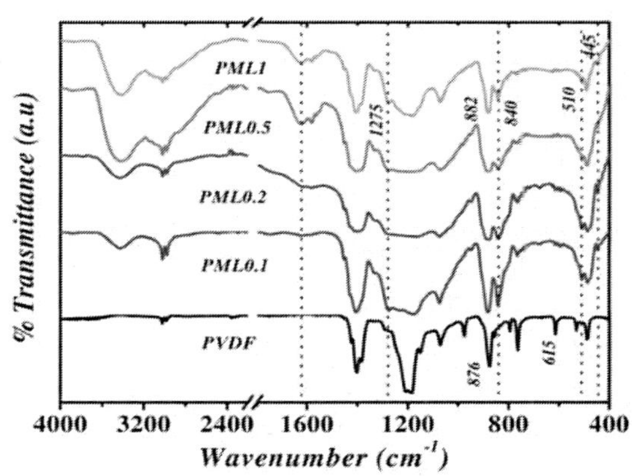

Fig. 4: FTIR Spectra of PML Nanocomposites at Indicated MWNT-IL Concentration

The β-phase formation is also identified from the XRD spectra (Fig. 5). PVDF show four characteristic diffraction peaks at 2θ = 17.8, 18.5 20.0 and 26.7^0 which characterizing the α-phase of PVDF.[18,27-29] The formation of piezoelectric b-polymorph is evident by the single peak at 2θ = 20.3^0 from the (110) and (200) crystal planes.[24-26] In PML1 samples no diffraction peaks of α crystals and a single peak at 2θ = 20.3^0 is observed indicating β-crystal formation and stabilization in the samples. This result suggests that at the immediate vicinity of MWNTs surface the nucleation of the β-crystal occurs and it grows to form the β-crystals.[21,22]

This means that the MWNT-IL integration in PVDF induce the β-crystal formation of PVDF in this nanocomposites. Here the IL acts as compatibilizers to improve the interaction between PVDF matrix and MWNTs. The MWNTs surface functionalization by the ionic liquid (IL) significantly enhances the crystal phase transitions and stabilization. The dipolar and specific interaction between the MWNT-IL with >CF_2 of PVDF renders the >CF_2 located along the one side of the molecular chain. Therefore, the all PVDF chains adopt the all trans conformation on the surface of ionic liquid coated MWNTs. This type of interaction can overcome the energy barrier between the α to β-phase transition and large amount of piezoelectric β-crystals are formed.[29]

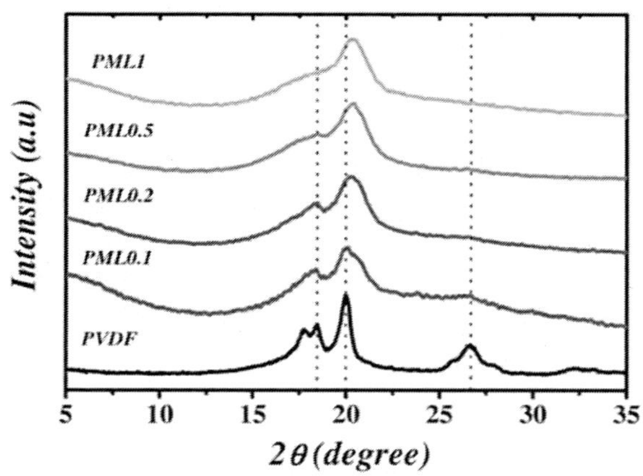

Fig. 5: XRD Spectra of Different PML Nanocomposites.

CONCLUSIONS

Here we functionalized multiwalled carbon nanotubes (MWNTs) with ionic liquid (IL). The functionalization of MWNTs is characterized by FTIR, HR-TEM micrographs and the TEM micrographs indicates the average coating thickness of IL on MWNTs surface is about 8-10 nm. We have successfully fabricated PVDF nanocomposites with β-crystal formation using MWNT-IL as the nanofillers confirmed by FTIR, XRD. The FE-SEM and TEM micrographs indicate that IL helps to disperse the MWNTs in the PVDF matrix and the most important piezoelectric β-phase was achieved and stabilized without any mechanical deformation.

ACKNOWLEDGEMENTS

We gratefully acknowledge the 'Unit of Nanoscience at IACS', DST, New Delhi for financial support. Amit Mandal acknowledges CSIR, New Delhi for the fellowship.

REFERENCES

[1] Iijima, S., Helical microtubules of graphitic carbon, *Nature*, 354, 56-58, 1991.

[2] (a) Ajayan, P. M., Nanotubes from Carbon, *Chem. Rev.*, 99, 1787-1799, 1999.
(b) Niyogi, S., Hamon, M. A., Hu, H., Zhao, B., Bhowmik, P., Sen, R., Itkis M. E., and Haddon, R. C., Chemistry of Single-Walled Carbon Nanotubes, *Acc. Chem. Res.*, 35, 1105-1113, 2002.

[3] Odom, T.W., Huang, J.L., Kim, P., and Lieber, C. M. Atomic Structure and Electronic Properties of Single-Walled Carbon Nanotubes, *Nature*, 391, 62-64, 1998.

[4] Ebbesen, T. W., Carbon Nanotubes: Preparation and Properties, CRC Press, Boca Raton, FL, 1997.

[5] Moniruzzaman, M., and Winey, K. I., Polymer Nanocomposites Containing Carbon Nanotubes, *Macromolecules*, 39, 5194-5205, 2006.

[6] Ajayan, P. M., Stephan, O., Colliex, C., and Trauth, D., Aligned Carbon Nanotube Arrays Formed by Cutting a Polymer Resin-Nanotube Composite, *Science*, 265, 1212-1214, 1994.

[7] Baughman, R. H., Zakhidov, A. A., and Heer, W. A., Carbon Nanotubes-the Route Toward Applications, *Science*, 297, 787-792, 2002.

[8] Cadek, M., Coleman, J.N., Ryan, K. P., Nicolosi, V., Bister, G., Fonseca, A., Nagy, J. B., Szostak, K., Beguin, F., and Blau, W. J., Reinforcement of Polymers with Carbon Nanotubes: The Role of Nanotube Surface Area, *Nano Lett.*, 4, 353-356, 2004.

[9] Kis, A., Csanyi, G., Salvetat, J.P., Lee, T. N., Couteau, E., Kulik, A. J., Benoit, W., Brugger, J., and Forro, L., Reinforcement of single-walled carbon nanotube bundles by intertube bridging, *Nat. Mater.*, 3, 153-157, 2004.

[10] Coleman, J. N., Khan, U., and Gun'ko, Y. K., Mechanical Reinforcement of Polymers Using Carbon Nanotubes, *Adv. Mater.*, 18, 689-706, 2006.

[11] Mandal, A., and Nandi, A. K., Physical properties of poly(vinylidene fluoride) composites with polymer functionalized multiwalled carbon nanotubes using nitrene chemistry, *J. Mater. Chem.*, 21, 15752-15764, 2011.

[12] Kong, H., Gao, C., and Yan, D.Y., Controlled Functionalization of Multiwalled Carbon Nanotubes by in Situ Atom Transfer Radical Polymerization, *J. Am. Chem.* Soc., 126, 412-413, 2004,

13] Shaffer, M.S.P., and Koziol, K., Polystyrene grafted multi-walled carbon nanotubes, *Chem. Comm.*, 2002, 2074-2075.

[14] Hong, C.Y., You, Y.Z., Wu, D.C., Liu, Y., and Pan, C.Y., Multiwalled Carbon Nanotubes Grafted with Hyperbranched Polymer Shell via SCVP, *Macromolecules*, 38, 2606-2611, 2005.

[15] Zhao, B., Hu, H., Yu, A.P., Perea, D., and Haddon, R.C., Synthesis and Characterization of Water Soluble Single-Walled Carbon Nanotube Graft Copolymers, *J. Am. Chem. Soc.*, 127, 8197-8203, 2005.

[16] Baskaran, D., Mays, J.W., and Bratcher, M.S., Polymer-Grafted Multiwalled Carbon Nanotubes through Surface-Initiated Polymerization, *Angew. Chem. Int. Ed.*, 43, 2138-2142. 2004.

[17] Lovinger, A.J., In Developments in crystalline polymers-1, Bassett, D.C., Ed., Applied Science Publishers, p 195, London, 1981.

[18] Hasegawa, R., Takahashi, Y., Chatani, Y., and Tadokaro, H., Crystal structure of three crystalline forms of Poly(vinylidene fluoride), *Polym. J.*, 3, 600-610, 1972,

[19] Priya, L., and Jog, J.P., Poly(vinylidene fluoride)/Clay Nanocomposites Prepared by Melt Intercalation: Crystallization and Dynamic Mechanical Behavior Studies, *J. Polym. Sci., Part B, Polym. Phys.*, 40, 1682-1689, 2002.

[20] Shah, D., Maiti, P., Gunn, E., Schmidt, D.F., Jiang, D.D., Batt, C.A., and Giannelis, E.P., Dramatic enhancements in toughness of Polyvinylidene fluoride nanocomposites via nanoclay-directed crystal structure and morphology, *Adv. Mater.*, 16, 1173-1177, 2004.

[21] Mandal, A., and Nandi, A.K., Noncovalent Functionalization of Multiwalled carbon nanotube by a Polythiophene-based Compatibilizer: Reinforcement and Conductivity improvement in Poly(vinylidene fluoride) Films, *J. Phys. Chem.* C 116, 9360-9371, 2012.

[22] Manna, S., and Nandi, A.K., Piezoelectric β Polymorph in Poly(vinylidene fluoride)-Functionalized Multiwalled Carbon Nanotube Nanocomposite Films, *J. Phys. Chem.* C 111, 14670-14680, 2007.

[23] Shang, S.M., Zeng, W., and Tao, X.M., High stretchable MWNTs/polyurethane conductive nanocomposites, *J. Mater. Chem.*, 21, 7274-7280, 2011.

[24] Roerdink, E., and Challa, G., Computerized infra-red Study of interaction of poly(vinylidene fluoride) with stereoregular poly(methyl methacryalate), *Polymer*, 21, 509-513, 1980.

[25] Belke, R.E., and Cabasso, I., Poly(vinylidene fluoride)/poly(vinyl acetate) miscible blends: 1. Thermal analysis and spectroscopic (FTir.) characterization, *Polymer*, 29, 1831-1842, 1988.

[26] Tashiro, K., Kobayashi, M., and Tadokoro, H., Vibrational Spectra and Disorder-Order Transition of Poly(vinylidene fluoride) Form I11, *Macromolecules*, 14, 1757-1764, 1981.

[27] Datta, J., and Nandi, A.K., Cocrystallization of poly(vinylidene fluoride) and vinylidene fluoride-tetrafluoro-ethylene copolymer blends: 3. Structural study, *Polymer*, 38, 2719-2724, 1997,

[28] Guerra, G, Karasz, F.E., and MacKnight, W.J., On blends of poly(vinylidene fluoride) and poly(vinyl fluoride), *Macromolecules*, 19, 1935-1938, 1986.

[29] Prest, W.M., and Luca, D.J., The morphology and thermal response of high temperature–crystallized poly(vinylidene fluoride), *J. Appl. Phys.*, 46, 4136-4143. 1975.

EFFECT OF IMPINGEMENT ANGLE ON EROSIVE WEAR BEHAVIOR OF GRAPHITE FILLED POLYAMIDE66/POLYPROPYLENE BLEND

VEENA PATIL, MOHAMMED ISMAIL AND B. SURESHA[1]

Department of Industrial & Production Engineering, The National Institute of Engineering, Mysore, Karnataka
[1]Department of Mechanical Engineering, The National Institute of Engineering, Mysore, Karnataka
E-mail: veenapatil777@gmail.com

ABSTRACT

The present work describes the processing and study of the erosive wear behaviour of 2.5wt% and 5wt% graphite (Gr) filled Polyamide66/Polypropylene (PA66/PP) blend. Composites were fabricated using co-twin extruder followed by injection moulding. The mechanical properties were investigated in accordance with ASTM standards. Experiments were carried out to study the effects of impingement angle at constant velocity and evaluated wear resistance on solid particle erosion behavior. The erosion wear test was conducted using Air jet type test rig as per ASTM G-76 standard, at different impingement angles from 15° to 90° at constant velocity of 30 m/s. The erodent used was silica sand of angular shape having grain size 212 μm. From the experimental investigation, it was found that the Polyamide66/Polypropylene (PA66/PP) and Graphite (Gr) filled PA66/PP showed ductile behavior with maximum erosion wear rate at 15° and 30° impingement angle respectively. It was observed that the 5% graphite filled PA66/PP improved the erosion wear resistance of PA66/PP blend. Scanning Electron Micrography (SEM) was used to analyze the surface morphologies of tested materials.

V. Rajendran, P. Manivasakan and K.E. Geckeler (eds.)
Advanced Nano Materials for Industrial Applications, pp. 255-262 (2012).

INTRODUCTION

Solid particle erosion is a typical erosion wear mode, where the loss of material results from repeated impact of small, solid particles. The effects of solid particle erosion have been recognized for a long time. In some cases solid particle erosion is a useful phenomenon, as in sandblasting and high-speed abrasive water jet cutting, but it is a serious problem in many engineering systems, including steam and jet turbines, pipelines and valves carrying particulate matter, and fluidized bed combustion systems.[1]

Composite materials provide an opportunity to combine different properties and design materials for applications requiring multiple functionalities. Polymeric matrices reinforced with hard and non dissipative fillers can possess high stiffness and damping, which is ideal for structural properties. Over the past decades, injection molded thermoplastic composites have been increasingly used for numerous mechanical and tribological purposes such as seals, gears and bearings. These materials are light in weight and are better alternatives to metallic components [2, 3]. The feature that makes polymer composites so promising in industrial applications is the possibility of tailoring their properties with functional fillers. It has been found that short fiber reinforcements can generally improve mechanical properties of the polymer composites [4, 5]. Filler-reinforced thermoplastic composites can improve the stiffness, decrease thermal expansion, improve long-term mechanical performance and reduce costs [6]. Many materials, including mineral and glass fillers, glass beads, carbon black, and wood flour, are currently used as fillers in thermoplastic composites [7, 8].

In recent years much research has been devoted to exploring the potential advantages of thermoplastic matrix for composite applications. Nylon is a good bearing material because of its high strength and good wear resistance. However, it has high friction coefficient under dry sliding motion. Various studies have demonstrated that the tribological properties of nylon can be further enhanced by the addition of inorganic fillers or fibers. Bahadur and Polineni [9] found that the wear rate of nylon 11 was reduced by a factor of five to six with the inclusion of glass fabric content up to 20.7 vol%. Bolvari et al. [10] also found that the optimal reinforcement for the minimum wear rate was 5% aramid fiber in nylon 66 composite. Tilly [11,12,13] has investigated the influence of velocity, impact angle, particle size, weight of abrasive quartz particles impacted and type of reinforcement (glass compared with carbon fiber) in plastic materials such as nylon, polypropylene and epoxy resins. The composite materials exhibited poor erosion resistance; the nylon exhibited a fourfold increase in erosion weight loss if reinforced with glass or with carbon fibers and the epoxy reinforced with steel powder showed a substantial improvement in erosion resistance compared with that of unreinforced material. Mathias et al. [14] and also Karasek et al. [15] have evaluated the erosion behavior of a graphite fiber reinforced bismaleimide polymer composite. These investigators observed the erosion rates of the PMC to be higher than the unreinforced polymer. Basava T. et al. [16] studied the development of graphite particles filled epoxy resin composite material and its erosive wear behavior. Graphite particles as a filler material in epoxy E-glass composite laminate resist the formation of crack growth and resists fiber matrix debonding which improves the resistance of erosive wear rate.

EXPERIMENTAL DETAILS

Material used

Polymer alloy of polyamide66 and polypropylene, and particulate graphite filled PA66/PP composites were prepared for this study with compatibilizer. The polymer alloy produced consists of 50% by weight. Maleic anhydride polypropylene (MAgPP) as compatibilizing agent was used in this study. The amount of compatibilizer added was 1 wt.% based on previous literature and this compatibilizer proportion was high enough for interaction with PA66/PP interface.

Compounding

Before compounding, the polymer granules and fillers were dried at 80^{0}C for 10 h in an air circulated oven and then dry mixed with polyamide66 and other additives. Composition shown in Table 2 was mixed and extrudated in a co-rotating twin extruder. The L/D ratio of the screw is 40:1. Mixing speed of 60 rpm was maintained for all the compositions. The extrudate from the die were quenched in a tank at 20–30^{0}C and then palletized. The temperatures from the feed zone to the die of the extruder were 205, 235, 245, 255 and 265 ^{0}C, respectively. The extrudate of the composition was palletized in palletizing machine.

Injection Molding

The granules of the extrudates were pre dried in an air circulated oven at 80^{0}C for 10 h and injection molded in a microprocessor based injection molding machine fitted with a master mould containing the cavity for tensile strength, flexural and impact specimens. After its ejection from the mould, specimens were cooled in ice-water. The temperatures maintained in three zones of the barrel were 200^{0}C, 235^{0}C and 260^{0}C, respectively. The details of the composites fabricated for the present investigation are given in Table 1.

Table 1: Materials Composition

Sl. No.	*Sample (designation)*	*Composition by Wt.%*		
		PP	*PA66*	*Particulate filler*
1	PA66/PP	50	50	–
2	PA66/PP+2.5%Gr	47.5	50	2.5(Gr)
3	PA66/PP+5%Gr	45	50	5(Gr)

Mechanical Characterization

Densities of the composites were determined by using a high precision electronic balance (Mettler Toledo, Model AX 205) using Archimede's principle. Shore hardness of the samples was measured as per ASTM D2240, by using a Hiroshima make Hardness Tester (Durometer). The tensile test was carried out using Universal tensile testing machine (JJ Lloyd, London,

United Kingdom, capacity 1–20 KN), according to ASTM D638. The tensile and flexure tests were performed at crosshead speed of 20 mm/min.

Erosion Test Method

Figure 1 shows the schematic diagram of erosion test rig confirming to ASTM G 76. The set up is capable of creating reproducible erosive situations for assessing erosion wear resistance of the prepared composite samples. It consists of an air compressor, an air particle mixing chamber and an accelerating chamber. Dry compressed air is mixed with the particles which are fed at constant rate from a sand flow control knob through the nozzle tube and then accelerated by passing the mixture through a convergent brass nozzle of 3 mm internal diameter. These particles impact the specimen which can be held at various angles with respect to the direction of erodent flow using a swivel and an adjustable sample clip. The velocity of the eroding particles is measured using double disc method. In the present study, dry silica sand (angular) of particle size 212μm is used as erodent. Each sample is cleaned in acetone, dried and weighed to an accuracy of ±0.1 mg using a precision electronic balance. It is then eroded in the test rig for 6min and weighed again to determine the weight loss. The process is repeated till the erosion rate attains a constant value called steady-state erosion rate. The ratio of this weight loss to the weight of the eroding particles causing the loss is then computed as a dimensionless incremental erosion rate. The erosion rate is defined as the weight loss of the specimen due to erosion divided by the weight of the erodent causing the loss. The conditions under which the erosion tests were carried out are listed in Table 2.

Table 2: Test Parameters

Erodent	*Silica Sand*
Erodent Size(μm)	212
Erodent Shape	irregular, slightly rounded
Impingement Angle(degree)	15,30,45,60,75 and 90
Impact Velocity(m/s)	30
Stand Off Distance(mm)	10

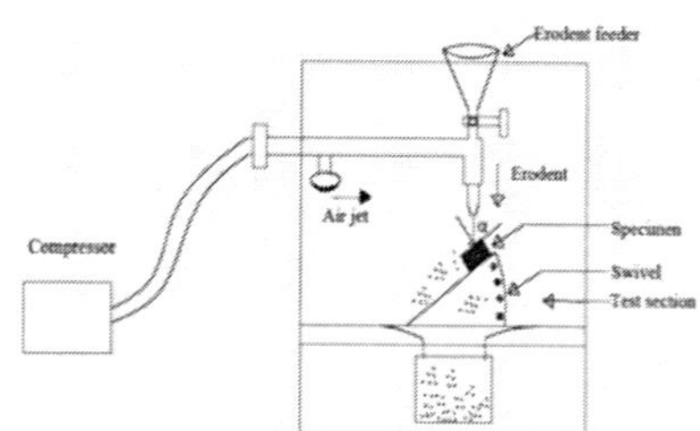

Fig. 1: A schematic Diagram of the Erosion Test Rig.

RESULTS AND DISCUSSION

Physico-Mechanical Properties

Table 3 shows the measured densities of PA66/PP blend, Gr-PA66/PP composites. The densities of graphite filled PA66/PP was higher than the density of PA66/PP blend. The density of 5% graphite filled PA66/PP composite is 0.9904 which is highest when compared to other composites. This is because the graphite has higher density than that of PA66/PP. By using Duro-hardness tester, the hardness of the composites is measured; the values recorded are given in Table 3. The hardness of PA66/PP blend increased with inclusion of graphite filler. From Table 3, it can be seen that the 5% graphite greatly increased the hardness of PA66/PP blend, which can be attributed to the higher hardness. The hardness of 5Gr-PA66/PP composite is 71, which is highest among all the composites tested. The improvement in hardness with incorporation of filler can be explained under the action of a compressive force, the thermoplastic matrix phase and the solid filler phase will be pressed together, touch each other and offer resistance. Thus the interface can transfer load more effectively although the interfacial bond may be poor. This results in enhancement of hardness of filled composites. The mechanical properties of composites are dependent on the characteristics of matrix, filler as well as interface phase. In the present work, Addition of graphite as filler in PA66/PP resulted in slight increase in the tensile strength and percentage elongation. Tensile properties are given in Table 3. Due to addition of PA66 to PP, the tensile strength increases (by 64%). However, the tensile strength improvement occurs sacrificing the ductility of the polymer blend similar to any engineering material. Micro filler inclusion though improves the mechanical strength imparts brittleness of the composites.

Table 3: Density, Hardness and Tensile Properties of Composites

Sl No	*Sample*	*Hardness (Shore- D)*	*Density(g/cc)*	*Tensile Strength (MPa)*	*Strain*
1	PA6/PP	64	0.9072	30.65	0.13
2	2.5Gr-PA66/PP	66	0.9843	30.65	0.11
3	5Gr-PA66/PP	71	0.9904	30.9	0.10

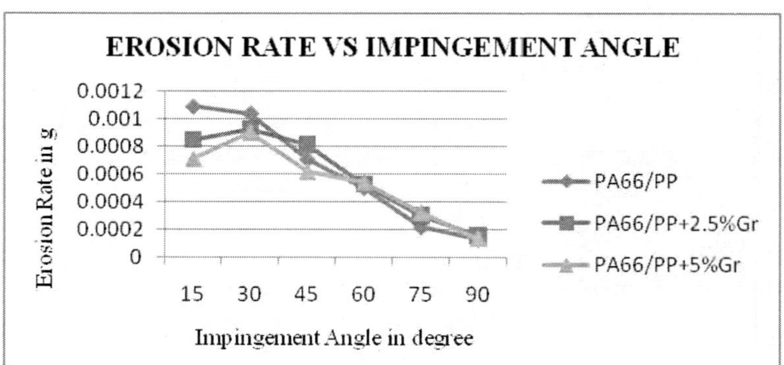

Fig. 2: Erosion Rate vs. Angle of Impingement for PA66/PP and with Graphite Filled

Erosion wear behavior of materials can be grouped as ductile and brittle categories although this grouping is not definitive. Thermoplastic matrix composites usually show ductile behavior and have the peak erosion rate at around 30^0 impingement angle because cutting mechanism is dominant in erosion. While the thermosetting ones erode in a brittle manner with the peak erosion occurring at normal impact. Figure 2 Shows erosion wear rate of unfilled and graphite filled PA66/PP composites tested as a function of the angle of impingement and at a constant impinging velocity of 30 m/s. Erosion rates were calculated by dividing the weight loss of specimen by the mass of erodent that impacted. The erosion rate of graphite filled PA66/PP increases with increase in the impingement angle and attains the peak value at 30^0. Whereas the weight loss of unfilled PA66/PP is initially very high and the peak value is at 15^0. This shows ductile erosion behavior of the composite. Sundararajan et al. [93] concluded that this behavior is attributed to the fact that the harder the material, larger is the fraction of the crater volume that is removed. In this investigation higher hardness values have been noted for composites with higher fiber loading and this is therefore the reason why the composites exhibit declining erosion resistance with the increase in fiber content. There is minimum weight loss for the filler content of 5% graphite. However, erosion wear weight loss is strongly affected by the variation of impingement angle of the particles.

Scanning Electron Microscope (SEM) Analysis

The worn surface of the composites was observed by SEM examination to search for a correlation of worn surface and erosion wear loss with increasing impingement angle. Figures (3a–3d) show morphology of the worn surface of the PA66/PP blend, 5%Gr-Pa66/PP composites at 15^0, $30^0$and 90^0 at a constant velocity of 30 m/s test conditions.

It is evident from the micrographs (Figure 3a) that the material removal in blend PA66/PP is dominated by microploughing, micro-cutting and plastic deformation. Due to mechanical action and the accumulated frictional heat the molecular structure of PA66/PP blend has been partially destroyed. The interfacial bonding strength is weaker for the blend thereby, producing severe plastic deformation. PP has straight molecular chain and relatively lower softening point. The poor adhesion between PA66 and PP results in higher erosion wear at angle 15^0; most material is lost when a maximum strain in the target is exceeded. Under normal impact (Figure 3b), formation of micro-cracks and embedment of fragments of sand particles is evident from the micrographs. This is due to largest part of the initial energy is converted in to heat and hence matrix is softened which resulted in embedment of sand particles. The embedded sand particles control the further erosion of the target surface. Figure 3c and 3d shows the eroded surface of 5% Gr filled PA66/PP composite. The micrograph 3c indicates Plastic flow of matrix material along the erosion direction for the composite eroded at 30^0 lower impact angle. When impacting at such low angle, the hard erodent particles penetrate the surface and cause material removal mostly by micro-ploughing. Under normal impact (Figure 3d), wear mechanism is dominated by extensive plastic deformation and propagation of multiple cracks in all the directions by repeated impacts. Such a damage process does not lead to effective removal of material. It is expected that several cumulative impacts are required for material removal in a ductile polymer matrix.

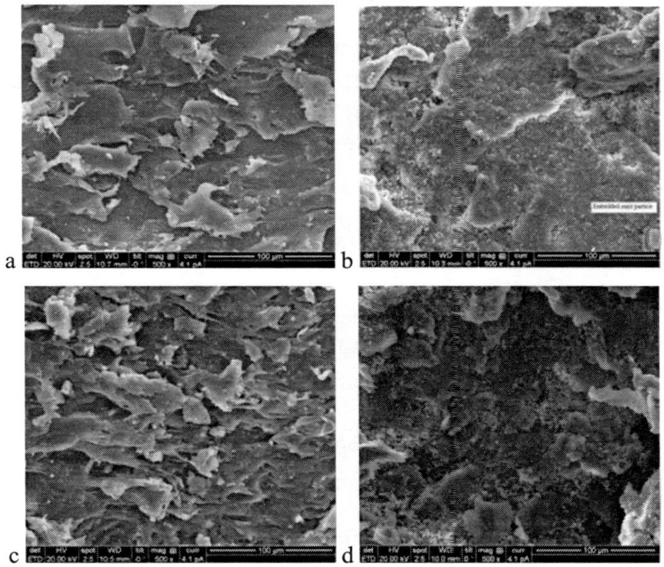

Fig. 3. SEM Micro-photographs of Blend PA66/PP at Impact Angle: (a) 15^0, (b) 90° and 5% Gr-PA66/PP at Impact Angle, (c) 30^0, (d) 90^0

CONCLUSIONS

Based on the solid particle erosion studies of unfilled and graphite filled PA66/PP composites, the following are the conclusions were drawn.

- The unfilled and filled Polyamide66/polypropylene composites were fabricated using Brabender co-twin extruder followed by injection moulding.
- Incorporation of Graphite (Gr) filler modifies the hardness, density, tensile properties of blend PA66/PP composite.
- Impingement angle and material properties such has matrix and particulate filler have strong influence on the erosive wear performance.
- The erosive wear behavior strongly depends on impingement angle.
- The PA66/PP blend has shown peak erosion rate at an impingement angle of 15^0. However, graphite filled PA66/PP composite showed peak erosion rate at 30^0 impingement angle, under the present experimental conditions studied.
- SEM was used to examine the worn surface. It was found that the main erosive wear mechanisms of both PA66/PP blend and graphite filled PA66/PP are microploughing and microcutting impacted under low angles and that plastic deformation and fatigue failure caused the mass loss for impacted under high angles.

ACKNOWLEDGEMENTS

It gives me immense pleasure to express my deep sense of gratitude to Dr. B. Suresha, Professor, Department of Mechanical Engineering, NIE, Mysore and to my guide Mohammed Ismail, Associate Professor, Department of Industrial and Production Engineering, NIE Mysore and to Dr. Y.S. Varadarajan Professor and Head Department of Industrial and Production Engineering, NIE, Mysore for providing me the necessary facilities in the department for conducting the present study.

REFERENCES

[1] Amar Patnaik, Solid particle erosion wear characteristics of fiber and particulate filled polymer composites: A review Wear 268 (2010) 249–263.

[2] Hutchings IM. Tribology friction and wear of engineering materials. London: CRC Press; 1992.

[3] Zhang SW. State of the art of polymer tribology. Tribology Int 1998, 31: 49–60.

[4] Vinson JR, Chou T. Composite materials and their uses in structures. London: Applied Science Publishing; 1975.

[5] Tsai SW. Strength characteristics of composite materials. NASA report: NASACR- 224: 1965.

[6] Palabiyik M, Bahadur S. Mechanical and tribological properties of polyamide 6 and high density polyethylene polyblends with and without compatibilizer. Wear 2000; 246 :149–58.

[7] Briscoe BJ, Wear of polymers: an easy on fundamental aspects. Tribology 1981;14:231–43.

[8] Huang R, Engineering plastic handbook. Beijing: Mechanical Industry Press; 2000.

[9] Bahadur .S, Polineni, V.K. Tribological studies of glass fabric-reinforced polyamide composites filled with CuO and PTFE, *Wear 200* (1996) 95–104.

[10] Bolvari A, Glenn S, Janssen R, Wear and friction of aramid fiber and polytetrafluoroethylene filled composites, *Wear 202–204* (1997) 697–702.

[11] Tilly G.P, Erosion caused by airborne particles, *Wear, 14* (1) (1969) 63 – 79.

[12] Tilly G.P, Sand erosion of metals and plastics: a brief review, *Wear, 14 (4) (1969) 241 - 248.*

[13] Tilly G.P and Sage W, The interaction of particle and material behaviour in erosion processes, *Wear,* 16 (6) (1970) 447 - 466.

[14] Mathias P-J., Wu W, Goretta K.C, Routbort L.J, Groppi D.P and Karasek K.R, "Solid particle erosion of a graphite-fiber reinforced bismaleimide polymer composite", *Wear,* Vol. 35, pp: 161-169, (1989).

[15] Karasek K.R., Goretta K.C., Helberg D.A. and Routbort J.L, "Erosion in Bismaleimide Polymers and Bismaleimide-polymer Composites", *Journal of Materials Science Letters,* Vol. 11, Issue 16, page: 1143–1144 (1992).

[16] Basava T, A N Harirao, "Development of Graphite Particles filled Epoxy Resin Composite Material and its Erosive wear Behavior" Proc. of Int. Joint Colloquium on Emerging Technologies in Computer Electrical and Mechanical 2011. © 2011 AMAE DOI: 02.CEM.2011.01.513

[17] Sundararajan G, Roy. M, Venkataraman B, Erosion efficiency a new parameter to characterize the dominant erosion micro-mechanism, Wear 140 (1990) 369.

PHYSICO-CHEMICAL, ELECTRICAL AND DIELECTRIC PROPERTIES OF THE SPRAY DEPOSITED WO_3 FILMS FOR GAS SENSOR APPLICATION

V.V. GANBAVLE, V.S. SAWANT AND K.Y. RAJPURE

Electrochemical materials laboratory, Department of Physics, Shivaji University, Kolhapur
E-mail: rajpure@yahoo.com

ABSTRACT

Synthesis and characterization of tungsten oxide films using cost effective spray pyrolysis technique is reported and effect of the substrate temperature on physicochemical properties is studied. Raman spectra of the films confirmed the W-O-W stretching and bending vibrations in concern with XRD results. With increase in substrate temperature from 400°, (30 nm) crystallite size decreased up to 425°C (28nm) and increased further at 475°C (43 nm). Scanning electron micrographs depict wired network of the tungsten oxide. Thermo-emf is linearly changing with temperature difference and decreases with increase in the deposition temperature. High roughness of the films suggests application of the films in gas sensor as more surface is provided for the analyte gas adsorption.

INTRODUCTION

Tungsten trioxide (WO_3) is a versatile material that finds variety of applications in different fields such as electrochromism, photocatalysis, gas sensor and thermochromism [1-3]. WO_3 is promising semiconducting metal oxide, for gas sensing devices due to its inherent electrical conductivity and excellent sensitivity coupled with selectivity towards various gases such as NO_2, NH_3, H_2S and various other gases [2]. Because of its properties WO_3 is found promising

V. Rajendran, P. Manivasakan and K.E. Geckeler (eds.)
Advanced Nano Materials for Industrial Applications, pp. 263-270 (2012).

especially for the oxidizing gases like NO, NO_2 among all other metal oxide semiconductors (MOS) such as ZnO, SnO_2, MoO_3 for the gas sensors [4]. MOS are known for their ability to detect trace concentrations of various gases in air via charge transfer interactions between the sensor and chemisorbed species that modify the electrical properties of the sensor [5]. With the help of modern technology of material synthesis manipulation of microstructure on a nanometer scale can produce new class of nanostructured materials with enhanced gas sensing properties to increase the performances of solid state gas sensors [5]. In the present work, we report the synthesis and characterization of WO_3 films by spray pyrolysis method and studies of the physico-chemical properties which are desirable for gas sensing application.

EXPERIMENTAL

WO_3 films on glass substrates were prepared by pyrolytic decomposition of ammonium tungstate by spray pyrolysis technique at various substrate temperatures, viz., 400°C to 475°C. Ammonium tungstate was prepared by dissolving tungstic acid powder (s.d. fine chemicals, Mumbai) in the liquor ammonia (Thomas Bakers) at 80°C with constant stirring for 20 minutes. Obtained solution is sprayed on to preheated substrates, spraying parameters were optimized by observing the uniformity and adherence of the films. The spray rate was kept constant at 4.5 ml/l and the spraying quantity of the solution was optimized to 75 ml. All the depositions were carried out in locally fabricated automated spray pyrolysis deposition system.

The structural information of deposited films was revealed by X-ray diffraction studies (λ = 1.5405 Å Cu Kα) and FT-Raman studies. Surface morphology was studied with the help of scanning electron micrographs and surface profiler (Ambious USA SP-I). Absorption spectra of the films were obtained from UV-Visible spectrophotometer (Hitachi). Electrical and dielectric measurements were performed using the Keithley 6514 electrometer and LCR meter (HP 4284-A). Thermoelectric power was measured to check the type of electrical conductivity of the material deposited.

RESULTS AND DISCUSSION

X-ray Diffraction Studies

Figure 1 shows the x-ray diffraction patterns of the films deposited at various temperatures which confirm the polycrystalline nature of the films. The WO_3 films are formed with good crystallinity at substrate temperature of 425°C while for higher as well as lower substrate temperature crystallinity is poor. Pronounced (113) preferred orientation is observed for all the substrate temperatures but with different peak intensities which indicate extent of crystallization along that plane is more. The intensity of the peaks for the film deposited at 475°C is very low which is due to the less thickness of the film (91.4 nm). The prominent triplet is observed at 2θ = 24.06°, 24.58° and 25.2° for the film deposited at substrate temperature 450°C. For the film deposited at 400°C and 425 °C, the doublet of XRD peaks is visibly prominent, presence of peaks ($\bar{1}016$) and (113) and a peaks at ($\bar{3}111$), ($\bar{2}116$) confirms typically monoclinic crystal structure of WO_3. All these peaks are indexed by using JCPDS card number 71-0070. Other planes at 2θ values 2θ = 29.96°, 42.96° and 48.32° are observed with relatively less intensity with (h k l) values ($\bar{4}$ 0 2), (0 0 28), (6 0 6) for the film deposited at substrate temperature of 425°C.

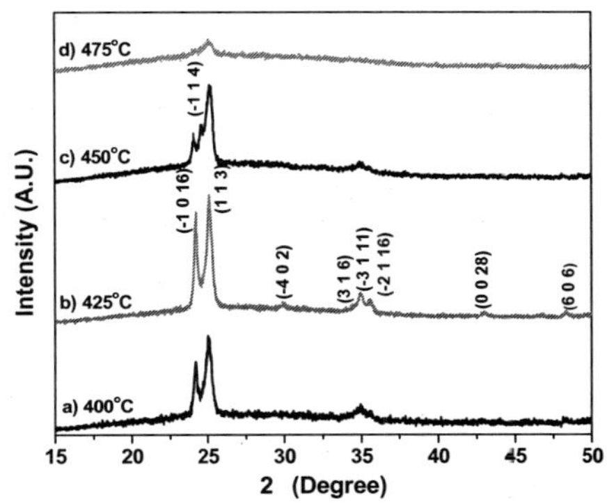

Fig. 1: X-ray Diffraction Patterns of WO_3 Thin Film for Different Substrate Temperature

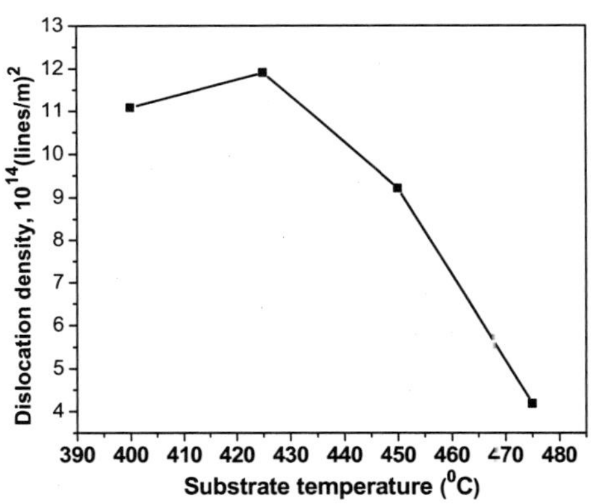

Fig. 2: Dislocation Density Variation with Substrate Temperature

Average crystallite size (D) was calculated using the Scherer formula. Smallest D value (28 nm) is observed for the film deposited at substrate temperature of 425°C. From the D values it is clear that material is nanocrystalline in nature. Microstrain in the films is calculated using equation [6]

$$\beta = \frac{\lambda}{D\cos(\theta)} - \varepsilon \tan(\theta) \quad \text{... (1)}$$

where ε is microstrain developed in the film. The micro strain was found to increase markedly with increase in the substrate temperature (not presented). The increase in the strain is due to the rapid growth of the film at higher temperature which may form rapid formation of the film. Dislocation density is calculated using formula [7],

$$\delta = \frac{1}{D^2} \quad \ldots (2)$$

where δ is the dislocation density. With increase in substrate temperature δ increases initially from 11×10^{14} to 12×10^{14} lines/m² and decreases there after to value of 4.1×10^{14} lines/m² (Fig. 2).

Fourier Transform Raman Spectroscopy studies

The FT-Raman spectroscopy is a powerful tool that combines enhanced visualization of chemical information which was used to investigate the chemical bonds present in film. This helps to get supplementary information concerning structural homogeneity of the film. FT-Raman spectrum of WO_3 film deposited at 425°C is shown in Fig. 3. Several characteristic peaks are obtained within the range 200 cm^{-1} to 1200 cm^{-1}. In this spectrum, peaks at around 268, 313, 780, and 950 cm^{-1} correspond to monoclinic phase, which is one of the stable forms of WO_3 [8]. Intensity of the bands is relatively low which is due to the high temperature synthesis of the films that causes lowering of thickness [9]. Peaks observed at 268 and 313 cm^{-1} correspond to O-W-O bending mode vibrations due to bridging oxygen. While the stretching vibrations of O-W-O lie at around 780 cm^{-1}. The band at 950-980 cm^{-1} can be assigned to the stretching mode of W=O terminal which are surface dangling bonds (terminal bonds) [8,10]. Intensity of

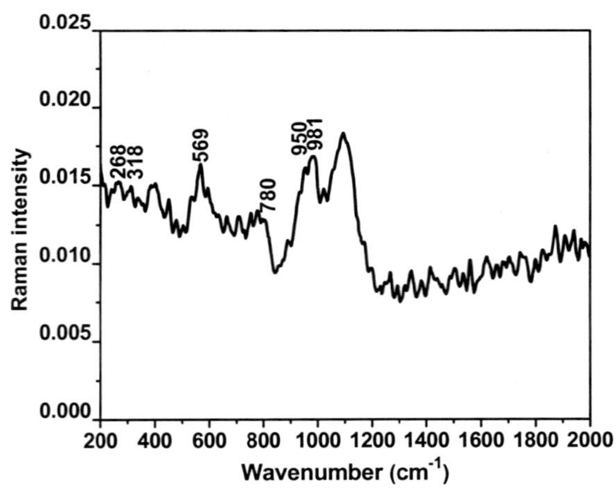

Fig. 3: Ft-Raman Spectrum of the Film Deposited at Substrate Temperature of 425 ºC

the peak at this band is relatively higher, indicating larger number W=O bonds present on the surface because termination of the O-W-O network is done by a double bond. Additional bands at 394 cm^{-1} and particularly at 1094 cm^{-1} indicates that the analyzed material still contains an unknown form of hydrated WO_3 [11]. This mode is common for all types of tungsten trioxides derived from WO_3 hydrates [11].

Scanning Electron Microscopy

Scanning electron micrograph (SEM) of WO_3 film deposited at 425°C is presented in Fig. 4. For all the samples uniform coverage and pin hole free deposition on the substrate with wires of around 250 nm thickness was observed (SEM not presented). At the optimized temperature of 425°C, well covered uniform wire network along with grains on the surface is observed. These wired network of the film form heavily disorientated net. In addition to wires, agglomerated very tiny grains of the material are also visible in the higher magnification scanning electron micrograph of 425°C (Fig. 4b).

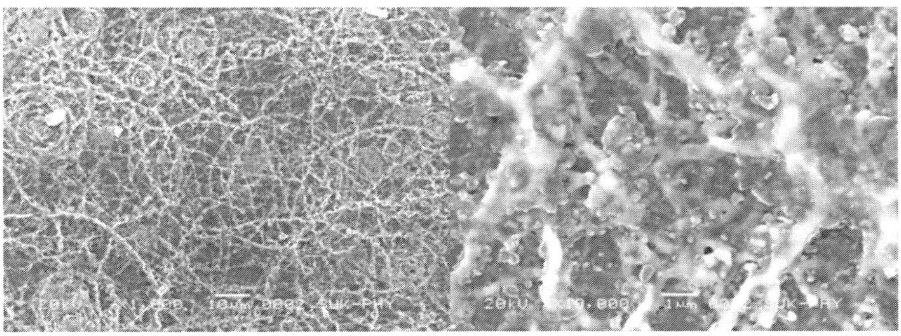

Fig. 4: SEM of WO_3 Deposited at Substrate Temperature 425°C (a) 1000X, (b) 10000X

Electrical Properties and Dielectric Properties

The resistivity decreases with increase in temperature suggesting semiconductor behavior of the films. Decrease in the resistivity with increase in temperature was due to the addition of thermal energy, electron could be set free from O^{2-} ions and from unbounded tungsten also. When an electron is introduced in the sample it might be associated with cations, which results in an unstable valence state [12]. The log of resistivity of the films is plotted with inverse of temperature in Fig. 5. Slight variation in resistivity with temperature variation might be due to the increase in crystallite size of the films from 28 nm to 43 nm. Activation energies of the films are calculated using formula [13],

$$\rho = \rho_0 \exp\left(\frac{\Delta E}{kT}\right) \qquad \text{... (3)}$$

where ΔE is the activation energy, ρ is the resistivity at temperature T, k is the Boltzmann constant and ρ_0 is temperature independent constant.

Table 1: Activation Energy of WO_3 Films Deposited at Various Substrate Temperatures

Substrate temperature (°C)	400	425	450	475
Activation energy (eV)	0.305	0.317	0.215	0.266

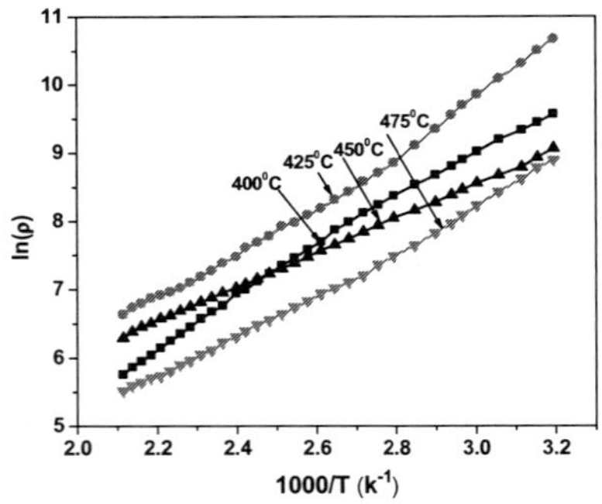

Fig. 5: Variation of Resistivity with Temperature for Different Substrate Temperatures

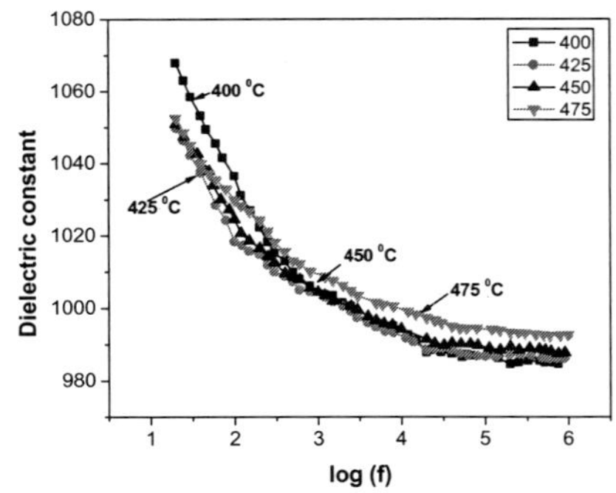

Fig. 6: Dielectric Constant Variation with Frequency for Various Substrate Temperatures

Dielectric measurements at room temperature with respect to change in frequency are shown in Fig. 6. The value of dielectric constant decreases suddenly at lower frequencies and remains nearly constant at higher value of frequencies, showing the dispersion of dielectric constant at lower frequencies. Decrease in dielectric constant takes place when jumping frequency of electric charge cannot follow the alternations of applied electric field beyond a certain critical frequency. At higher frequencies the losses are reduced and the dipoles contribute to the polarization [13]. The large value of dielectric constant is associated with space charge polarization and inhomogeneities like voids, wire like structure and pores dielectric structure.

Optical Properties

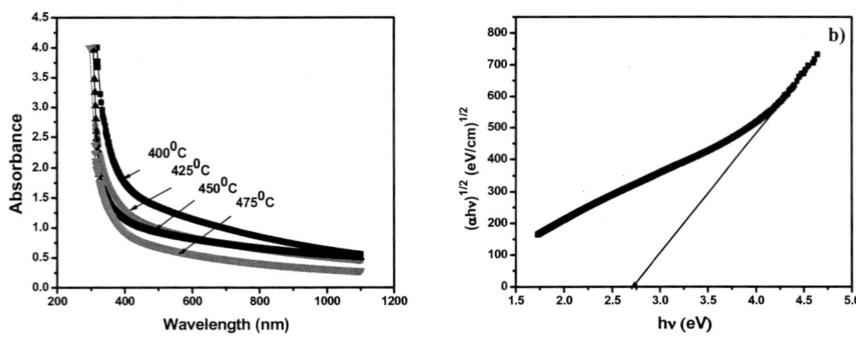

Fig. 7: (a) Optical Absorbance Spectra of WO_3 Thin Films Prepared at Different Substrate Temperatures (b) Plot of $(\alpha h\nu)^{1/2}$ Versus Photon Energy ($h\nu$) for the Film Deposited at 425°C

Generally for metal oxides absorption of the light is low and transmission of light is very high in the visible region of the spectrum [14], and metal reach films usually have large absorbance [15]. Absorbance spectra of the films showed decrease in absorbance with increase in temperature. For the film deposited at 400°C absorbance is larger while the film deposited at 475 °C absorbance is less. Absorbance at wavelength 550 nm is found to be less with increase in substrate temperature which is 0.6 and 1.2 for 400°C and 475°C, respectively. Graph of $(\alpha h\nu)^{1/2}$ versus hí for the film deposited at substrate temperature 425°C is plotted and indirect band gap of the films measured was 2.7 eV.

Thermoelectric Power Studies

The temperature of cold end was maintained at 0°C using ice and hot end was heated with the help of strip heater. The thermo-emf developed between two ends with cold end showing negative polarity pointing that the material has n-type electrical conductivity. N-type electrical conductivity can be attributed to non-stoichiometry due to oxygen-deficiency. From the TEP measurement for the films a smooth decrease in thermo-emf generation with increase in substrate temperature is seen and presented in Fig. 8.

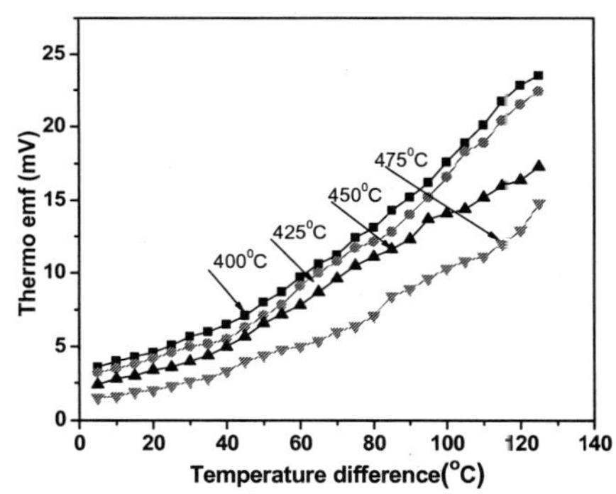

Fig. 8: Variation in the Thermo-emf Generated for the Films Deposited at Different Substrate Temperatures

CONCLUSIONS

The effect of substrate temperature on the physico-chemical and electrical properties of the films is studied. Oxide phase formation is confirmed through X-ray diffraction and FT-Raman spectroscopy studies. Polycrystalline, monoclinic, n-type tungsten oxide thin films with band gap of 2.7 eV were successfully synthesized by spray pyrolysis technique. Band gap matches well with that of the single crystal of WO_3. At the optimized temperature of 425°C a well covered, uniform, wire network along with densely packed grains on the film is observed. These films could find potential applications as gas sensor.

ACKNOWLEDGEMENTS

Authors are very much thankful to UGC-DSA, DST-FIST for their financial support through grants to Department of Physics, Shivaji University, Kolhapur.

REFERENCES

[1] Jarmo, K., Jani, M., Niina, H., Teemu, K., Geza, T., Maria, S., Andrey, S., Jyri-Pekka, M., Heli, J., Krisztian, K., Gas sensors based on anodic tungsten oxide, *Sensors and Actuators B* 153, 293–300, 2011.

[2] Korotcenkov, G., Metal oxides for solid-state gas sensors: What determines our choice?, *Mater. Sci. Eng., B*, 139, 1–23, 2007.

[3] Deepa, M., Singh, D.P., Shivaprasad, S.M., Agnihotry, S.A., A comparison of electrochromic properties of sol–gel derived amorphous and nanocrystalline tungsten oxide films, *Curr. Appl. Phys.*, 7, 220–229, 2007.

[4] Lee, D.S., Nam, K.H., Lee, D. D., Effect of substrate on NO_2 sensing properties of WO_3 thin film gas sensors, *Thin Solid Films*, 375, 142-146, 2000.

[5] Capone, S., Forleo, A., Francioso, L., Rella, R., Siciliano, P., Spadavecchia, J., Presicce, D.S., Taurino A.M., Solid state gas sensors: state of the art and future activities, 5, *J. Optoelectron. Adv. M.*, 5,1335-48, 2003.

[6] Thanikaikarasan, S., Mahalingam, T., Kathalinga, A., Kim, Y.D., Kim, T., Growth and characterization of electrosynthesized iron selenide thin films, *Vacuum* 83, 1066-1072, 2009.

[7] Thanikaikarasan, S., Mahalingam, T., Raja, M., Kim, T., Kim, Y. D., *J. Mater. Sci: Mater. El.* PP. 20 – 27, 2009.

[8] Gabrusenoks, J., Veispals, A., Czarnowski. A., Meiwes-Broer. K. H., Infrared and Raman spectroscopy of WO_3 and $CdWO_4$, *J. Electrochim. Acta.*, 46, 2229-2231, 2001.

[9] Santato C., Odziemkowski M., Ulmann M., Augustynski J., Crystallographically Oriented Mesoporous WO3 Films: Synthesis, Characterization, and Applications, *J. Am. Chem. Soc.* 123, 10639-10649, 2001.

[10] E Salje, *Acta. Cryst. B*, The orthorhombic phase of WO_3, 33, 574-77, 1977.

[11] Daniel M., Desbat B., Lassegues J., Gerand B., Figlarz M., Infrared and Raman Study of WO_3 Tungsten Trioxides and W03, xH_2O Tungsten Trioxide Hydrates, *J. Solid State Chem.* 67, 235-247, 1987.

[12] Rai R., Soni N.C., Sharma S., Choudhary R.N.P.,. Tandon A.P., Ferroelectrics and Dielectrics, Allied Publishers Pvt. Ltd., New Delhi, 177, 2004.

[13] Sawant, V.S., Shinde, S.S., Deokate, R.J., Bhosale, C.H., Chougule, B.K., Rajpure, K.Y., Effect of calcining tem- perature on electrical and dielectric properties of cadmium stannate, *Appl. Surf. Sci.,* 255, 6675-78, 2009.

[14] Rajpure K. Y., Lokhande C. D., Bhosale C. H., A comparative study of the properties of spray-deposited Sb_2Se_3 thin films prepared from aqueous and nonaqueous media, Mater. *Res. Bull.* 34, 1079-87, 1999.

[15] Hartnagel, H.L., Dawar, A.L., Jain, A.K., Jagadish, C., Semiconducting Transparent Thin Films, Bristol: Institute of Physics Publishing, P. 244, 1995.

SURFACE IMPREGNATION OF TRANSITION METAL ION ON POLYCRYSTALLINE TiO_2 USING Cu^{II} (en) $(im)_2^{2+}$ COMPLEX ION IN AQUEOUS MEDIUM

KANNIAH RAJKUMAR AND KRISHNAMOORTHY ANBALAGAN

Department of Chemistry, Pondicherry University, Kalapet, Pondicherry
E-mail: kanuniv@gmail.com; asgchem84@gmail.com

ABSTRACT

$Cu^{II}(en)(im)_2^{2+}$ ion was allowed to impregnate on the surface of polycrystalline TiO_2 in neat water at 27°C. The affinity of the Cu(II) complex on the surface leads to the formation of surface metal complex that is; copper(II) complex: surface compound. The uptake of the metal complex on the polycrystalline solid surface was followed spectrally. The uptake was continuous at definite time internals, however, limited due to thermodynamic considerations. Repetitive scan spectra illustrate the extent of copper(II) complex uptake by poly-TiO_2 in neat water system at definite time intervals. The electronic absorption band maximum shifted smoothly from 593 nm to 585 nm. It was observed that λ_{max} at 593 nm is decreased indicating the surface adherence of copper(II) complex on poly-TiO_2. The mechanistic aspects of the complex ion bound surface of the crystals were analyzed using DRS, PXRD, PL and SEM-EDX instrumental methods. EDX measurements characterize the surface species as implanted due to Cu(II) ion into poly-TiO_2 lattice to generate Cu^{II}: poly-TiO_2 (surface species). SEM images indicate particles are in spherical shape and each particle is agglomerated to form ~100 nm size. Influence of structure of the complex with a semiconductor metal oxide like TiO_2 is promising in waste water treatment. Moreover, metal implanted semiconductors are useful as efficient catalyst and in advanced electronics.

V. Rajendran, P. Manivasakan and K.E. Geckeler (eds.)
Advanced Nano Materials for Industrial Applications, pp. 271-278 (2012).

INTRODUCTION

The surface science of metal oxide is a relatively young field that enjoys a rapidly increasing interest. The general trend to take the 'next step' in surface science is to move onto more realistic and complex model system. This is motivated by the desire to contribute to the numerous applications were oxides surface play a role after most metals are oxidized immediately when exposed [1]. Heavy metals such as nickel, copper, etc., in aquatic system have harmful effect on human physiology and other biological systems when they exceed the tolerance level. Toxicity and persistence of metals in the environment causes a great environmental concern for heavy metal pollution [2,3]. Various methods have been proposed for the treatments of wastewater containing harmful ions, in which photochemical treatment is more convenient. This work reports on the study of copper complex ion on the surface of polycrystalline TiO_2 in neat water. Considerable concentration is adsorbed on the solid surface forming surface:complex adduct, simultaneously it becomes converted into surface species; TiO_2:Cu^{II} due to thermodynamically controlled surface chemistry. The products were properly investigated the interactions between copper complex ion on surface of polycrystalline TiO_2 in neat water. The formation of surface metal complex was characterized by DRS, FTIR, PL, PXRD and SEM-EDX analysis.

EXPERIMENTAL

Materials and Methods

$Cu(ClO_4)_2{\cdot}6H_2O$ (99%), spectral grade KBr, and polycrystalline TiO_2 were purchased from Sigma Aldrich and used as received. Imidazole (AR), ethylenediamine and all other chemicals were purchased from Himedia and SD Fine Chemicals (India). Solvents were distilled before use and water was triply distilled over alkaline potassium permanganate in an all glass apparatus. Analytically pure crystals of $[Cu(en)(im)_2](ClO_4)_2$ adsorbate was synthesized by a modified procedure [4] and recrystallized. The yield was 85%. Single crystals suitable for X-ray diffraction studies were obtained by recrystallization of the complex in methanol.

The FT-IR spectral investigations were carried out on a Thermo Nicollet-6700 FT-IR instrument (KBr pellet method). Electronic absorption spectral studies were undertaken on a double beam spectrophotometer (Shimadzu model 2450, Japan). Copper(II) complex was 1.65 $\times$ 10^{-3} M in neat water. Powder X-ray diffraction (PXRD) patterns were recorded X'PERT diffractometer employing Cu-Kα radiation. The morphology of the samples was investigated by scanning electron microscopy (SEM) (Hitachi, Model: S-3400N).

Adsorption Studies

Adsorption procedure is slightly modified from Fuerstenau method [5]. About 0.1 g (1.65 $\times$ 10^{-3} M) of the complex $[Cu^{II}(en)(im)_2](ClO_4)_2$ was taken in a 125 mL stoppered bottle. The complex was dissolved in neat water maintaining ionic strength with $NaClO_4$ (0.2 M). Then polycrystalline TiO_2 (200 mg) was added to the above solution and the entire solution was placed in a Technico cooling water bath shaker at different time intervals such as 0, 5, 10, 15, 20 and 240 min at 300 K. After definite time intervals the filtrate solutions were centrifuged and both the solid and solution were to instrumental analysis. The adsorbent (TiO_2) was removed by filtration through 0.45 ìm Whatman filter paper at various time intervals and dried over vacuum to understand the surface affinity of the complex ion.

RESULTS AND DISCUSSION

Adsorption of Copper(II) Complex ion on Poly-TiO_2

To study the surface affinity of the copper(II) complex ion on the surface of polycrystalline titanium dioxide, adsorption study was carried out in neat water. The complex binding was followed spectrally at various time intervals. Figure 1 is a repetitive scan spectra illustrating the extent of copper(II) complex uptake by poly-TiO_2 in neat water at definite time intervals. It was observed that λ_{max} at 593 nm is decreased indicating the surface adherence of copper(II) complex on poly-TiO_2. However, Figure 1 implies that the extent of adsorption is very high initially and after definite time interval, a decreasing trend was observed. When the complex concentrations are considered, it disappeared slowly from 0 to 240 min time interval and no new product or intermediate product causing absorption in the near UV region was observed.

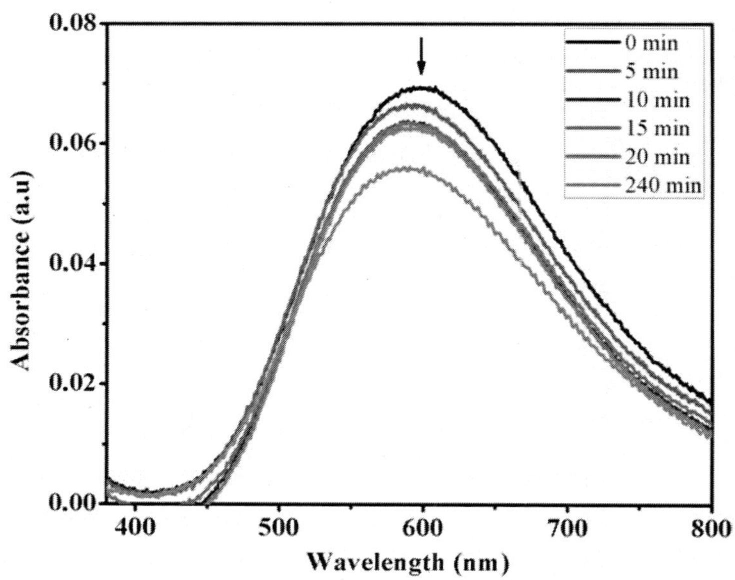

Fig. 1: Repetitive Scan Spectra Recorded for the $[Cu(en)(im)_2](ClO_4)_2$ Complex Adsorption on Polycrystalline TiO_2 in Neat Water at Various time Intervals. Complex Concentration = 1.65 x 10^{-3} M, Ionic Strength = 0.2 M $NaClO_4$, pH = 7.3 and at 300 K

FT-IR Spectral, DRS and Photoluminescence Analysis of Poly-TiO_2:Cu(II)-complex

IR analysis was used to determine the representative families of chemical intermediates adsorbed on the catalyst surface after the reuses [4]. The FT-IR spectrum of Cu^{II}/TiO_2 (Figure 2(a)) showed a large broad band at 3100-3500 cm^{-1} and rather narrow bands at 1638 and 1402 cm^{-1}, together with one weak absorption at 442 cm^{-1}. The broad one at 3100-3500 cm^{-1} can be assigned to both of νO-H (stretching mode) while those at 1638 and 1432 cm^{-1} can be assigned to the δO-H (bending mode) of hydroxyl (OH) group [5]. The absorption at 442 cm^{-1} can be assigned to the δTi-O (stretching mode) of Ti-O bond and this band tends to higher energy region at 538 cm^{-1} [6]. This Ti-O Therefore, the FT-IR spectra confirmed the presence of impurities such as H_2O and Cu^{2+} at the surface of poly-TiO_2 [7]. The amount of Cu^{2+} was, however, very small ca. 0.05-0.32% by elemental analysis for copper content. This small amount of copper also rendered it undetectable by SEM/EDX analysis. The λ_{max} at ~ 320 nm is due to $O^{2-} \rightarrow Ti^{4+}$ charge transfer transitions as shown in Figure 2(b). The diffuse reflectance spectra of all samples for 0-20 min and 240 min as shown in the Figures 2(b) reveal that wide absorption bands at the visible region wavelength of 400-800 nm, with λ_{max} at ~790 nm indicate the adsorption of copper on titanium(IV) dioxide surface [8]. It shows a shift in absorption to longer wavelength due to copper adsorption. Photoluminescence spectra of pure poly-TiO_2 and Cu^{II} complex adsorbed TiO_2 as shown in Figure 2(c), strong green emission at 553 nm is observed because of excitation using 312 nm. The emission characteristics of Cu^{2+}/TiO_2 (x = 0.05 to 0.32 at %) are quite similar to that of undoped nano-TiO_2, however, emission intensity drastically reduced for the former cases.

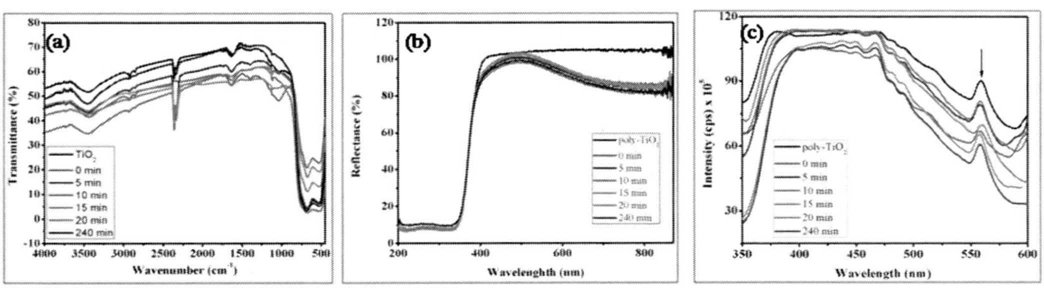

Fig. 2: (a) FT-IR (b) DRS and (c) Photoluminescence Spectra of the Polycrystalline TiO_2 Semiconductor Particles Collected from Adsorption of $Cu^{II}(en)(im)_2^{2+}$ on TiO_2 in neat Water and at Various Time Intervals. Medium: 1 M $NaClO_4$ at 300K

Powder X-ray Diffraction Analysis of TiO_2: Cu(II) Species

Pure polycrystalline TiO_2 and isolated surface adsorbed species TiO_2:$Cu^{II}(en)(im)_2^{2+}$ obtained at different time intervals were analyzed for their crystal structures. The PXRD patterns of both the samples basically belong to anatase form, therefore, copper(II) complex is chemically bonded either through surface species or through strong metal-supported interaction [9]. Both pure form of poly-TiO_2 and surface species; Cu^{II}:TiO_2 exhibit patterns (Figure 3) characteristic of anatase crystal phase (JCPDS file numbers 01-071-1166, 00-002-0387). The polycrystalline anatase structure was confirmed by (101), (004), (200), (105) and (211) diffraction peaks. The XRD patterns of anatase have a main peak at $2\theta = 25.2°$ corresponding to the (101) plane (JCPDS 21-1272), which confirms the phase purity of the crystalline solid. This suggests that the crystals do not contain impurity phase like rutile, hence transformation of crystal phase has not occurred [10].

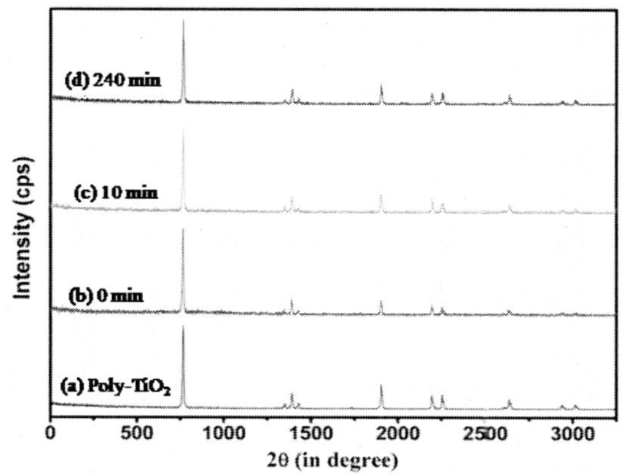

Fig. 3: Powder X-ray Diffraction Pattern of the Polycrystalline TiO_2 Semiconductor Particles Collected from Adsorption of $Cu^{II}(en)(im)_2^{2+}$ on TiO_2 in neat Water and at Various Time Intervals. Medium: 1 M $NaClO_4$ at 300K

Scanning Electron Microscopic Analysis

The SEM images of TiO_2:Cu^{II} particles are presented in Figure 4, which, reveals that the particle surface is relatively rough and apparently built from tiny particles. It can be seen that the particles are spherical with various size distributions. The copper(II) complex surface implanted particles are found to be highly agglomerated and some are non-spherical in characteristics [10].

These differences in morphology, particle shape and size indicate the presence of surface: Cu^{II} species [1]. From SEM images, it can be presumed that copper-deposited titania samples consist of very fine free individual grains plus other granular aggregates; very large distributions of shape and dimensions of the particles were observed, with average diameter of 160 nm. It is evident that the growth of particle is highly restricted due to Cu(II) deposition on polycrystalline TiO_2 surface.

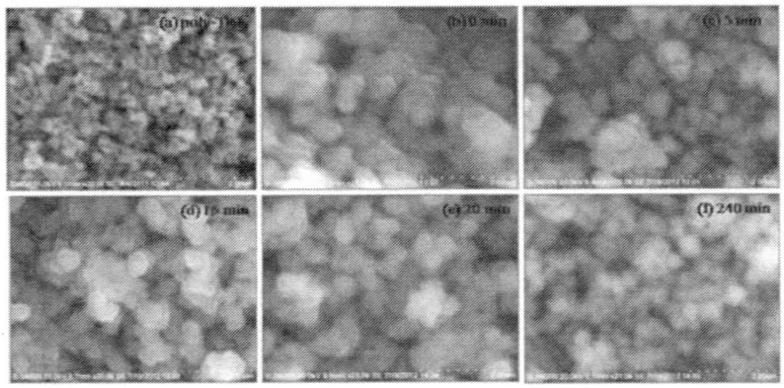

Fig. 4: SEM Images of the Polycrystalline TiO_2 Semiconductor Particles Collected from Adsorption of $Cu^{II}(en)(im)_2^{2+}$ on TiO_2 in Neat Water and at Various time Intervals. Medium: 1 M $NaClO_4$ at 300K

EDX Measurement of Poly-TiO_2:$Cu^{II}(en)(im)_2^{2+}$

Surface concentration of Cu(II) on poly-TiO_2 was calculated from energy diffuse X-ray studies. EDX spectra are shown in Figure 5 and Table 1 presents the concentration of copper loaded on the solid with respect to different time intervals. Table 1 exhibits the relationship between density of Cu(II) in atomic percent (*at.*%) with respect to time, in which Cu is enhanced from 0.05 to 0.32 %. EDX analysis shows the presence of titanium and oxygen in the TiO_2 sample at 0 min contact time in adsorption study, however, presence [2, 10], of a small quantity of copper was evident from the samples isolated at 5, 10, 15, 20 and 240 min contact time intervals. Figure 5 compares the EDX spectra for pure TiO_2 and complex adsorbed anatase (Cu-TiO_2). It could evidence that the copper atoms were introduced into the poly-TiO_2.

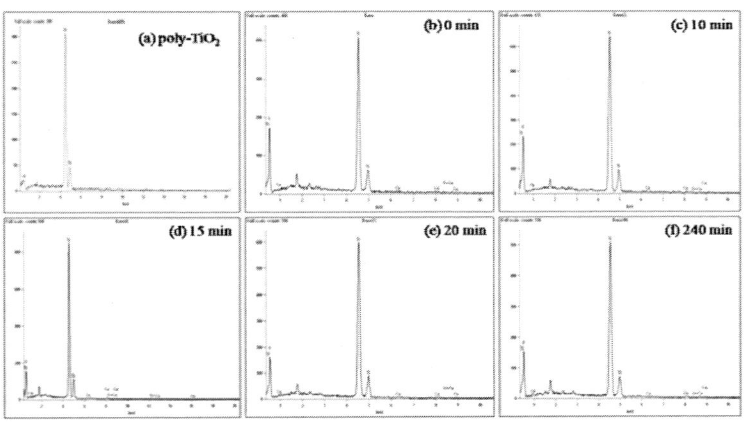

Fig. 5: EDX Spectra of the Polycrystalline TiO_2 Semiconductor Particles Collected from Adsorption of $Cu^{II}(en)(im)_2^{2+}$ on TiO_2 in Neat Water and at Various time Intervals. Medium: 1 M $NaClO_4$ at 300K

Table 1: The Chemical Content of the Polycrystalline TiO_2 Semiconductor Particles Collected from Adsorption of $Cu^{II}(en)(im)_2^{2+}$ on TiO_2 in neat Water and at Various time Intervals at 27°C According to the Energy Dispersive X-ray Analysis

Contact Time (min)	*Content Weight % (atm. %)*		
	Ti	*O*	*Cu*
0	40.99 (18.86)	58.88 (81.10)	0.13 (0.05)
5	36.34 (16.05)	63.42 (83.87)	0.24 (0.08)
10	43.53 (20.54)	56.16 (79.35)	0.30 (0.11)
15	38.00 (17.06)	61.59 (82.80)	0.41 (0.14)
20	42.52 (19.95)	56.81 (79.81)	0.68 (0.24)
240	49.95 (25.24)	49.20 (74.44)	0.85 (0.32)

CONCLUSIONS

$Cu^{II}(en)(im)_2^{2+}$ is preferably adsorbed on the solid surface forming surface: complex adduct, simultaneously it becomes converted into surface species; TiO_2:Cu^{II} due to thermodynamically controlled surface chemistry. The surface species was identified by EDX measurement, which infers the implantation of Cu in the lattice of poly-TiO_2. SEM images support the restriction of agglomeration of particles due to Cu(II) penetration into the lattice. In this investigation, it has been found that metal complex surface interaction leads to the formation of surface species that can be identified by DRS, FTIR, PL, PXRD and SEM-EDX analysis. Metal loaded semiconductor solids find applications as catalysts and in advanced electronics.

ACKNOWLEDGMENTS

KA is thankful to the Council of Scientific and Industrial Research (sanction order: No. 01(2570)/12/EMR-II/3.4.2012), New Delhi for financial support through major research project. The authors are thankful to CIF, Pondicherry University for instrumental facility.

REFERENCES

[1] (a) Diebold, U., *Surf. Sci. Reports,* 48, 53, 2003.

(b) Anbalagan, K., Ganeshraja, A.S., and Maharaja Mahalakshmi, C., *Met. Chem. Phys.* 134, 747, 2012.

[2] Bahadir, T., Bakan, G., Altas, L., and H. Buyukgungor, Enz. Microb. Technol. 41, 9, 2007.

[3] (a) Anbalagan, K., *J. Phys. Chem.*, C 115, 3821, 2011.

(b) Barkat, M.A., Chen, Y.T., and Huang, C.P., *Appl. Catal.*, 53, 13, 2004.

[4] Golchoubian, H., Nazaria, O., and Kariuki, B., J. Chin. Chem. Soc., 58, 60, 2011.

[5] Fuerstenau, D.W., and Asare, K.O., *J. Colloid. Inter. Sci.* 118, 524, 1987.

[6] Repo, E.H., Kurniawan, T.A., Warchol, J.K., and Sillanpaa, M.E. T., *J. Hazar. Mat.*, 171, 1071, 2009.

7] Khalil, K.M.S., and Zaki, I.Z., *Powder Technol.*, 92, 233, 1997.

[8] Zhang, J.Y., Boyd, I.W., O'Sullivan, B.J., Hurley, P.K., Kelly, P.V., and Senateur, J.P., *J. Non-Cryst. Solids*, 303, 134, 2002.

[9] Yoong, L.S., Chong, F.K., and Dutta, *B.K., Energy*, 34, 1652, 2009.

[10] Anbalagan, K., and Ganeshraja, A.S., *J. Indian Chem. Soc.*, (2012) in press.

OXYGEN SENSING PROPERTIES OF PLATINUM DOPED INDIUM OXIDE NANOPARTICLES PREPARED BY HYDROTHERMAL METHOD

D. SELVAKUMAR[1], N. DHARMARAJ[1], N.S. KUMAR AND V.C. PADAKI

Defence Bioengineering & Electromedical Laboratory, DRDO, Bangalore
[1]Department of Chemistry, Bharathiar University, Coimbatore
E-mail: dharmaraj67@yahoo.com

ABSTRACT

Nanostructured indium oxide (In_2O_3) and platinum doped indium oxide (Pt-In_2O_3) were prepared by surfactant assisted hydrothermal method followed by calcination at 400°C. Powder XRD and TEM results showed that the sizes of nanoparticle are around 15 nm. Oxygen sensors were fabricated by drop coating the nanopowders across gold electrodes and the sensing characteristics were studied by measuring the electrical resistance at 20% of oxygen partial pressure (pO_2) in the range of temperature between 80 to 375°C. The resistance of Pt-In_2O_3 was observed to be lower than In_2O_3 and the maximum response was obtained at 260°C. Further, the sensors showed a linear response of resistance with pO_2 ranging from 2.5% to 20%.

INTRODUCTION

Semiconducting metal oxides based oxygen sensors are being used in many industrial processes for quantitative detection and control of oxygen gas supply. The major disadvantage of metal oxide based sensors operating at higher temperature is their inadequate long term stability of sensor conductivity and thus the sensor signal [1]. Moreover, maintenance of sensor at high temperature increases power consumption and reduces sensor life and thus complicates the design of a sensor [2]. Among n-type metal oxides, In_2O_3 is a suitable candidate for oxygen

V. Rajendran, P. Manivasakan and K.E. Geckeler (eds.)
Advanced Nano Materials for Industrial Applications, pp. 279-286 (2012).

sensing at lower temperature due to its high conductivity. Recently platinum doped indium oxide (Pt-In_2O_3) nanopowder prepared by non-aqueous sol-gel route has shown promising performance for application in oxygen monitoring at lower temperature [3].

In the present study, we report the synthesis and oxygen sensing characteristics of In_2O_3 and Pt-In_2O_3 nanopowders prepared by cetyltrimethyammonium bromide (CTAB) assisted hydrothermal method. The sensitivity and response of the materials were compared by measuring their electrical resistance in oxygen gas with the range of pO_2 between 2.5 to 20% and temperature between 80 to 375°C.

EXPERIMENTAL

A stock solution of indium nitrate (0.08 $molL^{-1}$, pH 1) was prepared by dissolving indium metal in nitric acid and ultrapure water. To 12.5 mL of the above indium nitrate solution, 2.5 mL of ethylene glycol was added. To this solution, CTAB was added to a final concentration of 10 % (wt/vol). Then, 0.2804 g of HMT (1:2 mole ratio w.r.t indium metal) dissolved in 12.5 mL of water was added drop wise at a rate of 0.25 mL min^{-1} with vigorous stirring. The solution was stirred for further 5 h and the resultant solution was autoclaved in Teflon lined stainless steel autoclave at 180°C for 8 h. $In(OH)_3$ formed as a white precipitate was filtered, washed repeatedly with water followed by absolute ethanol and dried at 110°C. Finally, $In(OH)_3$ was calcined at 400°C for 3 h at atmospheric conditions to get bright yellow In_2O_3 NPs. For the preparation of Pt-In_2O_3, an aqueous solution of hexachloroplatinic acid (H_2PtCl_6) was added to the reactant solution keeping the atomic ratio of Pt to be 1% w.r.t indium metal, before autoclaving.

Sensing studies were performed on a device prepared by drop coating the nanoparticles on a pair of gold electrodes, screen printed on an alumina substrate. The size of the gold electrodes was 2 mm × 2 mm and the gap between the electrodes was 50 m. The sensor structure was mounted on a PID controlled heating block. The drop coated nanoparticles were dried in air and then treated in air for one hour at 380 °C in order to stabilize the film. Resistance measurements were carried out using Agilent 34970A data acquisition and switching unit linked with computer with a total gas flow rate of 200 sccm.The gas sensitivity S was determined by the relative resistance, $S = R_{oxygen}/R_{nitrogen}$, where R_{oxygen} and $R_{nitrogen}$ are the resistance of sensor in oxygen and in nitrogen, respectively.

RESULTS AND DISCUSSION

In this study, we employed a hydrothermal precipitation method to prepare nanosized In_2O_3 and Pt-In_2O_3, because it allows size control, morphology and crystallinity by tuning of the experimental variables [4]. As HMT decomposes into formaldehyde and ammonia with increasing temperature, OH^- species are released to the medium promoting the precipitation of the metal hydroxide and/or oxide [5]. H_2PtCl_6 was added into the hydrothermal reaction mixture for in-situ doping In_2O_3 with Pt metal. Though hexachloroplatinic acid is stable towards reduction by ethylene glycol at room temperature, it is reduced at 110 °C into elemental platinum [6]. Thus Pt metal was expected to dope with indium hydroxide, which is converted to Pt-In_2O_3 up on further calcinations.

Figure 1(a) shows the XRD pattern of In_2O_3 NPs after calcination at 400°C. All the diffraction peaks match well with that of bulk cubic bixbyite type In_2O_3 reflections according to JCPDS No. 0606416. The crystallite sizes estimated from the line-broadening using Debye-Scherrer equation [7] was 12 nm. The XRD pattern of Pt-In_2O_3 [Fig. 1(b)] did not show the diffraction peak due to platinum. This can presumably be ascribed to the very low content of doping, smaller size and homogenous distribution of platinum nanoparticles [8].

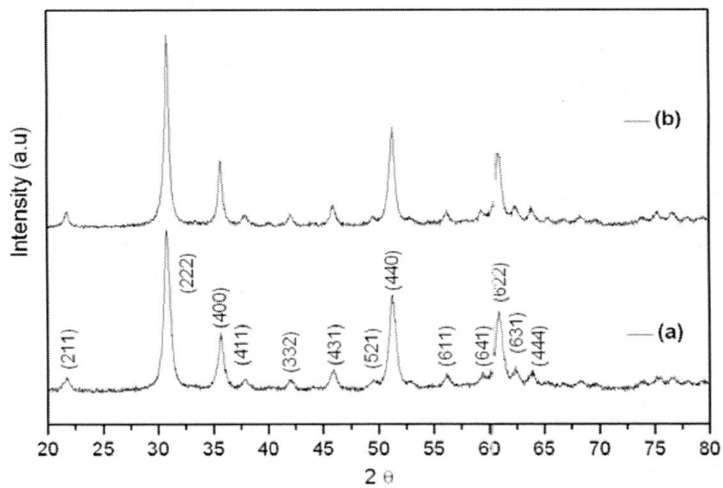

Fig. 1: XRD pattern of (a) In_2O_3 and (b) Pt-In_2O_3

The TEM micrographs of In_2O_3 and Pt-In_2O_3 nanoparticles are shown in Fig. 2 (a & b), which confirms that both the products are in nanometer size regime. The NP showed nearly monodispersed short rod-like particles with lengths in the range of 10-25 nm. The EDAX analysis of In_2O_3 matches with the theoretical values of indium and oxygen, while Pt-In_2O_3 showed the doping of platinum.

The experimental set up used for sensing oxygen using nanoparticles based sensor is depicted in Figure 3. When metal oxide is exposed to air, oxygen is adsorbed on the grains of oxide as anionic species (such as O_2^-, O_2^{2-} and O^-), inducing an electron-depletion layer to increase surface potential and work function. Therefore, a potential barrier for migration of electrons, often called double-Schottky barrier, is formed across each grain-boundary, which determines the resistance of the sensing material (R). When a reducing gas comes in contact the adsorbed oxygen species are consumed, leading to a diminishing of potential barrier and therefore R decreases with increasing partial pressure of the reducing gas. On the other hand,

R increases with increasing partial pressure of an oxidizing gas due to enhancement of potential barrier [9]. It has been established that the sensitivity of metal-oxide gas sensors can be substantially improved by doping of noble metal additives, such as Pt on surface of oxides or in its volume [10]. In case of Pt doped metal oxide nanoparticles, the gas molecules are first activated by the metal surface forming anionic gas species that then react via spillover process, increasing the surface coverage of oxides with gas.

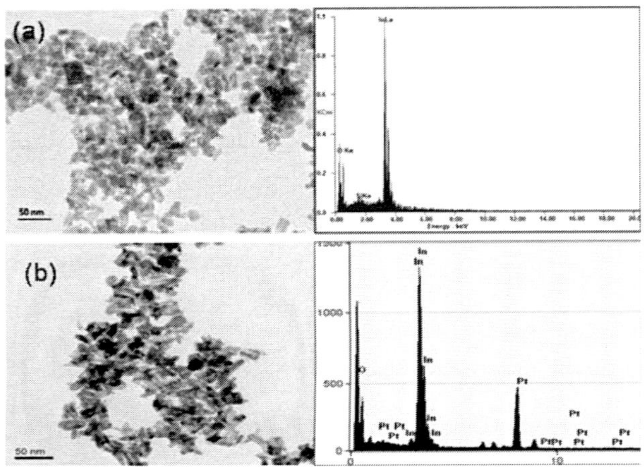

Fig. 2: TEM Micrograph and EDAX Pattern of (a) In_2O_3 and (b) Pt-In_2O_3

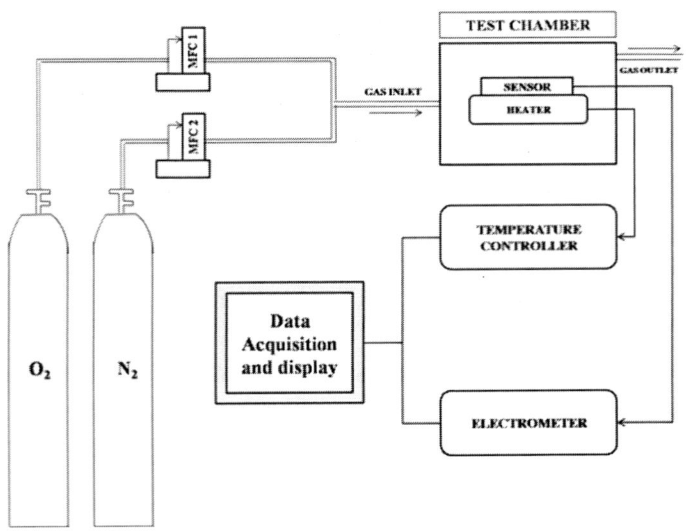

Fig. 3: Schematic Diagram of Experimental Set Up for Oxygen Sensing

Figure 4 shows the temperature dependant resistance of the sensors made from In_2O_3 and Pt-In_2O_3 measured at 20% oxygen partial pressure (pO_2). The resistance of Pt-In_2O_3 was less than In_2O_3 due to the presence of platinum metal. Moreover, the resistance of both the sensor decreased with increasing the operating temperature, following typical exponent law [11].

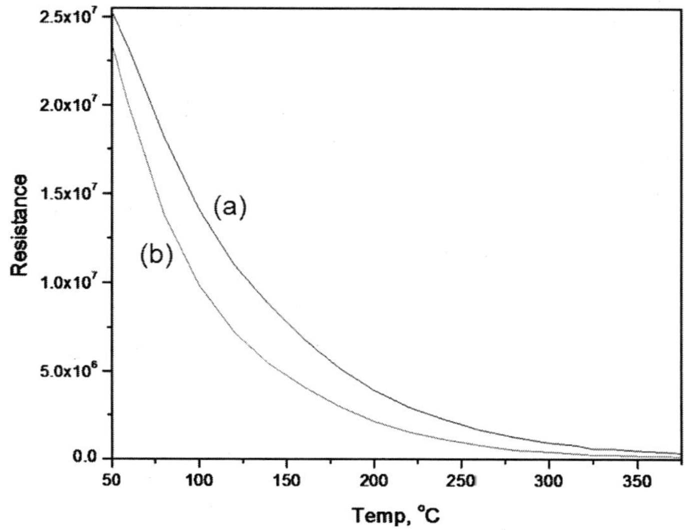

Fig. 4: Temperature Dependant Resistance (a) In_2O_3 and (b) Pt-In_2O_3

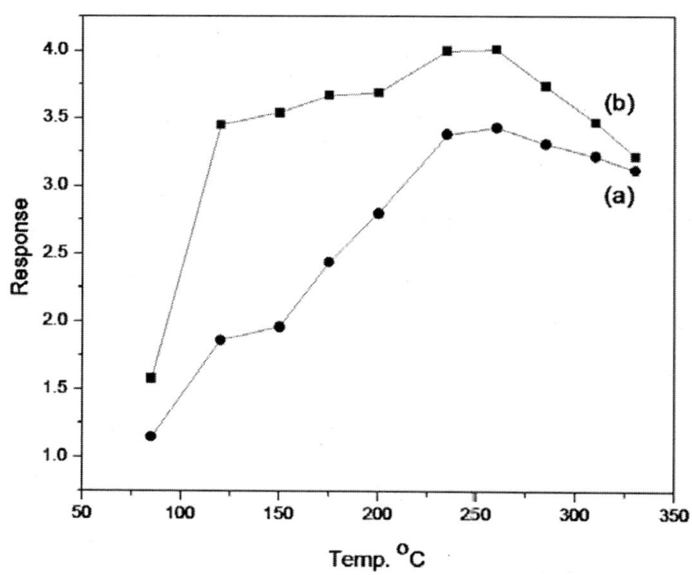

Fig. 5: Gas Response *vs.* Temperature (a) In_2O_3 and (b) Pt-In_2O_3

The plot of gas response *vs*. temperature, measured at 20% oxygen partial pressure (pO_2) is shown in Fig. 5. Initially the responses increase with operating temperature and reach a maximum at a certain temperature, and then decreased with further increasing the temperature. At lower temperature, the adsorption of oxygen species proceeds slowly, which is responsible for lower response. The response increases with increasing temperature due to faster adsorption kinetics. Beyond the optimum temperature, the lowering of response is due to enhanced desorption of ionic oxygen species as compared to adsorption [12]. In our studies the optimum response was obtained at 260°C for both from In_2O_3 and Pt-In_2O_3 sensors. This temperature is comparable to the results obtained in earlier reports [3]. Therefore, further studies on the sensing characteristics of the sensors were conducted at 260°C.

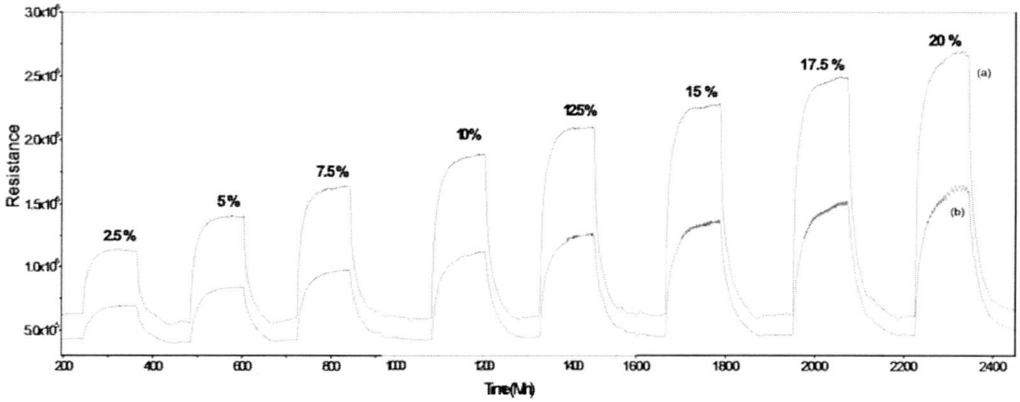

Fig. 6: Oxygen Sensing Response

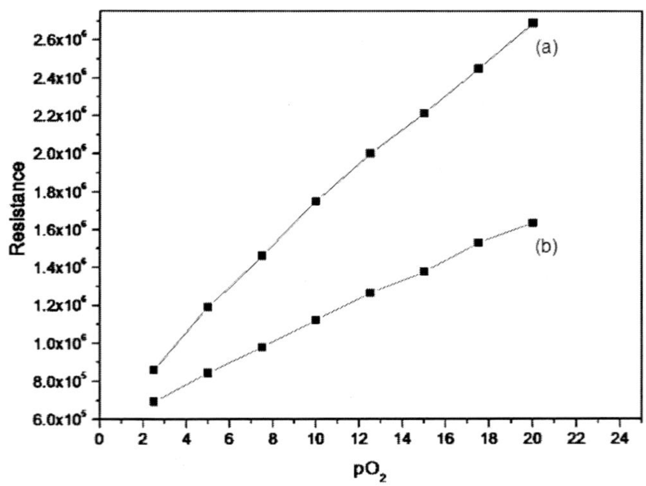

Fig. 7: Calibration Curve (a) In_2O_3 and (b) Pt-In_2O_3

The transient response of the undoped and Pt-doped In_2O_3 sensors to different oxygen concentrations at 260 °C is presented in Fig. 6. The sensors were exposed to successive pulses of oxygen ranging from 2.5 to 20 vol% in dry nitrogen. A reversible response was obtained as the sensors showed the same resistance value before and after each oxygen pulse. A plot of resistance *vs.* pO_2 showed a linear response (Fig. 7).

CONCLUSIONS

Platinum doped indium oxide nanoparticles of size around 15 nm were synthesized by surfactant assisted hydrothermal method. The resistance of Pt-In_2O_3 was less than In_2O_3 and the maximum response was observed at 260°C. Both the sensors showed a linear resistance value with oxygen partial pressure studied between 2.5 to 20%. The sensor fabricated from Pt-In_2O_3 showed higher response as compared to In_2O_3. The above results revealed that the Pt-In_2O_3 nanoparticles prepared by CTAB assisted hydrothermal route have potential applications for the fabrication of oxygen sensor operating at low temperature.

ACKNOWLEDGEMENTS

The authors are thankful to Dr. K. Kadirvelu, Coordinator, DRDO-BU CLS, Bharathiar University, Coimbatore, for XRD & EDAX facilities and Dr. D. Bharathimohan, School of Physics, Pondicherry University for TEM studies.

REFERENCES

[1] Meixner, H. and Lampe, U., Metal Oxide Sensors, *Sens. Actuat. B*, 33,198-202, 1996.

[2] Johnson, C.L., Schwank, J.W. and Wise, K.D., Integrated ultra-thin-film gas sensors, *Sens. Actuat.* B, 20, 55-62, 1994.

[3] Neri, G., Bonavita, A., Micali, G., Rizzo, G., Pinna N., and Niederberger, M., In_2O_3 and Pt-In_2O_3 nanopowders for low temperature oxygen sensors, *Sens. Actuat. B*, 127, 455-462, 2007.

[4] Aneesh, P.M., Vanaja, K.A., and Jayaraj, M.K., Synthesis of ZnO nanoparticles by hydrothermal method, *Nanophotonic materials* IV, 6639, 66390 J1 –J9, 2007.

[5] Govender, K., Boyle, D.S., Kenway, P.B., and O'Brien, P.J. Understanding the factors that govern the deposition and morphology of thin films of ZnO from aqueous solution, *Mater. Chem.*, 14, 2575-2591, 2004.

[6] Bonet, F., Delmas, V., Grugeon, S., Herrera Urbina R., Silvert, P.Y., and Tekaia, K.E., Synthesis of monodisperse Au, Pt, Pd, Ru and Ir nanoparticles in ethylene glycol, *Nano Structured Mat.*, 11, 8, 1277–1284, 1999.

[7] Patterson, A. The Scherrer Formula for X-Ray Particle Size Determination, *Phys. Rev.* 56 (10): 978–982, 1939.

[8] Ou, H.H. and Lo, S.L., Effect of Pt/Pd-doped TiO2 on the photocatalytic degradation of trichloroethylene, *J. Mol. Cat. A: Chem.,* 275, 200–205, 2007.

[9] Noboru, Y., and Kengo, S., Theory of power laws for semiconductor gas sensors, *Sens. Actuat.* B, 128, 566–573, 2008.

[10] Madler, L., Roessler, A., Pratsinis, S.F., Sahm, T., Gurlo, A., Barsan, N., and Weimar, U., Direct formation of highly porous gas-sensing films by in situ thermophoretic deposition of flame-made Pt/SnO_2 nanoparticles, *Sens. Actuat.* B, 114, 283-295, 2006.

[11] Bose, A.C., Balaya, P., Thangaduri, P., Ramasamy, S., Grain size effect on the universality of AC conductivity in SnO_2, *J. Phys. Chem. Solids.*, 64, 659-663, 2003.

[12] Kolmakov, A., Klenov, D.O, Lilach, Y., Stemmer, S., and Moskovits, M., Enhanced gas sensing by individual SnO_2 nanowires and nanobelts functionalized with Pd catalyst particles, *Nano Lett.*, 5, 667-673, 2005.

ELECTROFABRICATION OF NANOSTRUCTURED MULTILAYER COATINGS FOR BETTER CORROSION PROTECTION

VAISHAKA R. RAO AND A. CHITHARANJAN HEGDE

Electrochemistry Research Laboratory, Department of Chemistry, National Institute of Technology Karnataka, Surathkal Srinivasnagar
E-mail: hegdeac@rediffmail.com, achegde@yahoo.in

ABSTRACT

In an effort to improve the corrosion resistance of monolayer Zn-Ni alloy coating, the composition modulated multilayer alloy (CMMA) coatings have been electrofabricated using square current pulses. Successive layers of alloy having alternately changing compositions were accomplished on mild steel (MS) by making the cathode to cycle between two current densities. Multilayer coatings were electrofabricated with different combination current densities and number of layers, and their corrosion behaviors were studied by electrochemical AC and DC methods. The effect of cyclic cathode current densities (CCCD's), and number of layers on the corrosion performance of CMMA's have been studied, and results are discussed. The coating configurations have been optimized for the highest corrosion resistance of the coatings. The electrofabricated multilayers and their degradation pattern were confirmed by Scanning Electron Microscopy (SEM) analysis. The improved corrosion resistance of Zn-Ni CMMA coatings in relation to that of monolayer alloy deposited under identical bath condition has been explained in terms of increased number of interfaces due to layering.

INTRODUCTION

A new type of coating system so-called composition modulated multilayer alloy (CMMA) is gaining interest due to their improved functional properties, such as mechanical strength,

V. Rajendran, P. Manivasakan and K.E. Geckeler (eds.)
Advanced Nano Materials for Industrial Applications, pp. 287-294 (2012).

micro hardness, giant magnetoresistence and also the corrosion resistance [1]. The CMMA materials basically consists of alternating layers of metals/alloys in micro/nanometric scale deposited electrolyticaly by making the cathode to cycle between two current densities at regular time intervals. The multilayered coatings exhibit enhanced properties compared to respective, simple monolithic coatings like hardness, mechanical strength, wear and corrosion resistance [2]. Deposition techniques such as evaporation, molecular beam epitaxy and sputtering are some of the common methods still in use for development of multilayer coatings, its higher cost and practical difficulties made the way to electrolytic deposition process. Though Zn-Ni alloy coatings offer better corrosion protection than pure Zn, the electrofabrication of multilayer coating is more viable and is of distinct commercial interest.

The development CMMA coating with Zn and Ni as alternate layers was found to show best corrosion protection. Kalantary et al. have demonstrated presence of high % Ni in Zn-Ni alloy is responsible for enhanced corrosion protection [3]. The first multilayer deposition was reported by Blum as early as 1921 [4], and later by Brenner [5]. Many attempts have been made to understand the reasons for improved corrosion performance of multilayer coating [6,7]. Kirilova et al. [8, 9] have reported the deposition of CMMA coating of Zn-Co using single bath technique (SBT). Tench and White fabricated Cu/Ni metal multilayers and studied their mechanical properties [10]. It was proposed that the presence of Ni in the deposit is responsible for improved corrosion protection Recently, Thangaraj et al. optimized a chloride bath for the galvanostatic production of CMMA Zn-Ni coating, and showed that they are ~ 45 times more corrosion resistant than monolayer (bulk) Zn-Fe alloy coatings, deposited from same bath [11]. Though many reports are available with regard to the development of CMMA Zn-Ni coatings using different additives, no work has been reported for gelatin and glycerol, as additives. Hence the present looks at optimization of deposition condition for development of CMMA coatings of Zn-Ni for best protection of MS. The bath composition and operating parameters have been optimized by Hull cell method. The process and product of electrolysis were analyzed using different instrumental methods such as SEM and XRD, and results are discussed. All electroplating were carried out on pre-cleaned MS panels of 7.5 cm^2 active surface area at 303K and pH = 3.5. The computer controlled DC Power Analyzer (N6705, Agilent Technologies) was used as power source for electrodeposition. The modulation of composition in alternate layers was effected by periodic pulsing of current. The power pattern used for monolayer and multilayer coatings are shown in Fig. 1. The corrosion behaviors of the coatings were evaluated using 5% NaCl (a representative corrosion medium). All coatings were carried out for 10 min (~ 20 μm thickness) for comparison purpose.

EXPERIMENTAL

The optimal composition and operating parameters for deposition of monolayer bright, uniform and homogeneous Zn-Ni alloy on MS was arrived by Hull cell method, and is given in Table 1.

Table 1: The Composition and Deposition Conditions of Optimized Bath

Bath Ingredients	*Composition (g/L)*	*Operating Parameters*
$ZnCl_2.7H_2O$	27.2	Anode: Nickel
$NiCl_2.6H_2O$	94.9	Cathode: Mild Steel
Boric acid, BA	27.7	pH: 4
NH_4Cl	100	Temperature: 303K
Gelatin	5	
Glycerol	2.5	

The deposit characterizations were made by different analytical techniques, like spectrophotometry, XRD, SEM and electrochemical methods. A standard three electrode system was used for corrosion study, at scan rate of 1mV/s. The EIS measurements were made in frequency range of 100 kHz to 10 mHz in 10 mV perturbing voltage. The CMMA coatings with alternate layers of alloys are represented as: $(Zn\text{-}Ni)_{1/2/n}$ (where 1 and 2 indicate the first and second cathode current density (CCCD's) and 'n' represents the number of layers formed during total plating time. i.e. (10 min).

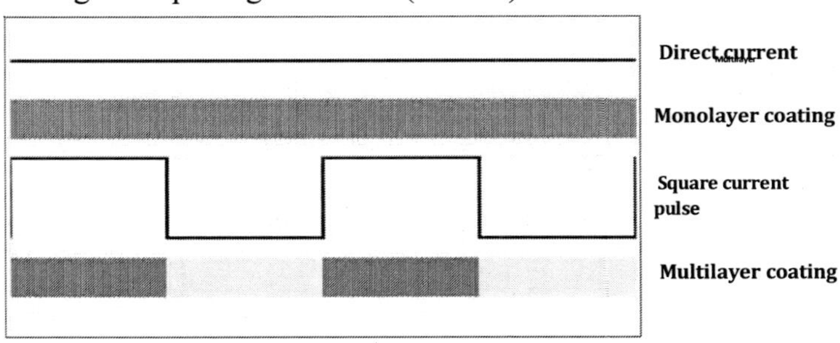

Fig. 1: Power Pattern Used Deposition of Monolayer (Direct Current) and Multiplayer (Pulsed Current) Coatings

RESULT AND DISCUSSION

Electrodeposition of Zn-Ni Alloy Coating

Zn-Ni CMA coatings were electrofabricated on polished MS with 7.5cm^2 exposed area. Monolayer coatings were developed at different c.d.'s, 1.0 to 4.0 Adm^{-2} and their corrosion rates (CR) were calculated by Tafel extrapolation method, and corresponding data are given in Table 2. It may be noted that at 3.0 Adm^{-2}, Zn-Ni alloy coating is found to be most corrosion resistant, and has been selected as optimal configuration, represented by $(Zn\text{-}Ni)_{3.0/mono}$.

Table 2: Corrosion Data for Zn-Ni Monolayer Coatings

c.d. (A/dm^2)	*Wt. % of Ni*	*Vickers Hardness* (V_{100})	$-E_0$ (*V vs. SCE*)	i_{corr} ($\mu A/cm^2$)	$CR \times 10^{-2}$ (mmy^{-1})
1.0	2.62	146	1.019	17.62	23.42
2.0	4.05	144	1.086	13.87	18.44
3.0	7.95	202	1.105	10.88	14.46
4.0	8.07	184	1.162	12.53	16.66

Optimization of Cyclic Cathode Current Densities (CCCD's)

The objective of improving the corrosion resistance of monolayer Zn-Ni alloy coating by CMMA technique is guided by two principles: i) Successive change in current density (c.d.) allows the growth of alloys having successive change in its chemical composition, ii) The layering of alloys having different noble metal content leads to the formation of distinct interface [7-9]. Hence multilayer coatings having 10 layers (arbitrarily chosen) have been accomplished at different CCCD's, namely at 1.0/3.0 and 2.0/4.0 A/dm^2. The coatings at 2.0 and 4.0 A/dm^2 difference between CCCD's were found to be more corrosion resistant, and corresponding corrosion data are reported in Table 3. These CCCD's were employed for optimizing the number of layers for better corrosion protection.

Table 3: Corrosion Rates of MNC are of Zn-Ni with 10 Layers to Set CCCD's

CCCD's (A/dm^2)	$-E_0$ (*V vs. SCE*)	$i_{corr.}$ ($\mu A/cm^2$)	$CR \times 10^{-2}$ (mmy^{-1})
$(Zn\text{-}Ni)_{1.0/3.0/10}$	1.085	8.416	11.18
$(Zn\text{-}Ni)_{2.0/4.0/10}$	1.108	6.606	8.78
$(Zn\text{-}Ni)_{4.0/mono}$	1.105	10.88	14.46

Optimization of Number of Layers in MNC's

The incredible claims of electroplating can be exploited for development of nanostructure multilayer coating for better corrosion protection. Hence, by choosing the above CCCD's multilayer coatings with higher degree of layering, i.e. with 20, 60, 120, 300 and 600 layers have been electrofabricated by proper setting up of the power source. The corrosion behavior of CMMA coating having different number of layers have been studied, and corresponding polarization behaviors are as shown in Fig. 3 (only representative). The corrosion data are reported in Table 4. It may be noted the CRs decreased drastically with increase in number of layers in both sets of CCCD's.

It indicates that the improvement in corrosion behavior of CMMA coating is not the unique property of individual layer composition; instead, the combined effect of composition modulation and number of layers. Further, the CR of CMMA Zn-Ni coating decreased only up to 300 layers and then started increasing, i.e. 600 layers as shown in Table 4. This decrease of CR at higher degree of layering is due to less relaxation time for deposition of metal ions on cathode. In other words, at very high degree of layering multilayer coating tend to become monolayer, showing more CR. Thus it may be stated that the amplitude of the composition modulation diminished rapidly when layer thicknesses below certain limit, i.e. less than 1000A°.

Table 4: Effect of Layering on Corrosion Behavior of CMMA $(Zn\text{-}Ni)_{1.0/3.0}$ and $(Zn\text{-}Ni)_{2.0/4.0}$ Coating in Comparison with Monolayer $(Zn\text{-}Ni)_{3.0/mono}$ Deposited from Same Bath at 303K

Coating Configuration	*Number of Layers*	*E_0 vs. SCE (volts)*	*i_c (μ A/cm^2)*	*CR × 10^{-2} (mmy^{-1})*
$(Zn\text{-}Ni)_{1.0/3.0}$	10	1.085	8.416	11.18
	20	1.037	7.125	9.47
	60	1.056	4.531	6.02
	120	1.067	2.357	3.13
	300	1.041	1.319	1.75
	600	1.050	3.577	4.75
$(Zn\text{-}Ni)_{2.0/4.0}$	10	1.108	6.606	8.78
	20	1.085	4.989	6.63
	60	1.050	2.125	2.82
	120	1.007	1.056	1.40
	300	1.006	0.295	0.39
	600	1.036	1.975	2.61
$(Zn\text{-}Ni)_{3.0/mono}$	-	1.105	10.88	14.46

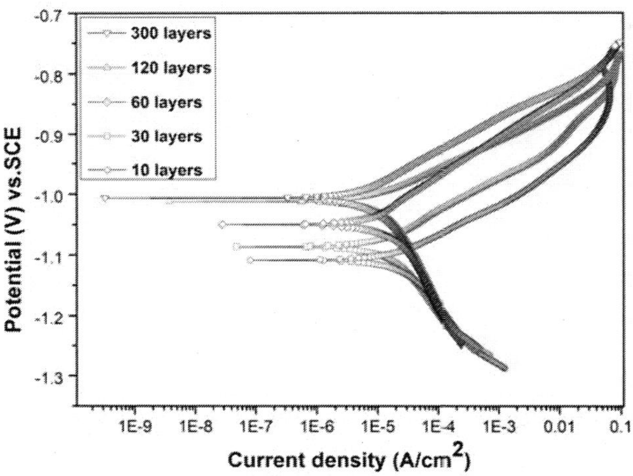

Fig. 2: Potentiodynamic Polarization Behaviors CMMA (Zn-Ni) Coatintgs having Different number of Layers, Deposited from Optimized at 303 K

Comparison of Monolayer and Multilayer Coatings

The corrosion behaviors of monolayer and multilayer Zn-Ni alloy coatings, electrofabricated from same bath for same time (under optimal conditions) are shown in terms of DC and AC polarizations as shown in Fig. 3. Corrosion data revealed that CMMA $(Zn\text{-}Ni)_{2.0/4.0/300}$ is about 37 (CR= 0.39 × 10^{-2} mm/y) times more corrosion resistant than $(Zn\text{-}Ni)_{3.0/mono}$ (CR =14.46 ×10^{-2} mm/y) coatings.

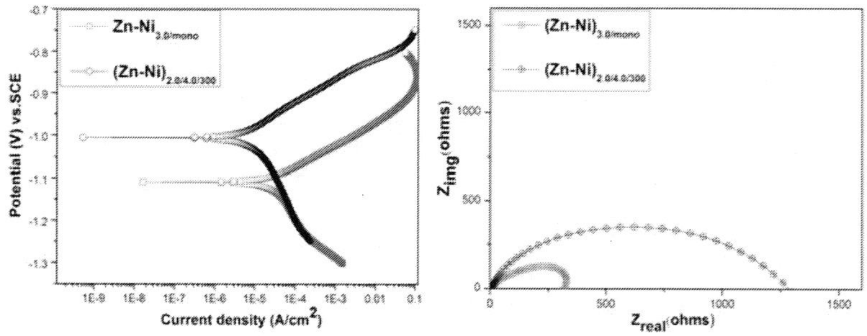

Fig. 3: Comparison of Corrosion Behaviors of Monolayer and Multilayer Coatings of (Zn-Ni) Alloy: (a) Potentiodynamic Polarization Behaviors and (b) Nyquist Response Under Same Frequency Limit

Scanning Electron Microscopic Study

The scanning electron microscopy (SEM) image of $(Zn\text{-}Ni)_{2.0/4.0/10}$ coating is shown in Fig. 4. The Figs. 4a and 4b present the surface morphology of the coating before and after bend test (electroplated coatings on thin foil of MS has bent by 135° repeatedly), respectively.

The fractured coating under SEM displays the layers as seen in Fig. 4b (shown by arrow mark). The high corrosion resistance of multilayer coatings can be envisaged due to formation of pores, crevices occurring in one layer is blocked or neutralized by the successively deposited coating layers, and thus the corrosion agent's path is delayed or blocked. The improved corrosion resistance afforded by multilayer coatings can be also explained in terms of formation of alternate layers of alloys with low and high wt. % of Ni as shown in Table 2.

Fig. 4: SEM Images of CMMA $(Zn\text{-}Ni)_{2.0/4.0/10}$ Coatings Electrofabricated from Optimized Bath: (a) Surface Structure, and (b) Cross-sectional View After Acid Test

Powder X-Ray Diffraction Study

The Fig. 5 shows the XRD patterns of Zn-Ni alloy deposited at different current densities from the same bath. The necessary requirement for better property of the multilayer coating is the compositional modulation, i.e. phase structure should vary with cyclic cathode current density. In present study, the deposits obtained at 1.0, 2.0, 3.0 and 4.0 Adm^{-2} were found to exhibit a marked difference in phase structure as shown in Fig. 5. Hence, layered coatings having distinct composition is responsible for modulation in composition, as evidenced by XRD study. The X-ray diffraction patterns, shown in Fig. 5 indicate that deposit at each c.d. has its own distinct phase structure.

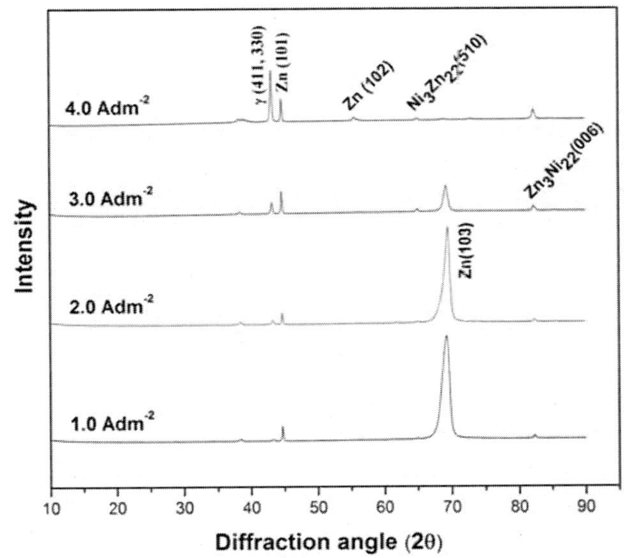

Fig. 5: XRD Patterns of Zn-Ni Alloy Coatings at Different c.d.s

CONCLUSIONS

- The proposed Zn-Ni bath followed anomalous type of deposition, i.e. wt. % Zn in deposit is more than that in the bath at all deposition c.d.'s.
- At different c.d., Zn-Ni coatings of different phase structures have formed, which enabled the compositional modulation in layered coating, evidenced by XRD and SEM analysis.
- The amplitude of the compositional modulation diminished rapidly when layer thicknesses below certain limit, i.e. less than 1000A°. Hence CMMA coatings fail to offer the improved properties at higher number of layers.
- Improved corrosion behavior at different set of CCCD's revealed that improvement is not specific to individual layers, instead it is a combined effect of compositional modulation and number of layers.

- The extended protection of CMMA coating is due to fact that the failure during the single layer deposition process is masked or neutralized by the successively electrofabricated layers. Hence, the path for the corrosive agents is extended or blocked.
- Under optimal conditions, multilayered Zn-Ni alloy coatings is about 37 times more corrosion resistant than corresponding monolayer alloy electrofabricated from same bath for same time.

ACKNOWLEDGMENTS

Mr. Vaishaka R. Rao acknowledge National Institute of Technology Karnataka (NITK), Surathkal for supporting this research in the form of Institute Fellowship.

REFERENCES

1] Jing-yin Fei and Wilcox, G.D., Electrodeposition of zinc–nickel compositionally modulated multilayer coatings and their corrosion behaviours. *Surf & Coat Tech.*, 200, 3533–3539, 2006.

[2] Artur Maciej, Ginter Nawrat, Wojciech Simka and Jerzy Piotrowski., Formation of compositionally modulated Zn–Ni alloy coatings on steel. *Mat Chem and Phy.*, 132, 1095-1102, 2012.

[3] Kalantary, M.R., Wilcox, G.D and Gabe, D.R, The production of compositionally modulated alloys by simulated high speed Electrodeposition from a single bath solution. *Electrochim Acta.,* 40, 1609-1616, 1995.

[4] Blum, W., The structure and properties of alternately deposited metals. *Trans Am Electrochem Soc.*, 40, 307–320,1921.

[5] Brenner, A, Electrodeposition of Alloy, 2nd vol., Academic Press, New York, 1963.

[6] Rahsepar, M and Bahrololoom, M.E., Corrosion study of Ni/Zn compositionally modulated multilayer coatings using electrochemical impedance spectroscopy. *Corr Science*. 51, 2537-2543, 2009.

[7] FEI Jing-yin, LIANG Guo-zheng, XIN Wen-li and WANG Wei-kang., Compositionally Modulated Multilayer Coatings. *J of Iron and Steel research*, 13(4), 61-67, 2006.

[8] Kirilova, I and Ivanov, I., Corrosion behaviour of Zn–Co compositionally modulated multilayers electrodeposited from single and dual baths. *J of App Electrochem.*, 29, 1133–1137, 1999.

[9] Kirilova, I., Ivanov, I and Rashkov, St., Anodic behaviour of composition modulated Zn-Co multilayers electrodeposited from single and dual baths. *J of App Electrochem.,* 28, 1359-1363, 1998.

[10] Tench, D and White, J., Tensile Properties of Nanostructured Ni-Cu Multilayered Materials Prepared by Electrodeposition, *J of Electrochem Soc.,* 138, 3757–3758, 1991.

[11] Thangaraj, V, Ravishankar, K and Chitharanjan Hegde, A., Surface modication by compositionally modulated multilayered Zn–Fe alloy coatings. *Chin. J. Chem.,* 26, 1–10, 2008.

SURFACE MORPHOLOGY, ELECTROCATALYTIC ACTIVITY, CORROSION RESISTANCE AND ELECTRO-ORGANIC APPLICATION OF Ti/TiO_2 ELECTRODE PREPARED BY THERMAL DECOMPOSITION OF TITANIUM TRICHLORIDE AND H_2O_2

K. Parameswari and C. Joseph Kennady

Department of Chemistry, School of Science and Humanities,
Karunya University, Coimbatore
E-mail: parameswari@karunya.edu

ABSTRACT

Ti/TiO_2 electrodes were prepared on metallic titanium substrate by the thermal decomposition of titanium trichloride and H_2O_2 at different temperatures. Surface morphology of TiO_2 film were studied by Scanning Electron Microscopy (SEM). The electrocatalytic activity and the corrosion resistance of the electrodes were studied by cyclic voltammetric studies and Tafel and impedance plots using electrochemical workstation (CHI660C). The electrode had been used for the carbonyl reduction with fairly good yield efficiency.

Keywords: Ti/TiO_2 Electrode, Structure, Phase, Organic synthesis

INTRODUCTION

Ti/TiO_2 cathode has been widely investigated due to the importance of its electrochemical, photochemical, catalytic and photo-electrochemical properties. Titanium dioxide coated titanium

V. Rajendran, P. Manivasakan and K.E. Geckeler (eds.)
Advanced Nano Materials for Industrial Applications, pp. 295-302 (2012).

(Ti/TiO_2) electrodes are used in a wide range of applications such as electrocatalysts, photocatalysis and dye-sensitized solar cells 1–4. Several authors have described the electrochemical and electroanalytical properties of this oxide electrode. Beck.et.al developed a TiO_2 cathode whose surface behaves as a stable redox electron carrier 5. Regarding the electrocatalytic property of a redox electrode many parameters involve which control the electrocatalytic property of the electrode such as surface area, particle size, phase formation etc. Apart from these, there are various parameters affecting the surface area of oxide electrodes which depends on the method of preparation. The parameters are the kind of metal complexes, alcoholic solvents, concentration of metal complexes, drying temperature, calcination temperature, calcination time, and decomposition speed of the salts or complexes coated on the titanium substrate and the number of dip-coating cycles, and so on. These preparation variables affecting the nature of TiO_2 on titanium substrate and its electrocatalytic properties are not well studied. Controlling the morphology of TiO_2 is a highly active area of research. One of the fundamental concepts in the development of these controlled shaped TiO_2 film is that select growth planes can be promoted or by changing the surface chemistry of the growth nuclei, by the oxidizing agent and the pH of the precursor. As it was discussed about the preparation of Ti/TiO_2 electrode from the thermal decomposition of $TiCl_3$ with HNO_3 the same preparation method 6, was carried out but by changing the oxidizing reagent HNO_3 to H_2O_2, to study the possible changes in the surface morphology, microstructure, stability and its electrocatalytic activity of the film TiO_2 on titanium substrate. This change is often easily accomplished by altering the pH of the growth media, and the nature of oxidizing agent which led to investigate the morphological variations of film TiO_2 at these conditions. And the preparation of titania is very much interesting since titania exhibits polymorphism (anatase, rutile and brookite) and the phase formation is highly dependent on the preparation process.

EXPERIMENTAL

Preparation of Ti/TiO_2 Electrode

In general the experimental procedure recommended by Beck and Co- workers 5 was adopted. The Ti/TiO_2 electrode was prepared by the following procedure. $TiCl_3$ solutions were prepared in different concentrations 0.05, 0.10, 0.15 and 0.20N using $TiCl_3$, H_2O_2 and isopropyl alcohol (Merck). In this solution the concentration of H_2O_2 and isopropyl alcohol were kept constant. With the above given concentrations of $TiCl_3$, coating solutions were prepared by adding H_2O_2 and isopropyl alcohol slowly and cooled to 10°C and maintained at 10°C for 20 minutes and then allowed to raise to room temperature. A stable clear light brownish yellow solution is obtained.

These solutions were sprayed over pretreated titanium substrate strips. And the adherent film was dried in an oven at 100-110°C. The metal strip was placed in a muffle furnace for 25 minutes. This procedure was repeated at least six times to get a good coating of TiO_2 on titanium metal. These electrodes were essentially prepared in two different standard sizes, viz.

1 × 5 cm and 9 × 12 cm for cyclic voltammetric and preparative scale experiments respectively. For cyclic voltammetric experiments a punched tape was usually bound around the electrode strip just exposing $0.0785cm^2$ for electrochemical activity. In preparative scale experiments 9 × 8 cm electrode surface was immersed in the electrolyte.

Surface Morphology and Electrocatalytic Studies

These electrodes were characterized for surface morphology by the scanning electron microscope (SEM) using JEOL 6390 instrument. Tafel and impedance plots were recorded using electrochemical workstation (CHI660C). From the above investigations optimized conditions were obtained. Cyclic voltammetric studies were carried out using electrochemical workstation (CHI660C). Pretreated titanium mesh was coated using the same procedure under optimum conditions. This Ti/TiO_2 electrode was used for galvanostatic electrolysis in which the electro reduction of fumaric to succinic acid was carried out. The product formed was isolated and confirmed by melting point measurement. The product was also confirmed by NMR spectroscopic technique using a T-60 Varian 60 MHz spectrometer.

RESULTS AND DISCUSSION

Surface Morphology Studies

The micrographs of Ti/TiO_2 electrodes prepared from different concentrations of $TiCl_3$ and at different temperatures by thermal decomposition method with H_2O_2 as oxidizing reagent are given in Fig. 1. Figure 1a is the micrograph of TiO_2 electrode at 500°C with 0.05N $TiCl_3$. TiO_2 film is formed evenly, but mostly amorphous and few particles with crystalline size in the range of 225-374nm. When the concentration of $TiCl_3$ increased from 0.05-0.10N, agglomeration of particle had been observed (Fig. 1b), still at increased concentration of $TiCl_3$ more of agglomeration of TiO_2 particles and the particle size increased as the concentration of $TiCl_3$ increased (Fig. 1c). Thermal treatment induces the formation of cracks and of small particles scattered on the surface of the titanium substrate. But further increase of concentration of $TiCl_3$ (Fig. 1d), there was no agglomeration but TiO_2 film formed evenly with the crystalline size in the range of 230-238nm. The film TiO_2 which was formed, of amorphous compared to other micrographs. In the electrodes prepared at 600°C it was found that agglomeration of TiO_2 had been reduced (Fig.1 e, f, g, h).

At 600°C the micrograph with the concentration 0.05N of $TiCl_3$ the particle size of TiO_2 on the substrate titanium was found to be in the range of 70-380nm. (Fig. 1e).The particle size of the TiO_2 with 0.10N concentration of $TiCl_3$ was in the range of 40-54 nm (Fig. 1f) and with 0.15N was 74-148 nm (Fig. 1g). The temperature factor plays an important role in controlling the particle size and the oxidizing agent H_2O_2 used along with $TiCl_3$. But it was observed that at higher concentration of $TiCl_3$ (0.20N) the TiO_2 particle size was in the range of 477-961nm.

Porous nature of the TiO_2 film was totally absent in the electrodes prepared at 500°C, but it is observed in the electrodes prepared at 600°C.

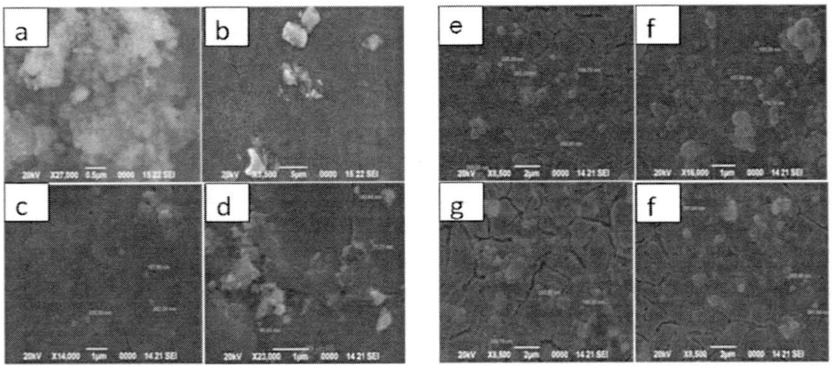

Fig. 1: SEM Micrographs of TiO_2 Film on Titanium Prepared at 500°C (a) 0.05N, (b) 0.10N, (c) 0.15N, (d) 0.20N of $TiCl_3$ with H_2O_2 and at 600°C, (e) 0.05N, (f) 0.10N, (g) 0.15N, (h) 0.20N of $TiCl_3$ with H_2O_2

From these observations it has been found that the small particle size and porosity of the film TiO_2 can be obtained at higher temperature (600°C) and low concentration of $TiCl_3$ which gives better electrocatalytic activity of the electrode by thermal decomposition method 7. Generally in these micrographs less or no cracks observed, so these electrodes may have less stress bearing ability and less surface area as these electrodes have even surfaces 8.

Corrosion Studies

Tafel Polarization Technique

The corrosion potential E_{corr} (mv vs. SCE), corrosion rate CR (mmpy), passivation range E_p and corrosion potential I_p (A) of the electrode and the uncoated titanium substrate in sulphuric was calculated and are presented in Table 1. The passivation range for the coated and uncoated from the Table 1 it shows that the uncoated has less corrosion resistance than the coated in 0.1M H_2SO_4. The E_{corr} (mv vs. SCE) also supports this finding that the coated one has comparatively good resistance than the uncoated one. The corrosion rate CR (mmpy) calculated from the Tafel polarization plots also confirms the results that the Ti/TiO_2 electrode has good corrosion resistance than the uncoated.

Table 1: Results Obtained from Tafel Polarization Techniques

System Studied	*Passivation Range (mA)*	*E_{corr} (mv vs SCE)*	*I_p (A)*	*CR (mm/year)*
Ti/TiO_2	(-0.0871) to (+0.9909)	-0.156	-3.3856	2.31
Titanium	(0.0955) to (+0.9943)	-0.556	-2.8173	1.3

The polarization curves of the electrode and substrate in sulphuric acid are presented in Fig. 1. It was previously reported that the corrosion of titanium or titanium alloys in sulphuric acid solutions is due to the dissolution of the material as Ti^{3+}. Krysa et al. confirmed the formation of Ti^{3+} analytically using spectrophotometric methods 9. The presence of oxygen in the electrolyte oxidizes Ti^{3+} into Ti^{4+}. As the concentration of Ti^{4+} increases the corrosion rate increases 10. As Ti^{4+} increases due to the film TiO_2 the corrosion rate of Ti/TiO_2 increased.

Electrochemical Impedance Spectroscopic Studies

Impedance curves for Ti/TiO_2 and the titanium substrate are given in Fig.3. The electrochemical parameters measured from these curves are given in Table. 2. The charge transfer resistance of the coated titanium Ti/TiO_2 R_{ct}=3.4 Ω shows the corrosion resistance of the film compared to the uncoated Rct = 1.1 Ω. And the I_{corr} (mA/cm^2) for the coated 1.7519 mA/cm^2, uncoated 11.85 mA/cm^2 clearly show, because of the thin and the dense layer of film TiO_2 on titanium substrate, the corrosion resistance of the electrode Ti/TiO_2 increased. From the calculated values, the anodic current density (I_{corr}), it is evident that coated titanium has good tendency of corrosion resistance behavior. By applying TiO_2 film coating on titanium, the corrosion resistance in 0.1M H_2SO_4 has improved.

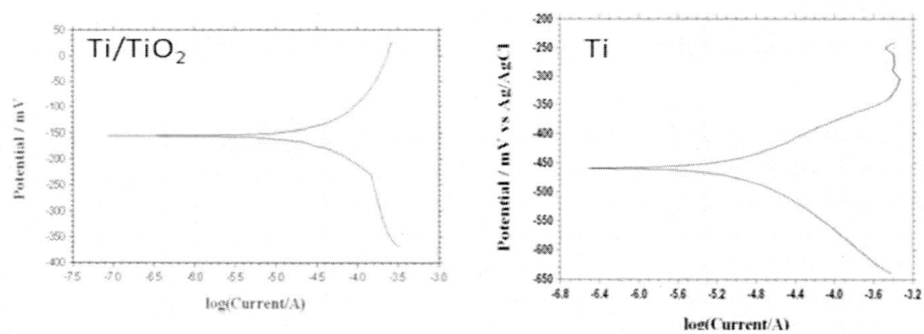

Fig. 2: Tafel Polarization Plot of Ti/TiO_2 Electrode in 0.1M H_2SO_4 Prepared at 600°C from 0.15N of $TiCl_3$ with H_2O_2 & Titanium Substrate in 0.1M H_2SO_4

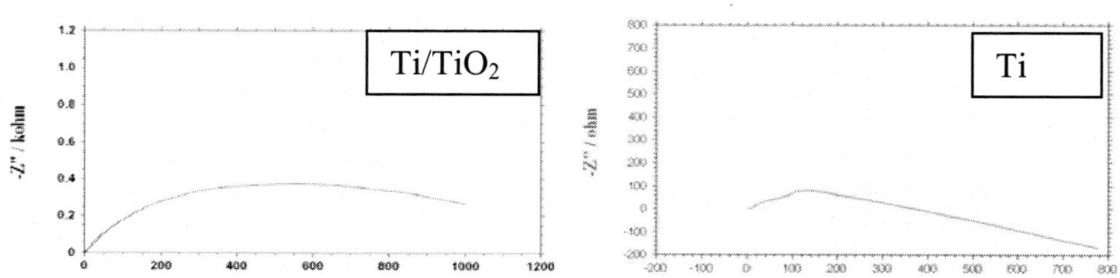

Fig. 3: AC-Impedance Spectroscopic Plot of Ti/TiO_2 Electrode in 0.1M H_2SO_4 Prepared at 600°C from 0.15N of $TiCl_3$ with H_2O_2 and Titanium

Table 2: Results Obtained from Electrochemical Impedance Spectroscopic Techniques

System Studied	*Phase Angle (θ)*	R_{ct} *(Ω)*	R_F*(Ω)*	I_{corr} *(mA/cm²)*
Ti/TiO_2	20	3.4	1.6	1.7519
Titanium	14	1.1	-	11.85

From the Table 2, it is found that the phase angle value obtained is 24 and the range of values for phase angle from 40 to70 indicates that the impedance has a significant contribution in the corrosion resistance properties of the electrode TiO_2 (phase angle 24) compared to the uncoated titanium substrate (phase angle 14) 11,12. A decrease of the maximum absolute value of phase angle of the electrode is attributed to the decrease of the corrosion resistance of the metal 13-16.

The calculated parameters from Tafel polarization techniques like passivation range E_p (mV), corrosion potential E_{corr} (mv vs. SCE), passivation current I_p (A) and corrosion resistance CR (mmpy) are in good agreement with the parameters calculated from AC-impedance spectroscopic techniques like phase angle (θ), charge transfer resistance R_{ct} (Ω), film resistance R_F (Ω) and corrosion current I_{corr} (mA/cm²) for both coated titanium and uncoated titanium substrate in 0.1M H_2SO_4.

CYCLIC VOLTAMMETRIC ANALYSIS FOR THE ELECTRODE Ti/TiO_2

The cyclic voltammogram recorded (sweep rate 50mVS^{-1}) for the electrode Ti/TiO_2 prepared by the thermal decomposition method from $TiCl_3$ and H_2O_2 mixture is presented in Fig. 4a. From the voltammogram it is evident that this Ti/TiO_2 electrode has very low redox electrocatalytic activity.

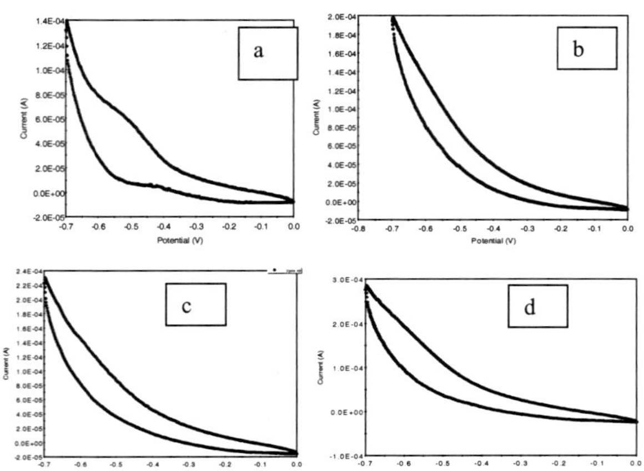

Fig. 4: (a) Cyclic Voltammogram of Ti/TiO_2 Electrode in 1.0 M H_2SO_4 Prepared at 600°C from 0.15N of $TiCl_3$ with H_2O_2 (b) with 0.1mM, (c) with 0.2mM, (d) with 0.5mM Fumaric Acid

In the case of Ti/TiO_2 electrode from $TiCl_3$ and H_2O_2, the XRD pattern clearly reveals that the crystalline structure is rutile type TiO_2. This might be the cause for the absence or very poor redox behavior of the electrode.

CYCLIC VOLTAMMETRIC ANALYSIS OF TI/TIO$_2$ IN H$_2$SO$_4$ WITH FUMARIC ACID

Cyclic voltammograms obtained with fumaric acid are presented in Fig. 4b-d as there was no redox activity at the base electrode, addition of fumaric acid does not change the cyclic voltammetric pattern appreciably. Even at higher concentration of fumaric acid the pattern remains same. The above point was well established with the results of preparative scale electrolysis. The yield efficiency is low with this electrode.

PREPARATIVE ELECTROLYSIS

A number of preparative scale experiments were carried out under galvanostatic scale experimental conditions to optimize the conditions for preparative electrolysis and achieve good yield efficiency in the preparation of succinic acid from fumaric acid. Some of these results are summarized in Table 3. The current was varied from 1.0-5.0 A/dm^2 and varied temperature. At lower current densities particularly at 4.0 A/dm^2 fumaric acid reduced to succinic acid. The temperature of the electrolysis was varied from 45-65°C as it is given in the experiments at 55°C the yield is fairly good and by increasing the electrolyte temperature beyond 65°C causes some destruction to the electrode by means of dissolving the TiO_2 coating. The optimum temperature for getting better yield was taken as 55°C. From the results it was found that the optimum conditions for the reduction of fumaric acid to succinic acid with fairly good yield (47.0%) are at 4.0 A/dm^2 current density and temperature 55°C.

Table. 3: Galvanostatic Reduction of Fumaric acid at a Ti/TiO_2 Electrode

Expt.. No	*Current Density (A/dm^2)*	*Temperature (^{0}C)*	*Weight of Succinic Acid (g)*	*Yield Efficiency (%)*
1	1	55	3.15	31.5
2	2	55	3.50	35.0
3	3	55	3.90	39.0
4	4	55	4.70	47.0
5	5	55	4.50	45.0
6	4	45	2.48	24.8
7	4	65	4.40	44.0

Conditions: Amount of fumaric acid taken: 10g, Electrolyte: 1M H_2SO_4 Anode: Stainless steel

PRODUCT ANALYSIS

The product was separated and analyzed by NMR spectroscopic method. The peak at 2.6 confirms the ethylene proton in succinic acid.

CONCLUSION

From SEM studies it has been inferred that the temperature factor plays an important role in controlling the particle size and the oxidizing agent H_2O_2 used along with $TiCl_3$. From the corrosion studies, the coated titanium substrate has improved corrosion resistance than the uncoated one. From the CV studies, it is clear that this electrode can be used as a redox electrode for surface processes with very low efficiency. Further work can be carried out to improve the yield.

REFERENCES

[1] Hagfeldt.A and Grätzel, M., Light-Induced redox reactions in nanocrystalline systems. *Chem. Rev*, 95, 49 1995.
[2] Hoffmann M.R, Martin S.T, Choi. W and D.W. Bahnemann. D.W, Photovoltaic Performance of DSSCs based on AN and AN-Br Films, *Chem. Rev*, 95, 69, 1995..
[3] Linsebigler A.L, Lu G and Yates J.T., Photocatalysis on TiOn Surfaces: Principles, Mechanisms, and. Selected Results. *Chem. Rev*, 95, 735, 1995.
[4] Chen. X and Mao. S.S, Titanium Dioxide Nanomaterials: Synthesis, Properties, Modifications, and Applications, *Chem. Rev*, 107, 2891, 2007.
[5] Beck F and Gabriel W, Angew. Heterogeneous Redox Catalysis on Ti/TiO_2 Cathodes—Reduction of Nitrobenzene, *Chem. Int. Ed. Engl*, 24, 771, 1985.
[6] K.Parameswari, C Joseph Kennady, Karunya Journal of Research, Crystalline structure and electro-organic applications of Ti/TiO_2 electrode prepared by thermal decomposition of Titanium Tetraisopropoxide. Vol. 3, Iss 1, 55-60, 2011.
[7] Ronconi.C.M, Pereira.C.M, Immobilised particulate TiO_2 catalysts for degradation of organic pollutants, Effect of layer thickness, *Journal of Applied Elec.Chem*, 31, 319, 2001.
[8] Patsi M.E and Hanutanimei J.A., Bonding strengths of titania sol-gel derived coatings on titanium, *Journal of Sol-Gel Science and Techcnology*, 11, 55, 1998.
[9] Krysa. J, Mraz. R. and Rousar. I., Mater. Chem. Phys, 48, 64, 1997.
[10] Kelly. E.J, The influence of fluoride ions on the passive dissolution of titanium, *J. Electrochem*. Soc, 123, 162, 1976; 126, 2064, 1979.
[11] Ping Gu, Yan Fu, Pig Xie, Deaudoin. J.J, Characteristics of surface corrosion of reinforcing steel in cement paste by low frequency impedance spectroscopy, *Cement and Concrete Research*, 24, No. 2 231, 1994.
[12] Rozenfeld I.C., Corrosion inhibitors, Mc Graw- Hill, International Book Company' part 2, Ch. 5, 145, 1981.
[13] Mansfeld. F., Shih. H, "Detection of pittig with electrochemical impedance spectroscopy, *J. Electrchem Soc*, 135, No. 4,171, 1988.
[14] Shih.H, Mansfeld. F, A pittig procedure for impedance spectroscopy obtained for cases of localized corrosion, 45, No 8, 610, 1989.
[15] Mannnsfeld. F, Lin. S, Kim. S, Shih. H, Corrosion protection of Al alloys & Al based metal matrix composites by chemical passivation, corrosion, 45, No.8, 615, 1989 .
[16] Macdonald. D.D, Tantawry. Y.A. EL, Rocha. R.C, Filho, Urquidi Macdonald. M., Evaluation of Electrochemical impedance techniques for detecting corrosion on rebar in reinforced concrete, SHRP-10/UFR-91-524, Strategic highway research programme report, National Research Council Washington, DC 1994.

COMPRESSIVE HYSTERESIS IN CNT REINFORCED RUBBER NANOCOMPOSITES

K. Sasikumar, Manoj N.R., T. Mukundan and D. Khastgir[1]

Materials Science Division, Naval Physical and Oceanographic Laboratory, DRDO, Kochi
[1]Rubber Technology Centre, Indian Institute of Technology, Kharagpur
E-mail: nrmanoj@indiatimes.com

ABSTRACT

Carbon nano tube (CNT) based nanocomposites are expected to bring out exceptional and revolutionary design changes in the realm of specialty composites for various applications. Significant challenges still remain the elucidation of full potential of CNT composites due to the lack of a multilevel analysis of composites with respect to the area of application. In this study, efforts are made to develop an understanding of the strain induced changes in the nanocomposites at different strains and CNT contents. The compressive hysteresis of carboxylated nitrile rubber and its composites, reinforced with CNT up to 7 parts per hundred rubber (phr), were investigated using universal testing machine for high strains (up to 25%) and dynamic mechanical analyzer for low strains (up to 3%). In both cases, Mullins effect was observed, i.e. the stress-strain curve follows a significantly softer path during unloading and the loop area of the first cycle is much higher than the rest. The loop area becomes constant by the 5th cycle in both cases. The hysteresis loop area was used to calculate the percentage damping and loss factor. For high strains, it was observed that the loop area of all the cycles increases proportionately with increasing CNT concentration and the percentage damping increases from 15% in unfilled to 35% in filled system. At low strains, as found from DMA, the hysteresis loop area of first cycle decreases from unfilled rubber up to nanocomposites with 3 phr and then increases, while the loop area of unfilled rubber is higher than even that of the nanocomposite containing 7 phr CNT for the 5th cycle. The damping increases from 22% to 30% for nanocomposite with 3 phr CNT and then decreases to 26% at 7 phr

V. Rajendran, P. Manivasakan and K.E. Geckeler (eds.)
Advanced Nano Materials for Industrial Applications, pp. 303-310 (2012).

level. The morphology of the composites was studied using Transmission Electron Microscope (TEM), which shows aggregate formation as well as breakage of nanotubes as the CNT concentration increases in the composite. The results would be of great assistance in the design of nanocomposites for effective damping applications.

INTRODUCTION

Reinforcement using fillers is an important area of rubber research and development. The property enhancement is found to be a function of content, size, aspect ratio of the fillers and the physics and chemistry of dispersion [1]. Rubber nanocomposites have gained great importance in recent times on account of their wide range of potential applications in many areas [2]. Due to the high surface area of nanofillers, one can achieve enhanced filler-matrix interaction at a significantly lower level of filler loading compared to the conventional fillers. Among the variety of nano materials available to function as fillers in rubber compounds, Multi wall carbon nanotubes (MWNT), due to their very high mechanical properties, low production cost and high performance have become a prime candidate for many engineering applications [3, 4]. Filler-reinforced rubber shows many interesting nonlinear effects such as the Mullins effect and Payne effect [5]. The hysteresis loop provides an idea about the energy loss in the cyclic deformation of a material. The amount of energy lost per cycle of deformation is known as the hysteresis. Mullins effect is a strain induced softening behavior observed in filled as well as gum rubber systems [6]. This is associated with a reduction in stress (softening) at a given strain level during the unloading path as compared to the stress on initial loading in cyclic tests. Also, when previously unstrained rubber is subjected to strain cycles at constant peak value, the first few oscillations result in reducing peak stress values. This phenomenon is attributed to the mechanical hysteresis from filler particles de-bonding from each other or from the polymer chains, caused by the stretching. As a result, highly filled systems suffer a prominent reduction in stiffness as compared to less filled systems [7-8].

In this work, MWNT was used as the nano filler and the matrix consisted of carboxylated nitrile rubber (XNBR), a high performance specialty rubber having pendent carboxyl groups at random along the chain. The hysteresis behavior of CNT-filled rubber is thoroughly analyzed in compression mode in universal testing machine and dynamic mechanical analyzer with the intention of developing an understanding of interaction of nanofillers with rubber matrix at low and high strains.

EXPERIMENTAL

The XNBR used was Chemigum NX-146 with a Mooney viscosity of ML 1+4[100°C] 45±5, acrylonitrile content 33%, carboxylic acid content 1.0% and obtained from M/s Eliochem, India. The MWNT, having a purity of 95% and an average diameter of 20-30 nm was supplied by M/s Chemapol Mumbai. The rubber was compounded in a 6" two-roll mill with various loadings (0, 1, 3, 5 and 7 phr) of MWNT along with zinc oxide, stearic acid, n-cyclohexyl-2-benzothiazole-sulfenamide (CBS) and sulfur. The stocks were cured to 90% (T_{90}) under pressure at 150°C, based on a rheometric study.

Compression studies of neat XNBR and nanocomposites (20 mm dia × 10 mm thick) were carried out according to ASTM standard D 575-91 in a Zwick 1476 (Germany) universal testing machine. Dynamic mechanical analysis (DMA) of composites (13 mm dia × 2 mm thick) was carried out on DMA Q 800 (TA instruments, USA) in compression mode. The dispersion characteristics of CNT in the cryo microtomed nanocomposites of 100 nm thickness were analyzed using JEOL 3010 Transmission Electron Microscope (TEM).

RESULTS AND DISCUSSION

The essential feature for a good composite is the filler-matrix interactions, which promotes the stress transfer from the matrix to the filler [9]. The higher surface area and aspect ratio of CNT would provide unique filler-matrix interactions at a significantly lower level of filler loading. The compressive stress–strain plots of neat XNBR and nanocomposites are shown in Fig. 1. Unlike the general observation of not getting sufficient dispersion of nanofillers in mill mixing and leading subsequently to a reduction in properties with increasing filler content, the modulus and compressive strength are seen to increase with increasing CNT content in this case. Also, the slopes of the successive plots indicate that reinforcement is higher at higher filler loading. Overall, up to a strain of 25%, there is 50% increase in compressive strength and 35% increase in compressive modulus with the incorporation of CNT up to 7 phr.

Hysteresis is a measure of energy density dissipated by a material during cyclic deformation. Fillers are known to cause an increase in hysteresis loss in rubbers, and hysteresis measurements can also be used to estimate the reinforcing ability of fillers [10]. Figs. 2a & 2b illustrates the representative stress-strain cycles of XNBR and the nanocomposites showing the typical hysteresis loops obtained on compression. The nonlinearity is clearly displayed in the bending of the ellipse from the linear long axis. It is found that on the initial loading, the virgin material exhibit a relatively stiff response. When it is subsequently unloaded, then reloaded, the stress-strain curve follows a significantly softer path. The stress-strain response is seen to stabilize by the fourth cycle, and hence the area of the fifth cycle is taken as the constant area (mentioned hereafter as last cycle). This typical stress softening is the Mullins effect - the stress and the area of the hysteresis loop decrease significantly between the first two cycles. Table 1 summarizes the hysteresis loop areas of first, second and last cycles (constant area) of all the composites.

The loop area directly gives the energy loss of deformation. Similarly the difference in loop area between the first and second cycles, indicating Mullins effect, also increases with increasing CNT content. It may be seen that the difference rises from 50 J/m² to almost 200 J/m² from unfilled to the nanocomposite with 7 phr CNT content. This indicates that in addition to the increasing reinforcement by CNT, there is agglomerate formation at the microscopic level that breaks down with the strain. The enhanced energy loss per cycle may be attributed to the high degree of internal friction that would result from poor interactions of CNT with polymeric chains, unlike the carbon black that has defects and structural irregularities for both physical and chemical interactions [11]. The formation of more aggregates and increased dynamic friction of CNT with polymer chains at increased levels of CNT causes the observed effects. The hysteresis loop area can be used to express the easily perceivable percentage damping as

$$\% \text{ damping} = \frac{U_1\text{-}U_2}{U_2} \times 100 = \frac{\frac{1}{2}\pi\sin\varphi \times 100}{1+\frac{1}{4}\pi\sin\varphi} \quad \text{... (1)}$$

where U_1 is total strain energy, U_1-U_2 is the area inside the loop and j is the mechanical loss angle [12]. The % damping and calculated loss factor are tabulated in Table 1.

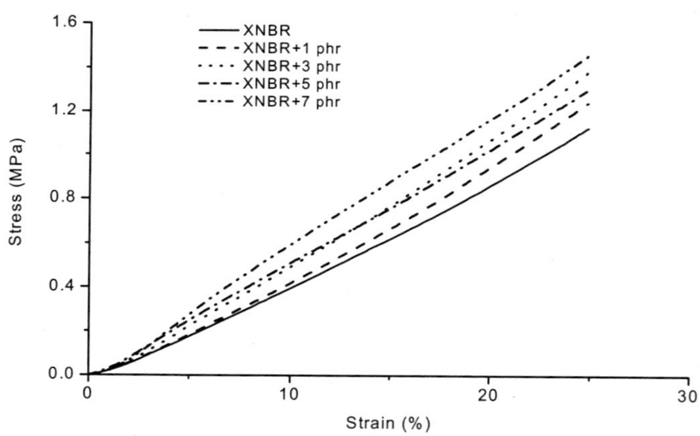

Fig. 1: The Compressive Stress–Strain Plots of Neat XNBR and Nanocomposites from UTM

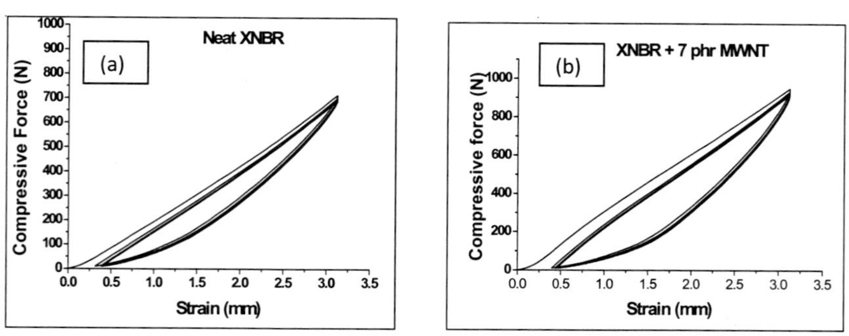

Fig. 2: Stress-Strain Cycles of XNBR and Nanocomposites Under Compression in UTM

Table 1: Hysteresis loop Area, % Damping and Loss Factor of XNBR and Nanocomposites from UTM

CNT Content (phr)	*Hysteresis Loop Area (J/m²)*			*% Damping*		*Calculated Loss Factor*	
	1st Cycle	*2nd Cycle*	*Last Cycle*	*1st Cycle*	*Last Cycle*	*1st Cycle*	*Last Cycle*
0	290	241	236	16.8	15.4	0.12	0.10
1	448	310	231	24.2	16.1	0.18	0.11
3	580	423	352	27.7	20.4	0.21	0.14
5	757	630	521	38.5	30.3	0.31	0.23
7	921	734	678	40.7	35.5	0.34	0.28

It can be seen that the incorporation of CNT into XNBR increases the % damping from 15% in unfilled to 35% in filled system, because of increasing friction at the polymer-CNT interface. The trend is the same for the calculated values of % damping from first cycle loop area also, although the values are 18-25% higher than other loop areas. In order to find out the impact of low strains on rubber nanocomposites, the experiments were repeated on dynamic mechanical analyzer (DMA). As a standard experimental approach, the DMA technique is quite useful to study these effects at low amplitudes of strain [13]. Figures 3a &3b show the representative stress-strain cycles generated by the rubber and the nanocomposites with a strain up to 55 μm (equivalent to 3%). The behavior of the nanocomposites is entirely different at low strains, as compared to the results from the UTM. The loop area is considerably less apparently due to the low strain in DMA. However, it may be seen that the loop areas do not increase with increasing CNT content. Mullins effect is clearly visible from the change in the area of first, second and last cycle by which the area becomes constant.

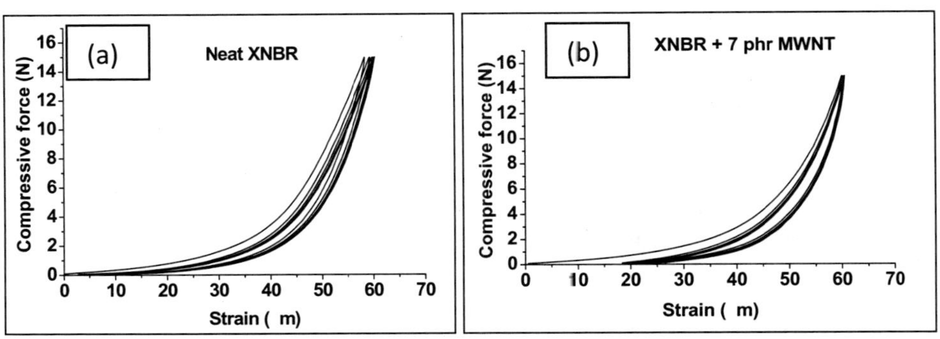

Fig. 3: Stress-Strain Cycles of XNBR and Nanocomposites Under Compression in DMA

The difference in the area of first and second cycles decreases from 117 J/m² to 72 for the last cycle decreases from 153 J/m² to 93 J/m² and then increases to 249 J/m² for the composite with 7 phr filler. The effect is less pronounced than unfilled in the case of 1 phr filled system. This may be due to the fact that the CNT particles or their aggregates are sufficiently separated from each other and hence cannot contribute to the energy loss due to friction and break down of structures at low strains. The hysteresis loop area of first cycles decreases from unfilled rubber up to nanocomposites with 3 phr and then increases. Another interesting observation is that the loop area of unfilled rubber is higher than even that of the nanocomposite containing 7 phr CNT for the last cycle. The % damping and calculated loss factor are tabulated in Table 2. The table also shows the % damping of all composites, where the damping decreases for composites with1 phr filler and then increases till 5 phr and shows a decreasing trend. The damping, calculated from the area of the last cycle, is the highest for 3 phr. The loss factor also shows the same behaviour. This may be attributed to the low level of strain imparted to the samples in DMA which may not be sufficient to break down any aggregates because of the higher interaction between aggregates at higher loading of CNT. At the low strain (~ 55μ) as well as low rate of strain (~ 15 μm/min) experienced in DMA, it is possible that the crosslinks in the rubber matrix stores the energy but the strong filler aggregates are not broken down and any small aggregates if broken may be reformed using the stored energy.

Table 2: Hysteresis Loop areas, % Damping and Loss Factor of XNBR and Nanocomposites from DMA

CNT Content (phr)	*Hysteresis Loop Area (J/m^2)*			*% Damping*		*Calculated Loss Factor*	
	1st Cycle	*2nd Cycle*	*Last Cycle*	*1st Cycle*	*Last Cycle*	*1st Cycle*	*Last Cycle*
0	0.445	0.328	0.292	30.7	22.6	0.24	0.16
1	0.302	0.230	0.209	25.3	19.7	0.19	0.14
3	0.429	0.296	0.275	38.6	30.7	0.32	0.23
5	0.457	0.301	0.256	39.5	28.1	0.33	0.21
7	0.533	0.336	0.284	38.7	26.5	0.32	0.20

It may also be explained based on a two-phase filler-elastomer structure where reinforcement occurs due to long filler formed chains and intermolecular rubber-filler bonds. Small deformations are always associated with straightening of macromolecular segments, which fails the filler formed chains at relatively low strains and filler-rubber bonds only at larger strains [14]. In order to understand the morphology of nanocomposites and the possibility of aggregate formation by CNT, TEM micrographs were taken for nanocomposites containing 1, 3 and 7 phr CNT and are shown in Figs. 4a-c. In all cases there is uniform distribution of CNT.

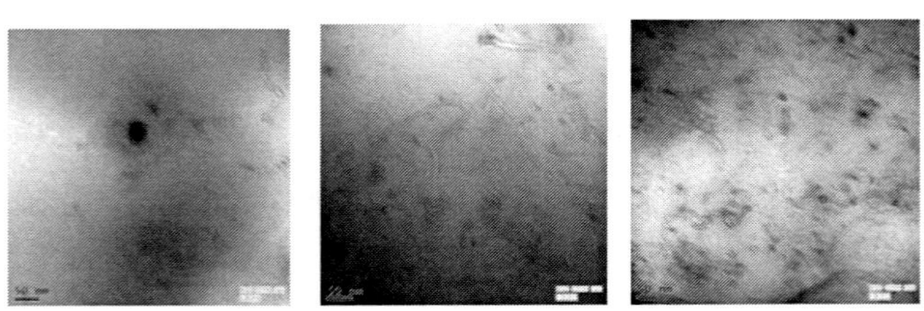

Fig. 4: TEM Micrographs of Nanocomposites Containing (a) 1, (b) 3, and (c) 7 phr CNT

The nanocomposite with 1 phr CNT does not show much of aggregate formation whereas that with 7 phr clearly shows clusters of CNT. Another interesting observation is in the change in the length of nanotubes in the composites. The composite with 1 phr CNT shows more or less uniformly long CNTs whereas that with 7 phr shows a good variation in the length of tubes, clearly due the breakage that happens while the nanocomposite is mixed in the mill due to the increased shearing force. The results indicate an important inference that the same nanocomposite behave differently at low and high strains when damping is considered. For high strain applications, the higher the loading of CNT, the higher is the damping and loss factor. For low strains however, damping increases up to 3 phr and effectively decreases with further addition of CNT. In many of the practical applications where rubber is used in mounts, bushes and baffles, the rubber layer may undergo large strains. However, in vibration damping or coating applications using free layer or constrained layer methods, viscoelastic layers usually undergo small deformations [15].

CONCLUSIONS

This study describes the nanocomposite based on XNBR and CNT particularly focusing on the compressive hysteresis and damping. In order to find the effect of strain, cylindrical samples were cyclically deformed up to 25% in a universal testing machine for high strains and up to 3 % in a dynamic mechanical analysis for low strains. Mullins effect was observed in both cases. The loop area becomes constant by the 5th cycle in both cases. The main difference in the behavior is that for high strains, the % of damping and loss factor continuously increases with CNT content, whereas for low strains, the highest damping was observed in 3 phr loaded composite after which there is a decrease in damping. The behavior is explained based on the nature and role of strain on the filler-rubber interaction and cluster formation of CNT. TEM micrographs clearly show the uniform distribution of filler, breakage of CNT and formation of clusters at higher loading. The results indicate that the extent of damping in a nanocomposite is determined by the strain experienced as well as the CNT content.

REFERENCES

[1] Bueche F., Reinforcement of Elastomer, G. Kraus Ed, Wiley Inter-Science, New York, 1965.

[2] Thomas S. and Stephen R., Rubber Nanocomposites: Preparation, Properties and Applications, John Wiley and Sons, Singapore, 2010.

[3} Meyyappan M., Carbon Nanotube: Science and Applications, CRC Press, Florida, 2005.

[4] O'Connell M.J., Carbon Nanotubes – Properties and Applications, Taylor and Francis, Boca Raton, 2006.

[5} Freakley P.K. and Payne A.R., Theory and Practice of Engineering with Rubber, pp. 3-23, Applied Science Publishers, London, 1978.

[6] Mullins L., Softening of rubber by deformation, *Rubber Chem Technol*, 42, 339–362, 1969.

[7] Dorfmann A. and Ogden R.W., A constitutive model for the Mullins effect with permanent set in particle-reinforced rubber, *Int. J. Solids Struct.*, 41, 1855–1878, 2004.

[8] Diani J., Fayolleand B., Gilormini P., A Review on the Mullins Effect, *Eur. Polym. Journal*, 45, 601–612, 2009.

[9] Blanchard A.F. and Parkinson D., Breakage of Carbon-Rubber Networks by Applied Stress, *Ind. Eng. Chem.* 44, 799, 1952.

[10] Abu-Abdeen, M, Degradation of the mechanical properties of composite vulcanizates loaded with paraffin wax. *J. Appl. Polym. Sci.,* 81, 2265–2270, 2001.

[11} Cataldo F., The impact of fullerene-like concept in carbon black science, *Carbon*, 40, 157–162, 2002.

[12] Gobel E.F and Brichta A.M., "Rubber Spring Design", Butterworths, London, 1974.

[13] Menard H.P., "Dynamic Mechanical Analysis – A Practical Introduction", CRC Press, London, 2008.

[14] Liebowitz H, Fracture: An Advanced Treatise, Vol 7, Fracture of Nonmetals and Composites, Academic press, New York, 1972.

[15] Sperling L.H. (Ed), "Sound and Vibration Damping with Polymers", American Chemical Society, Washington D.C., 1989.

STUDY OF FUNCTIONAL AND PHYSICAL PROPERTIES OF AMORPHOUS NANO SILICA COATED ON COTTON FABRIC

R. Palanivelu, V. Rajendran G. Thirunavukkarasu and N. Gobi

Department of Chemistry, Centre for Nano Science and Technology, K.S. Rangasamy College of Technology, Tiruchengode & Anna University, Tamil Nadu
E-mail: veerajendran@gmail.com

ABSTRACT

Silica nanoparticles were synthesized from rice husk and characterized from Transmission electron microscopy, X-ray diffraction and thermal gravimetric analysis. The particles were found to be amorphous in nature, ranging in size from 30 to 100 nm. The concentrations of silica nanoparticles, curing temperature and curing time have been taken as independent variables to design the experiment. Box-Behnken method has been used to derive the experimental plan and nine experiments were conducted. Based on the experimental plan, cotton fabric has been treated and analyzed the functional and physical properties. From this study, it has been observed that the increase in concentration of silica nanoparticles improve the flame retardancy due to its high LOI value at optimal conditions.

INTRODUCTION

Cotton is a flammable material and it is one of the most dangerous textile materials in the case of fire which will burn vigorously if not properly treated. Flame retardant materials are added or applied in textile product to increase its flame resistance time[1]. Primary flame retardants have either phosphorous or halogens or both. The compounds of phosphorous as flame retardant materials release formaldehyde during curing and finishing which is harmful to human being.

V. Rajendran, P. Manivasakan and K.E. Geckeler (eds.)
Advanced Nano Materials for Industrial Applications, pp. 311-318 (2012).

All halogens are effective for eliminating free radicals and the trapping efficiency increases with the size of the halogens (i.e. I > Br > Cl > F)[1]. During the process and usage, current popular flame retardants have disadvantages like release of environmental pollutants and persistent organic pollutants. Nanoparticles can provide high durability for fabrics, because of its larger surface area-to-volume ratio and high surface energy[2]. This provides better affinity and durability for fabric.[3,4] Some of nanoparticles such as SiO_2, TiO_2, ZnO and Fe_2O_3 are used as UV blocking agents, antimicrobial agents, water repellents, crease recovery, etc.[3] Of these, SiO_2 improves thermal and thermal-oxidative stability of polymers by the formation of thermally stable chemical bonds between polymer and the nanoparticles. Because of formation of cross-linking bonds and of a three-dimensional polymer- nanoparticles network at elevated temperatures, there is increase in the thermal stability of the polymer[5]. The higher LOI values observed might be due to formation of higher performance of carbonaceous-silica char on the surface.[6,7,8]

Mostly silica nanoparticles are synthesized using sol-gel method which is not economic[3]. In this investigation, silica nanoparticles are synthesized from natural resource such as rice husk by the thermal degradation method and its flame retardancy over coated fabric is characterized.[9,10,11,12]

MATERIALS AND METHODS

Materials

A bleached and mercerized 100% cotton fabric, having the specifications of plain weave, 120 gsm, warp count 40 Ne and 30 ends/cm, weft count 40 Ne and 28 picks/cm was used in this study. The solution contains silica nanoparticles, acrylic binder as binding agent, Lissopal-N as wetting agent and silicone emulsion as softening agent.

Methods

Synthesis of Silica Nanoparticles

Silica nanoparticles were prepared from natural resource of rice husk by thermal degradation method. The rice husk was rinsed, dried and aged at 80°C for 1 hour. The particles were refluxed with 10% HCl for 45 minutes and dried at 80°C. The residue was kept in a muffle furnace at 750°C for 6 hours and was powdered well homogeneously.

Experimental Plan

To analyze the effect of process parameters, an experimental plan was designed using Box-Behnken method. In this experiment, three different variables with three different levels in each variable were selected (Table 1) and nine trials were conducted (Table 2).

Table 1: Details of Process Parameters and Different Levels for Klame Retardant Finish

Process Parameters	*Different Levels*		
	-1	0	-1
Conc. of silica nanoparticles (X1), % (owf)	2.5	5	7.5
Curing temperature (X2), °C	140	160	180
Curing time (X3), min	1	2	3

Table 2: Functional and Physical Properties of Coated Fabrics

Sample No.	*Conc. of SiO_2*	*Curing Temp. (°C)*	*Curing Time (min)*	*Flame Retardant (s)*	*Drape Co-efficient*	*Air Permeability $cm^3/cm^2/s$*	*Bending Modulus (kg/cm^2)*
T1	2.5	140	2	17.80	0.72	4.000	189.04
T2	7.5	140	2	33.50	0.79	2.200	210.84
T3	2.5	180	2	18.20	0.74	4.000	195.01
T4	7.5	180	2	37.30	0.77	2.850	240.54
T5	2.5	160	1	19.30	0.75	4.100	193.15
T6	7.5	160	1	36.30	0.79	2.900	301.48
T7	2.5	160	3	17.10	0.70	4.250	193.01
T8	7.5	160	3	35.20	0.73	2.250	267.16
T9	5	140	1	39.70	0.70	3.250	220.32

Coating of Fabrics

In each experiment, the pad solution contains different concentrations of silica nanoparticles (2.5, 5 and 7.5%), silicone emulsion (1%), acrylic binder (5%) and wetting agent (0.1%, Lissapol-N). The padded fabrics were dried at 80°C and then, the curing process was carried out (Table 1). To remove non-binded silica nanoparticles, the cured fabrics were treated with soap solution for 5 min, rinsed for 10 min at room temperature and then dried.

Linking Agent Technique

The free hydroxyl groups of cellulose structures can be cross linked by polycarboxylic acids which depends on the number of free OH groups in the cellulose and it forms ester cross links. The remaining free OH groups in the polycarboxylic acids can be anchored by nano particles.[13] Cotton cellulose is a polysaccharide with many free hydroxyl groups on the surface. In this study, BTCA (1, 2, 3, 4 butane tetra carboxylic acid) has been used as an anchoring agent which has four free carboxylic acid groups. Ester bonds are formed between the hydroxyl groups of cellulose and BTCA. The free carboxylic groups of the linking agent are able to anchor SiO_2 by electrostatic interaction.

Evaluation of Functional and Physical Properties

The functional and physical properties of coated fabrics were tested as per ASTM and AATCC test methods such as flame retardant (ASTM D 1230), air permeability (ASTM D 737-04), bending rigidity and bending modulus (ASTM D 1388-08) and drape coefficient (ASTM E 21), the results are given in Table 2.

RESULTS AND DISCUSSIONS

Characterization of Silica Nanoparticles

The synthesized silica nanoparticles were characterized by X-ray Diffractometer (XRD) using CuKα (λ = 1.54 Å) as a radiation source. The powder diffraction pattern indicates a broad peak at 2θ = 22°, which reveals the amorphous nature of the silica nanoparticles.[3,6] The XRD patterns of the silica nanoparticles obtained from rice husk is shown in Fig. 1.

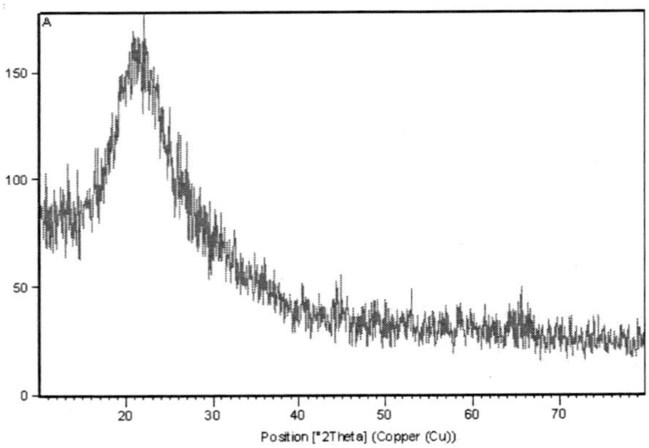

Fig. 1: XRD Pattern of Amorphous Nano Silica Particles

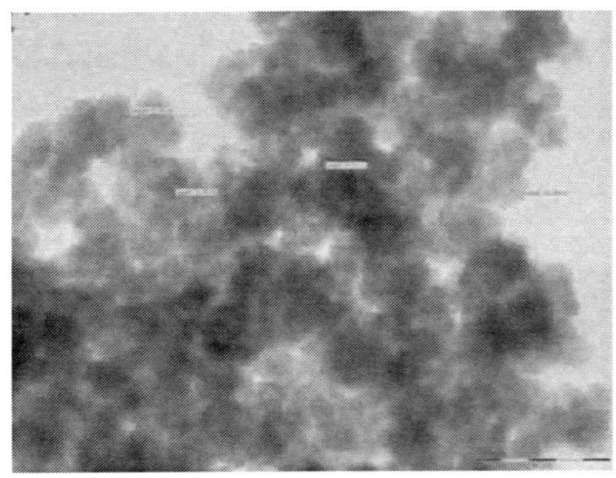

Fig. 2: TEM Pattern of Amorphous Nano Silica Particles

High Resolution Transmission Electron Microscope (HRTEM) was used to characterize the particle size and the morphological structure of nano particles is shown in Fig. 2. It shows that the particle size is around 100 nm.

Thermo gravimetric analysis (TGA) was used to analyze the weight loss of silica nano particles while increasing temperature are shown in Fig. 3. It shows that the weight loss of silica nano particles at 700°C is 5.5%. Hence, it withstands heat for long duration which is important for flame retardant materials.

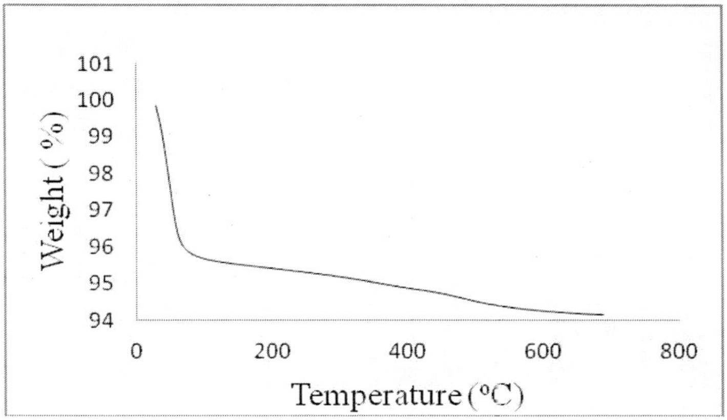

Fig. 3: TGA of Amorphous Nano Silica Particles

Effect of Process Variables on Flame Retardancy

Figure 4a & 4b show that, the increase in the concentration of silica nano particles, the flame retardancy of the fabric is increased due to the higher limiting oxygen value of silica nanoparticles.[6,7,8]

Fig. 4c shows that the initial increase of curing temperature and curing time increases the flame retardancy of the fabric. The increase in curing temperature upto 160°C helps to form cross links between the cellulose and silica nano particles, but further increase in temperature and time leads to degrade the cellulose material as the result of reduction in flame resistance.

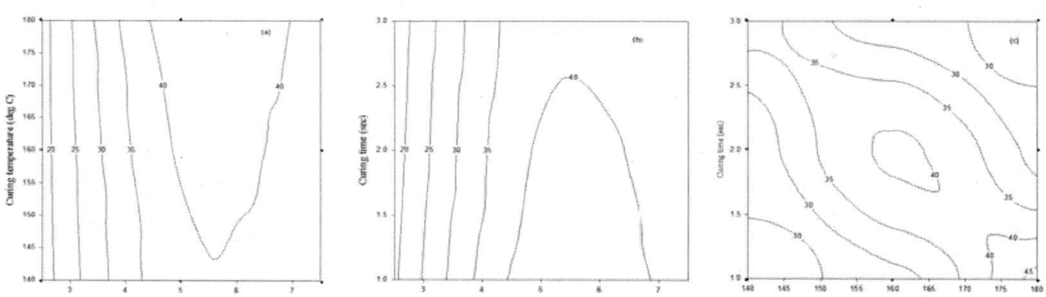

Fig. 4: Effect of Silica Nano Particles, Curing Temperature and Curing Time on Flame Retardancy (a) Silica Nano Particles and Curing Temperature (b) Silica Nano Particles and Curing Time (c) Curing Temperature and Curing Time

Effect of Process Variables on Air Permeability

Figure 5a & 5b show that, with the increase in the concentration of silica nano particles, the air permeability is reduced because cellulose material pore size is arrested but there is no influence on temperature and time.[21]

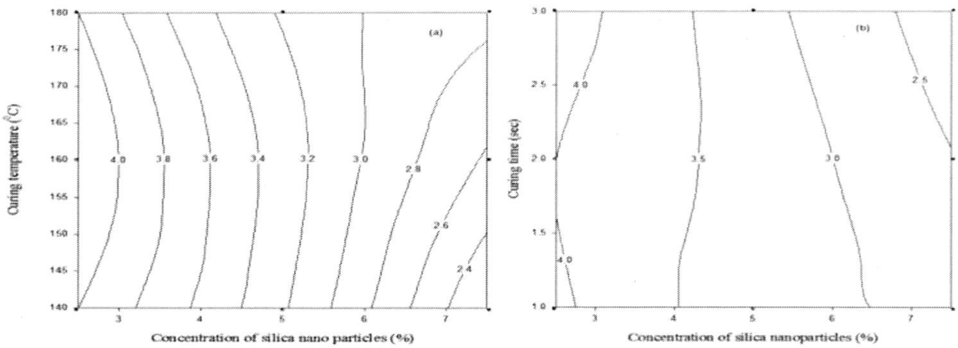

Fig. 5: Effect of Silica Nano Particles, Curing Temperature and Curing Time on Air Permeability (a) Silica Nano Particles and Curing Temperature (b) Silica Nano Particles and Curing Time

Effect of Process Variables on Drape Properties

Figure 6a & 6b show that, the increase in the silica nano particle, the binding action between the yarns happens so the drape ability is reduced, hence the drape co-efficient is increased but at higher curing time, the drape co-efficient is decreased because of material losses the bond between them.

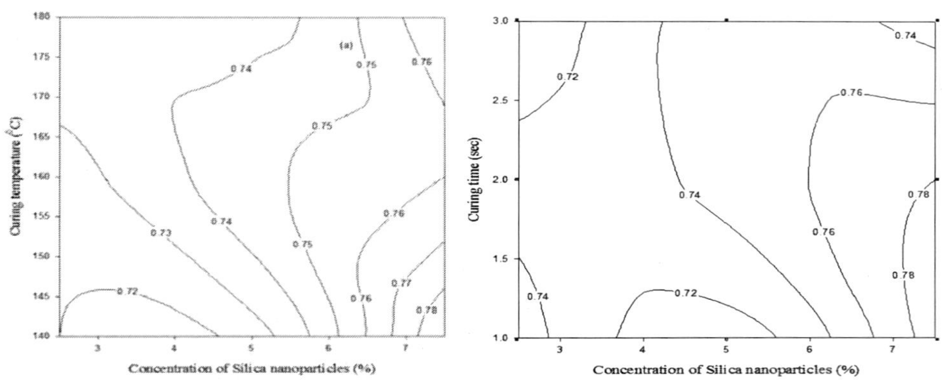

Fig. 6: Effect of Silica Nano Particles, Curing Temperature and Curing Time on Drape Property (a) Silica Nano Particles and Curing Temperature (b) Silica Nano Particles and Curing Time

Effect of Process Variables on Bending Properties

Fig. 7a & 7b show that increasing the concentration of silica nano particles, the material become stiffer because of increase in material weight (gsm) and thickness by coating, so bending modulus is increased. But the increase in curing temperature and curing time, then bending modulus is increased initially decreased.

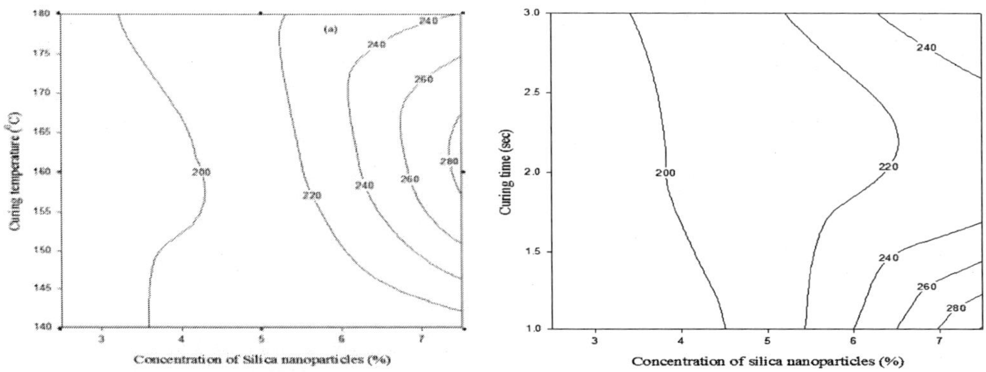

Fig. 7: Effect of Silica Nano Particles, Curing Temperature and Curing time on Bending Property (a) Silica Nano Particles and Curing Temperature (b) Silica Nano Particles and Curing Time

CONCLUSIONS

The silica nano particles were synthesized by thermal degradation method from rice husk and the particles were characterized. The XRD result of 2θ = 22° showed amorphous nature and the particle size was characterized through HR TEM with the size range from 30 to 100 nm. The TGA analysis of silica nano particles showed the reduction in weight loss is 5.5% even at 700°C. The combination of binding of cellulose and trapping of silica nano particles significantly increased the flame retardancy. The increase in curing time and temperature strength of the fabric were reduced. Drape ability, bending modulus and air permeability values were reduced while increasing the process parameters.

REFERENCES

[1] Mehran Alaeea, Pedro Ariasb, Andreas Sjo Dinc & Ake Bergmand, *J Environment International*, 29 (2003) 683.

[2] Jasiorski M, Bakardijeva S, Doroszkiewicz W Brzeziñski S, Malinowska, Marcinkowska D, Ornat M,Strêk W & Maruszewski K, *Materials Science* (2004) 22.

[3] Gobi Nallathambi, Thangavelu Ramachandran, Venkatachalam Rajendran & Rajagoundar Palanivelu, *Materials Research*, 14(4) (2011) 552.

[4] Wong Y.W.H., Yuen C.W.M., Leung M.Y.S., Ku S.K.A., & Lam H.L.I., *J AUTEX*, 6(11) (2006) 1.

[5} Prabir K. Patra, Steven B. Warner, Paul Calvert, Yong K Kim, Qinguo Fan & Sabit Adanur, *NTC Project*, 2002-2008.

[6] Biplab K Deka & Tarun K Maji, *Polymer Engineering & Science* (2012) 1.

[7] Periadurai T, Vijayakumar C.T. & Balasubramanian M, *J Analytical and Applied Pyrolysis*, 89 (2010) 244.

[8] Reshmi Rekha Devi & Tarun K Maji, *Wood Sci.Technol,* Spring - Verlag, 2012.

[9] Davinder Mittal, *Resonance*, 1997, 64.

[10] Kalapathy U, Proctor A & Shultz J., *J Bioresource Technology*, 73 (2000) 257.

[11] Kalapathy U, Proctor A & Shultz J., *J Bioresource Technology*, 85 (2002) 285.

[12] Nittaya Thuadaij & Apinon Nuntiya & Chiang Mai, *J Science*, 35 (1) (2008) 206.

[13] Klaus Hoffmann, MD; Jan Laperre, PhD; Annelies Avermaete, MD; Peter Altmeyer, MD; Thilo Gambichler, MD, *Arch Dermatol,* 137, Aug 2001.

[14] Yadav A, Virendra Prasad, Kathe A, Sheela Raj, Deepti Yadav, Sundramoorthy C & Vigneshwaran, *Mater. Sci.*, 29(2006) 641.

ALUMINA–ZIRCONIA NANOPARTICLES FOR CORROSION PROTECTION OF MILD STEEL (EN3) SPECIMEN

A. Karthik, S. Arunmetha, P. Manivasakan, R. Yuvakumar and V. Rajendran

Center for Nanoscience and Technology, K.S. Rangasamy College of Technology, Tiruchengode, Tamil Nadu
E-mail: veerajendran@gmail.com

ABSTRACT

In this investigation, α-Al_2O_3-stabilized t-ZrO_2 binary nanoparticles were produced at high temperature through eco-friendly approach using spray pyrolysis technique. The size of the produced nanoparticles was 50 nm with surface area of 60 m^2g^{-1}. The production of binary powder was investigated with respect to size and morphology, crystalline phase, and compositions using the particle size analysis, scanning electron microscopy, X-ray fluorescence spectroscopy, X-ray diffraction, and transmission electron microscopy. A multilayer thermal barrier coating of nanostructured Al_2O_3–ZrO_2 embedded silicate was applied to mild steel (EN3) specimen using spin-coating technique. The Al_2O_3-stabilized ZrO_2 with silicate matrix facilitated the formation of zirconium silicate nanostructured surface-protective coating on EN3 specimen. The Al_2O_3–ZrO_2/SiO_2 matrix-based hybrid inorganic coating shows effective thermal barrier for EN3 at 600°C.

INTRODUCTION

Al_2O_3–ZrO_2 nanocomposite is used in industrial applications such as thermal barrier and corrosion resistance due to their thermal shock resistance, room temperature strength, lower thermal conductivity, and thermal diffusivity [1]. The composite materials enhance

V. Rajendran, P. Manivasakan and K.E. Geckeler (eds.)
Advanced Nano Materials for Industrial Applications, pp. 319-326 (2012).

physicochemical properties when compared with individual components [2]. The ZrO_2-based materials are concentrated mainly on the stabilization of tetragonal (t)-ZrO_2 phase by addition of Al_2O_3, Y_2O_2, CeO_2, CaO, and MgO [3-4]. To prevent the oxidation, delamination, cracking, spallation, and surface deterioration in high-temperature atmosphere, the nanostructured multilayer coating is applied to various alloy substrates using Al_2O_3 and ZrO_2. Generally, in multilayer coating, each layer functions as heat dissipater across the coating. During this process, thermal conductivity tends to decrease linearly with increase in the number of structural layers [5]. Moreover, in comparison to ZrO_2, Al_2O_3–ZrO_2 nanocomposite has been used extensively in thermal barrier coatings (TBCs). This is because of the Al_2O_3 content as it functions effectively against the high-temperature oxidation with respect to high hardness, low porosity, and also improves the performance of composite [6]. In addition, Al_2O_3 enhances the high-temperature corrosion prevention rate of metal substrate due to its low oxygen diffusivity [7]. The strength and toughness of Al_2O_3–ZrO_2 composite increase due to the stress-induced transformation of t-ZrO_2 to m-ZrO_2 and mechanism of microcrack [8]. It is clear from the studies carried out so far that the volume of m-ZrO_2 decreases with increase in the Al_2O_3 content in Al_2O_3–ZrO_2 matrix. In addition, the resultant Al_2O_3 reduces the size of ZrO_2 and increases the phase volume of t-ZrO_2 in the matrix [9]. It is evident that the Al_2O_3–ZrO_2 nanocomposite shows increased toughening behavior than that shown by ZrO_2. Highly toughened ZrO_2-based ceramics (Al_2O_3–ZrO_2 nanocomposite) are required for industrial coating applications. The processing of sol–gel coating is one of the most widely used structural coatings that can modify the surface properties of substrates without altering the nature of materials at low temperature [10]. The deposition of sol–gel based Al_2O_3-stabilized ZrO_2 coating provides better protection over working mild and stainless-steel specimen in harsh environments [11].

This work reports on the mass production of high surface area Al_2O_3–ZrO_2 nanoparticles with free-flowing structure. The binary nanoparticles with spherical morphology and high surface area, and quite reduced grain size, were produced in large quantity. The nano-Al_2O_3–ZrO_2 embedded silicate sol coating was applied over the specimen, and the thermal stability of coating was evaluated through fire treatment.

EXPERIMENTAL PROCEDURE

Alumina-stabilized zirconia binary nanoparticles (50 wt% Al_2O_3/50 wt% ZrO_2) were prepared by spray pyrolysis technique, using precursors such as aluminum nitrate ($Al(NO_3)_3 \times 9H_2O$; 98.5%; Merck GR, India) and zirconium oxynitrate ($ZrO(NO_3)_2 \times 2H_2O$; Loba Chemie, India). The metal nitrate precursor solution was prepared by mixing $Al(NO_3)_3$ and $ZrO(NO_3)_2$ in 500 ml double-distilled water and stirring it vigorously until homogeneous solution was obtained at an ambient temperature. Then, the nitrate precursor was allowed to decompose within the hot-air zone at 400°C. The spray pyrolysis is a thermal decomposition-dependent process that breaks the atomized droplets (nanoentities) of the nitrate precursor. It can be used to produce large quantities of Al_2O_3–ZrO_2 binary spherical nanoparticles with high surface area. The homogeneous solution of mixed metal nitrate precursor, which was fed through a two-way nozzle into the reaction chamber with the help of a feed pump (of 2–3 rpm), was used as the starting phase. Then, compressed air was sprayed for atomization of the precursor solution.

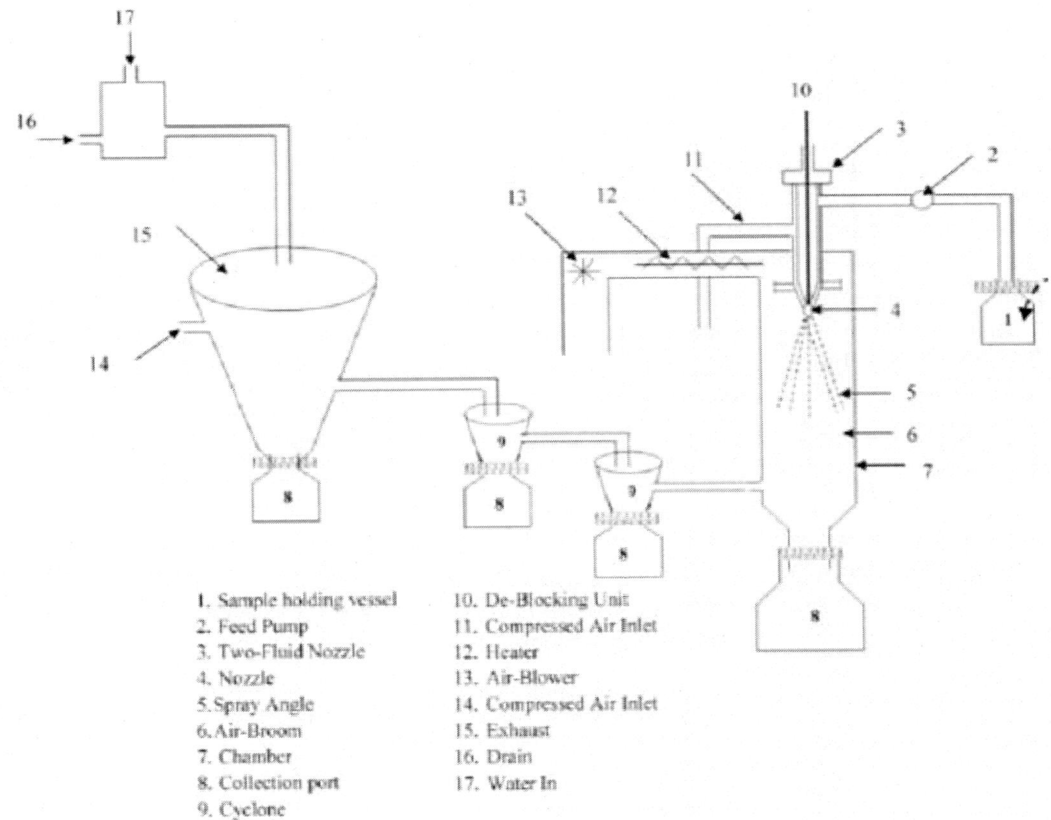

Fig. 1: Schematic Diagram of Spray Pyrolysis Experimental Set-up

The hot air was injected into the reaction chamber followed by spraying the precursor in the hot zone of the reaction chamber. Atomized droplets of the precursor were decomposed at 400°C to generate nanoparticles with free-flowing structure. The complete process of spray pyrolysis is described in our previous study [12]. The primary experimental setup for spray pyrolysis is described in Fig. 1.

CHARACTERIZATION STUDIES

The peak positions and phase structure of Al_2O_3–ZrO_2 binary particles were determined by X-ray diffractometer (X'Pert Pro; PANalytical, the Netherlands) using CuKα as a radiation source (λ = 1.54060 Å). The diffractometer was operated at 40 kV (30 mA) with an angle (2θ) of 10°–80° at a scanning rate of 10° sec^{-1}. The Fourier transform infrared (FTIR) spectra of binary particles were identified by an infrared spectrometer (Spectrum 100; PerkinElmer, USA) in the range of 400–4000 cm^{-1} using the KBr pellet method (90 wt% KBr). The purity and stoichiometry of Al_2O_3–ZrO_2 powder were determined using chemical composition analysis (X-ray fluorescence (XRF); XGT-2700; Horiba, Japan). The particle size distribution (PSD) of prepared samples was determined using dynamic light-scattering principle (Nanophox; Sympatec, Germany) in the range of 1–1000 nm with a scattering angle of 90°. Surface area of binary nanoparticles was measured using the surface area analyzer (BET Autosorb AS-1MP; Quantachrome, USA). The microstructure and morphology of Al_2O_3–ZrO_2 particles were examined using scanning electron microscopy (SEM; JSM-6390LV; JEOL, Japan). Grain size and microstructures of binary nanoparticles were examined by transmission electron microscopy (TEM; CM200; Philips, USA) operated at a potential of 120 kV. Further, the surface features of specimen and compositions of coating were studied using SEM.

Silica sol was prepared using 2.5 ml tetraethyl orthosilicate (TEOS; 99%; Merck), HNO_3 (69%), and ethanol. TEOS (2.5 ml, 0.1 M) and ethanol (90 ml) were mixed thoroughly for 10 min and then 5 ml polyethylene glycol was slowly added to reduce the solvent (ethanol) evaporation. Further, 2.5 ml HNO_3 was drop-wise added with continuous stirring for 20 min and was maintained for 1 h. Then, 0.25 g as-prepared Al_2O_3–ZrO_2 binary nanoparticles (fired and decomposed at 400°C) were incorporated in silicate sol under sonication for 30 min to get homogeneous mixture. The prepared homogeneous nano-Al_2O_3–ZrO_2 embedded silicate sol was deposited on the surface of the mild steel (EN3) specimen using spin-coating technique (SM 108 BT; Sawatec, Germany). The sol was slowly (drop-by-drop) loaded up to 30 sec in air while maintaining the spin speed at 3000 rpm [10] under vacuum pressure of -0.6 bar. The coated specimen was allowed to dry in air at ambient temperature for 1 h to remove its surface hydration. Further, the coated specimen was exposed to heat treatment for 30 min at 400°C and was allowed to cool to promote the intersplat fine creep that enhances the densification of coating. To deposit relatively dense and crack-free coating, we individually applied six-layer coating (layer by layer) on both sides of mild steel (EN3). The deposition of each layer was stabilized through the thermal treatment that improves the quality of coatings.

The XRD pattern of produced Al_2O_3-ZrO_2 binary particles calcination at different temperatures for 3 h is shown in Fig. 2. The obtained result shows that at 600°C, the powder exhibits an amorphous nature. It shows that the characteristics of as-prepared Al_2O_3-ZrO_2 amorphous binary particles. However, the powder sintered at 800 and 1000°C reveals the strong crystalline phase of t-ZrO_2 and α-Al_2O_3 with an average grain size of 6 and 12 nm respectively. The existence of controlled grain growth at high temperature is due to surface enthalpy reduction of zirconia by influence of alumina.The controlled grain growth at high temperature is due to surface enthalpy reduction of zirconia by alumina [13]. FTIR spectra of Al_2O_3–ZrO_2 binary particles at different temperatures are shown in Fig.2. However, the O–H stretching vibration and bending frequency of ZrO_2 molecule appeared at 1637 and 1386 cm^{-1} [14]. In addition, the characteristic active mode shows peak at 950–808 and 526–513cm^{-1}correspond to vibration spectra of Al–O and Zr–O, respectively [14,15].

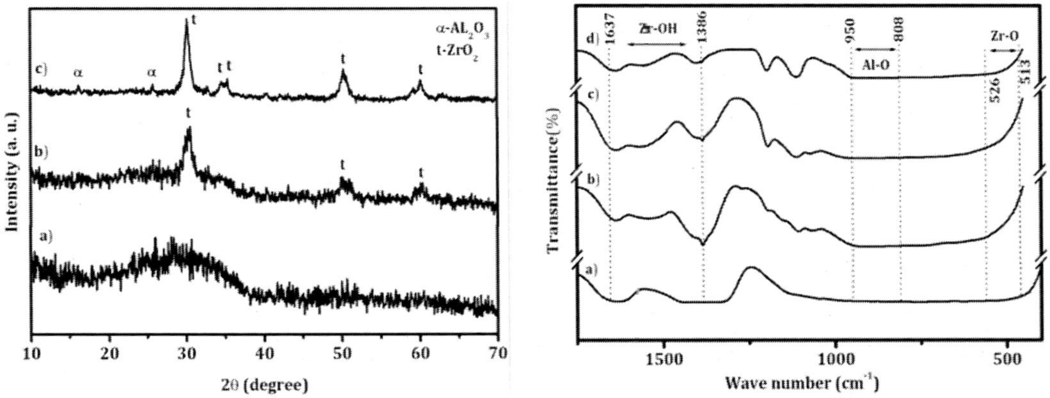

(a) XRD (b) FTIR

Fig. 2: Al_2O_3–ZrO_2 Binary Particles: (a) XRD Pattern, and (b) FTIR Spectrum

Figure 3(a) shows the PSD of as-prepared Al_2O_3–ZrO_2 nanoparticles. It can be seen from PSD analysis that the mean diameter of particle is 50 nm. The surface area of Al_2O_3–ZrO_2 as-prepared powder is 60 m^2g^{-1}. The XRF spectra of as-prepared Al_2O_3–ZrO_2 binary particles are shown in Fig. 3(b). The chemical composition of as-prepared binary particles (at 400°C) consists of 55 wt% Al_2O_3 and 45 wt% ZrO_2, which is in agreement to an experiment precursor ratio (50 wt% Al_2O_3/50 wt% ZrO_2). The microstructure and surface morphology of as-prepared Al_2O_3–ZrO_2 binary particles are shown in Fig. 3(c). The surface image seems to be spherical and uniformly distributed porous nature. It can be seen that clusters of particles are closely packed to form weaker agglomeration. TEM micrograph of $Al_2O_3ZrO_2$ binary particles is shown in Fig. 3(d). It shows uniform distribution of nanoparticles with order of 50 nm.

The microstructure of TBC at 600°C for 6 h is shown in Fig. 4. After heat treatment, the uncoated specimen surface (Fig. 4(a)) is shuttered with an increased roughness. Fig. 4(b) shows six-layer coating of nano-Al_2O_3–ZrO_2 filled hybrid silicate on EN3 mild steel after heat treatment at 600°C for 6 h. The SEM image (Fig. 4(b)) shows that the nanostructure

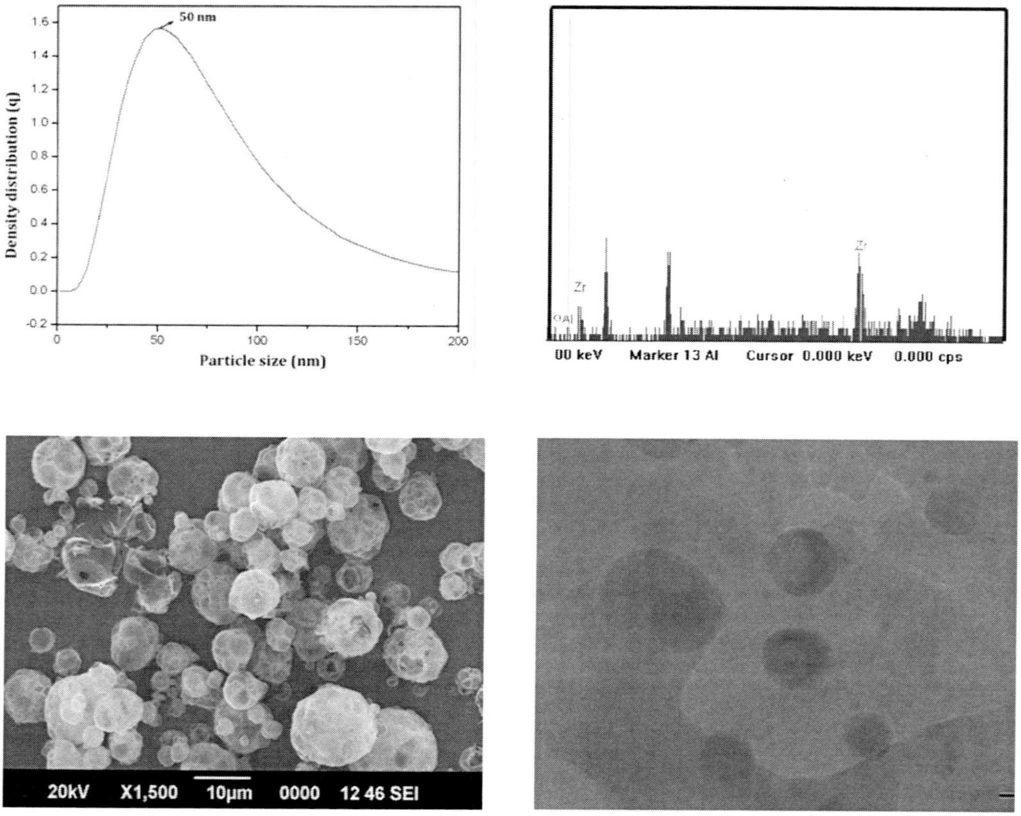

Fig. 3: Al_2O_3–ZrO_2 Binary Particles: (a) Particle Size Analyser, (b) XRF-spectrum (c) Micrograph of SEM and (d) TEM

coating is structurally uniform on the specimen. The filling of Al_2O_3–ZrO_2 nanoparticles in silicate features a smooth-surface nanostructured coating covered with superior adhesion to EN3 base materials. The thermal barrier of Al_2O_3–ZrO_2 filled silicate coating indicates the better uniformity throughout the surface of EN3 specimen. The superimposed nanostructure coating provides a better thermal-insulating barrier and effectively suppresses the corrosion-oxidizing species on the specimen at high temperature with minimum surface deterioration. Thus, it is clear from this study (Fig. 4(b)) that the high-temperature TBC is not linked with volume shrinkage and spallation. Moreover, the complex Al_2O_3–ZrO_2 filled silica sol matrix is used as thermal barrier in high-temperature applications.

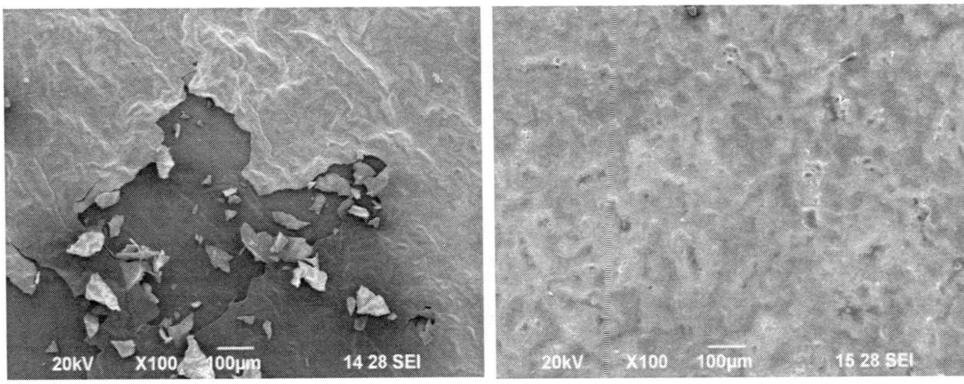

Fig. 4: TBC on Mild steel Specimen fired at 600°C for 6 h: (a) Uncoated, and (b) Six Layer Coated

CONCLUSIONS

α-Al_2O_3-stabilized t-ZrO_2 binary nanoparticles are produced by hot-air spray pyrolysis using metal nitrate precursors. The toughening effect of t-ZrO_2 is achieved at high temperature with critical crystallite size less than 12 nm. This work identified the novel process for large-scale production of high surface area (60 m^2g^{-1}) α-Al_2O_3-t-ZrO_2 nanoparticles. Particles produced by this process have the capability of free-flowing structure which is used as nanofiller for silicate-based inorganic–inorganic hybrid coating. The nanostructured Al_2O_3–ZrO_2 filled silicate multilayer coating is achieved on the mild steel (EN3) specimen using spin-coating technique. After high-temperature exposure, the Al_2O_3–ZrO_2 embedded silicate coating shows better compatibility with improved surface damage resistance as compared with that of the uncoated specimen. The reinforced nanostructured hybrid TBC is observed as a better high-temperature barrier by means of their morphological and structural analyses. This study shows that the nano-Al_2O_3–ZrO_2 filled silicate coating is best suited to obtain the required TBC properties to protect the mild steel (EN3) specimen against high temperatures.

REFERENCES

[1] Zhang J., Kobayashi A., *Vacuum* 2009, 83, 1, 92–97.

[2] Heidly L., Pulgarin C., Garrido L.B., Albano M.P., *Ceram. Int.* 2012, 38, 3, 1843–1849.

[3] Zukerman I., Zhitomirsky V.N., Beit-Ya'akov G., Boxman R.L., Raveh A., Kim S.K., *J. Mater. Sci.* 2010, 45, 23, 6379–6388.

[4] Tuan W.H., Chen R.Z., Wang T.C., Cheng C.H., Kuo P.S., *J. Eur. Ceram. Soc.* 2002, 22, 16, 2827–2833.

[5] Jang B.K., *Surf. Coat. Technol.* 2008, 202, 8, 1568–1573.

[6] Shanmugavelayutham G., Kobayashi A., *Mater. Chem. Phys.* 2007, 103, 2-3, 283–289.

[7] Limarga A.M., Widjajab S., Yipc T.H., *Surf. Coat. Technol.* 2005, 197, 1, 93–102.

[8] Sarkar D., Adak S., Chu M.C., Cho S.J., Mitra N.K., *Ceram. Int.* 2007, 33, 2, 255–261.

[9] Grabis J., Steins I., Rasmane D., Krumina A., Berzins M., *Proc. Estonian Acad. Sci. Eng.* 2006, 12, 4, 349–357.

[10] Sakamoto H., Qiu J., Makishima A., *Sci. Technol. Adv. Mater.* 2003, 4, 1, 69–76.

[11] Nazeri A., Qadri S.B., *Surf. Coat. Technol.* 1996, 86–87, 166–169.

[12] Rajendran V., Manivasakan P., Saravanakumar B., Rauta P.R., Shau B.B., Panda B.K., Renukadevi P., Kolandaivel P., *CiiT Int. J. Autom. Auton. Syst.* 2009, DOI: AA052009005.

[13] Beitollahi A., Hosseini-Bay H., Sarpoolaki H., *J. Mater. Sci.-Mater. Electron.* 2010, 21, 2, 130–136.

[14] Sarkar D., Mohapatra D., Ray S., Bhattacharyya S., Adak S., Mitra N., *Ceram. Int.* 2007, 33, 7, 1275–1282.

[15] Chandradass J., Jun B., Bae D.-s., *J. Non-Cryst. Solids* 2008, 354, 26, 3085–3087.

STRUCTURAL AND TEMPERATURE DEPENDENT ON OPTICAL PROPERTIES OF NANO SCALE $BaTiO_3$ THIN FILMS PREPARED FROM $BaTiO_3$ NANOPARTICLES SYNTHESISED BY WET CHEMICAL METHOD

R. SENGODAN, B. CHANDAR SHEKAR[1] AND S. SATHISH[1]

Department of Physics, Kumaraguru College of Technology, Coimbatore, Tamil Nadu.
[2]Departments of Physics, Kongunadu Arts and Science College, G.N-Mills, Coimbatore, Tamil Nadu
E-mail: sengodan78@gmail.com

ABSTRACT

$BaTiO_3$ nanoparticles prepared by wet chemical method were thermally evaporated onto a well cleaned glass substrate under a vacuum of 2 x 10^{-5} Torr, using a Hindu High vacuum coating unit. Nano thin films of $BaTiO_3$ were prepared on well cleaned glass plates for the first time by vacuum evaporation method. The thickness of the coated films was measured by using quartz crystal thickness monitor. Optical properties of coated films in the spectral range 300–2500 nm were carried out using a UV-VIS-NIR Spectrophotometer. The band gaps of the films were found to be 3.95eV to 3.85eV. It was found to be direct allowed transition and decrease little with the increase of thickness and annealed temperature. The extinction coefficient and absorption coefficient decrease with increase of film thickness and annealed temperature. Energy dispersive Spectrum (EDS) were used to identify the composition of the nano powders and films. X- ray diffraction (XRD) spectrum indicated that the powder has tetragonal structure whereas the film was predominantly amorphous. Surface morphology of the powders and coated films were studied by scanning

V. Rajendran, P. Manivasakan and K.E. Geckeler (eds.)
Advanced Nano Materials for Industrial Applications, pp. 327-336 (2012).

electron microscope (SEM). No pits and pin holes were found in the surface of the film. The nano–scale, impurity–free defect–free microstructure was observed on the surface of the powder.

INTRODUCTION

Barium Titanate (BT) has been one of the best known and widely used materials for electric ceramics due to its excellent dielectric, piezoelectric, pyroelectric and ferroelectric properties [1–3]. $BaTiO_3$ is mainly used in multilayer ceramic capacitors (MLCCs), sensors and actuators, electro-optic devices, thermistors, integral capacitors in printed circuit boards (PCBs), temperature–humidity gas sensors, memory applications, etc. However, it is difficult to develop a dielectric layer of less than 10 nm with large capacitance, which is a major requirement for MLCC miniaturization and electronic/microelectronic devices. Therefore, in order to get further compactness is strongly needed to synthesize BT powders with particle sizes smaller than 100nm and with high dielectric constant. The controllable preparation of crystalline BT particles and thin films has been extensively investigated since BT dielectric and ferroelectric properties strongly depend on its microstructure (crystallinity and grain size) [4]. Indeed, it is known that the improved BT microstructure has a direct impact on the increasing of electronic devices performance. The traditional preparation methods of perovskite-type powders are performed by solid state reactions at high temperatures, usually higher than 1000°C, and under specific pressure and pH conditions [5]. Several techniques are available for synthesizing fine BT powders; these include sol–gel [6], microwave [7], hydrothermal methods and sputtering [8]. In recent years some new methods have been introduced to make porous $BaTiO_3$ too. These methods include the development of soft-chemistry routes to produce nano particles or specially shaped materials, such as one dimensional nanowire [9] and the introduction of biosynthesis methods to prepare $BaTiO_3$ nano powders [10]. Among these synthetic methods, wet chemical methods are of great interest, because they are safe and environmentally friendly. Thin films of $BaTiO_3$ could be prepared by different technique like r.f sputtering, pulsed laser ablation and metal- organic chemical vapour deposition [11] etc. Despite that several technique have been explored to deposit thin film of $BaTiO_3$ less attention has been devoted to thermal evaporation.

In the present study, we have reported about the applying of thermal evaporation technique for nano scale thin film preparation of $BaTiO_3$ nanopowder synthesised by using wet chemical method. Microstructure, surface morphology and optical properties of thin film prepared on well cleaned glass substrate have also been investigated.

EXPERIMENTAL

Synthesis of $BaTiO_3$ Nano Powder

$BaTiO_3$ nano powders were synthesized using wet chemical method. The starting materials used were barium chloride ($BaCl_2.2H_2O$), Titanium dioxide (TiO_2) powder and Oxalic acid. A solution of Ba: Ti: Oxalic acid having mole ratio1: 1: 1 was stirred and evaporated at 70°C till a clear, viscous resin was obtained and then dried at 110°C for 20 hours. The precursor formed was heated at 900° C for 2 hours to form $BaTiO_3$ nano powder.

$BaTiO_3$ thin Film Preparation

Thin films were thermally grown onto well cleaned glass substrate under a vacuum of 2×10^{-5} torr, using a Hind High vacuum coating unit. The growth rate and thickness were measured during growth using a quartz crystal oscillator thickness monitor. The growth rate was adjusted to be as low as 1 Å/sec to avoid the differential evaporation of elements of the alloy.

Characteristics of $BaTiO_3$ nano Powder and $BaTiO_3$ Thin Films

Energy Dispersive X-ray micro analyser (EDS) were used for the identification of the chemical composition. The XRD patterns of the resulting products were obtained from X-ray powder diffraction with CuKα radiation. The micrograph of $BaTiO_3$ was examined by direct observation via scanning electron microscope (SEM). The optical transmission measurements were made in the UV and visible regions at near normal incidence for all the film samples deposited on glass substrates. In the investigation, JASCO-UV/VISIBLE spectrophotometer (model UVIDEC 610, Japan) was used. It is also provided with CRT (Cathode Ray Tube) monitor having graphic display at high resolution.

RESULT AND DISCUSSION

EDS Analysis

Figure 1 and 2. shows the EDS spectrum of as prepared $BaTiO_3$ nano powder and as deposited $BaTiO_3$nano scale thin film on glass plate respectively. Elemental composition analysis indicated the presence of Ba, Ti and O in the synthesized $BaTiO_3$ nano powder as well as in the $BaTiO_3$ nano scale thin films.

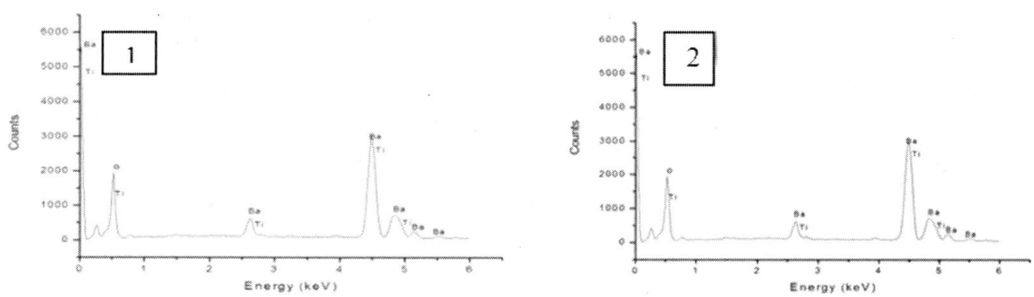

Fig. 1 and 2: EDS Spectrum of $BaTiO_3$ Nano Powder and Nano Film

X–Ray Diffraction Analysis

Organic acid precursor method was used to prepare $BaTiO_3$ powder in nanosize. Figure 3 shows XRD patterns of precursor sample produced from solution mixture of barium chloride, titanium dioxide and oxalic acid mole ratio 1:1:1 evaporated at 80^0C and thermally treated at 900^0C to a well crystalline barium titanate $BaTiO_3$phase (JCPDS Card # 79-2263). The crystallite size of $BaTiO_3$ powder was determined from the X-ray diffraction data using the Debye-Scherrer formula

$$d = k\lambda/\beta\cos\theta$$

where d is the crystallite size, k = 0.9 is a correction factor to account for particle shapes, β is the full width half maximum (FWHM) of the most intense diffraction peaks, λ is the wave length of Cu target = 1.5406 Å, and θ is the Bragg angle. The average crystallite size of the prepared $BaTiO_3$ powder was 55 nm. The system of the nano powder has Tetragonal nature. In the XRD plot, the peak positions from $2\theta = 44^0$ to 46^0 play important role in determining the tetragonal nature of the system, the peak splitting was found clearly in the (002) plane which shows its high degree of tetragonality [12].

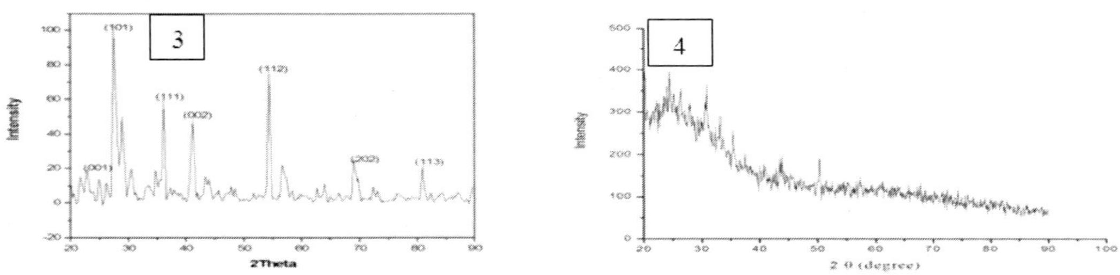

Fig. 3 and 4: XRD Spectrums of $BaTIO_3$ Nano Powder and Nano film of Thickness 200 nm

A typical X-ray diffraction (XRD) pattern of $BaTiO_3$ film thickness (200 nm) is as shown in the Fig. 4. The absence of sharp peaks in the spectrum indicated the predominantly amorphous nature of the film

SEM Analysis

Typical scanning electron micrographs of $BaTiO_3$ nano powder and $BaTiO_3$ film of 200 nm are as shown on Figs. 5 and 6. The micrograph indicates that the average grain size of $BaTiO_3$ nano powder varies from 90.68 nm to 133 nm. In the case of $BaTiO_3$ thin film the average grain size varies from 195 nm to 1270 nm. The grain size of vacuum evaporated $BaTiO_3$ thin films are relatively large compared to as prepared $BaTiO_3$ nano powder. Beside, good distribution and complete surface coverage of glass by vacuum evaporated particles conformed that dense and void free deposit is produced by thermal evaporation of Barium titanate nano powder. No pits, pin holes and dendritic features are found on the surface in SEM micrographs.

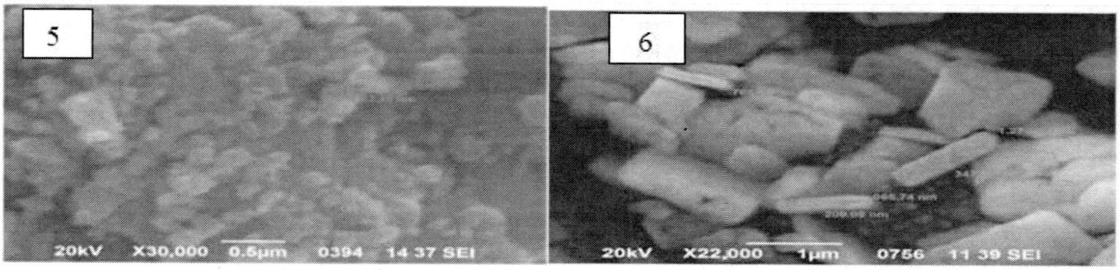

Fig. 5 and 6: SEM Micrograph of $BaTiO_3$ Nano Powder and Nano Film of Thickness 200 nm

Optical Properties

Figure 7 and 8 shows the transmittance spectra for the different thickness of as deposited and annealed nano thin films. It was found that the transmittance of the film decreases as the increase of film thickness, which leads to decrease a decrease in light scattering losses [13]. High transmittance in the higher wavelength region and a sharp absorption edge were observed in the films. The average transmittance value for the $BaTiO_3$ films is over 90%. The transmittance falls steeply with decreasing wavelength. It can be positively concluded that the material is of highly absorbing nature in the visible region. On the other hand with increase of annealing temperature no change in the transmittance it is reveals that minimum imperfections in the films.

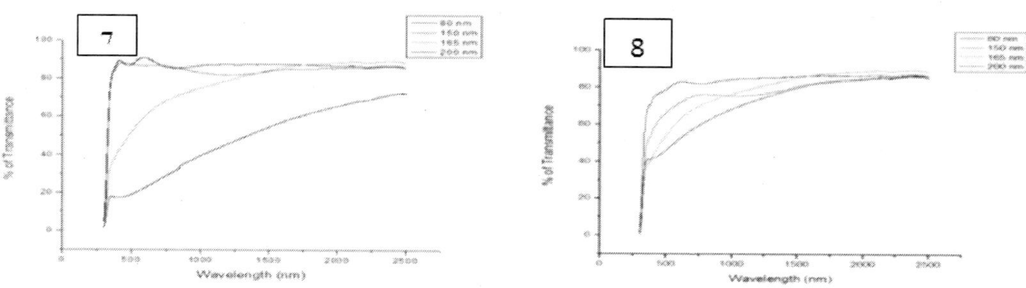

Fig. 7 and 8: Transmittance Spectra of $BaTiO_3$ Films as Deposited and Annealed at 373K

The Absorption co-efficient (α) can be calculated from the relation,

$$\alpha = 4\pi k/\lambda$$

where k is the Extinction coefficient of the films and λ wavelength of the incident light radiation and the Extinction co-efficient *(k)* can be calculated from the relation

$$k = 2.303\ \lambda \log (1/T_0)/4\pi d$$

Figure 9 and 10 represents the variation of extinction coefficient (k) with wavelength for various thicknesses as deposited and annealed films. It is seen that in all the films the extinction coefficient increases as the wavelength decreases [14]. In addition, the extinction coefficient is fall at lower wavelength. The value of extinction coefficient decreases with increasing film thickness and decrease with increase of annealing temperature, which may be due to the improvement in the crystallinity with the increase in film thickness leading to the minimum imperfections

Figurs 11 and 12 represents the variation of absorption coefficient (α) with wavelength for various thicknesses of as deposited and annealed $BaTiO_3$ nano thin films. The absorption coefficient increases as wavelength decreases [15]. In addition, the absorption coefficient decreases with increase in film thickness and decrease with increase of annealing temperature, which is an expected inverse nature of the transmittance spectra. The absorption Coefficient is one of the very important factors for ferroelectric materials in the IR detectors and arrays, which make use of their pyroelectric properties [16].

By determining α from that transmittance curves. For direct allowed transition the absorption coefficient and photon energy are related by expression.

$$\alpha h\nu = A\,(h\nu - E_g)^n$$

where A is a constant and E_g the optical band gap energy. The $(\alpha h\nu)^2$ is plotted against the photon energy ($h\nu$) for $BaTiO_3$ nano thin films of various thickness and the energy gap for these films are obtained by extrapolating the linear straight line portion of the curve to the energy axis. All the plots show straight line portions supporting the interpretation of direct band gap for all the films. From the plot, it is concluded that the optical transition in these films is direct and allowed. The variation of $(\alpha h\nu)^2$ versus photon energy ($h\nu$) for various thickness of as deposited and annealed nano thin films is shown in Figure. 13, 14, 15 and 16 and 17, 18, 19 and 20 respectively.

It is observed that the band gap (E_g) decreases with increase in thickness and annealing temperature. It is good agreement with the earlier investigations on $BaTiO_3$ films [13]. The decreases in the band gap may be due to many reasons. It may be due to the increase of grain size, a decrease in strain, dislocation density and an increase of lattice constant. The estimated optical band gap values of both the as deposited and annealed films are shown in Table 1.

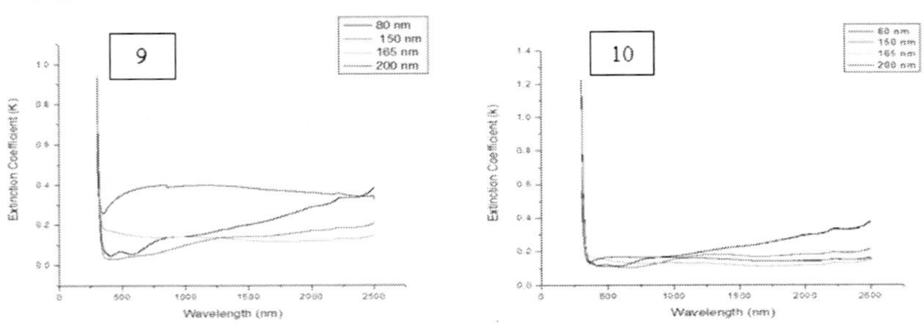

Fig. 9 and 10: Variation of Extinction Co-efficient (k) with Wavelength for $BaTiO_3$ Nano Film at Different Thickness as Deposited and Annealed at 373K for 2 Hours

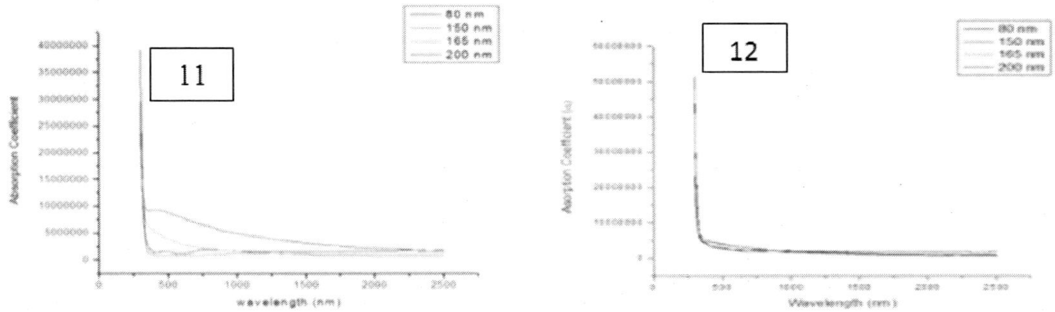

Fig. 11 and 12: Variation of Absorption Co-efficient (α) with Wavelength for $BaTiO_3$ Film at Different Thickness as Deposited and Annealed at 373K for 2 Hours

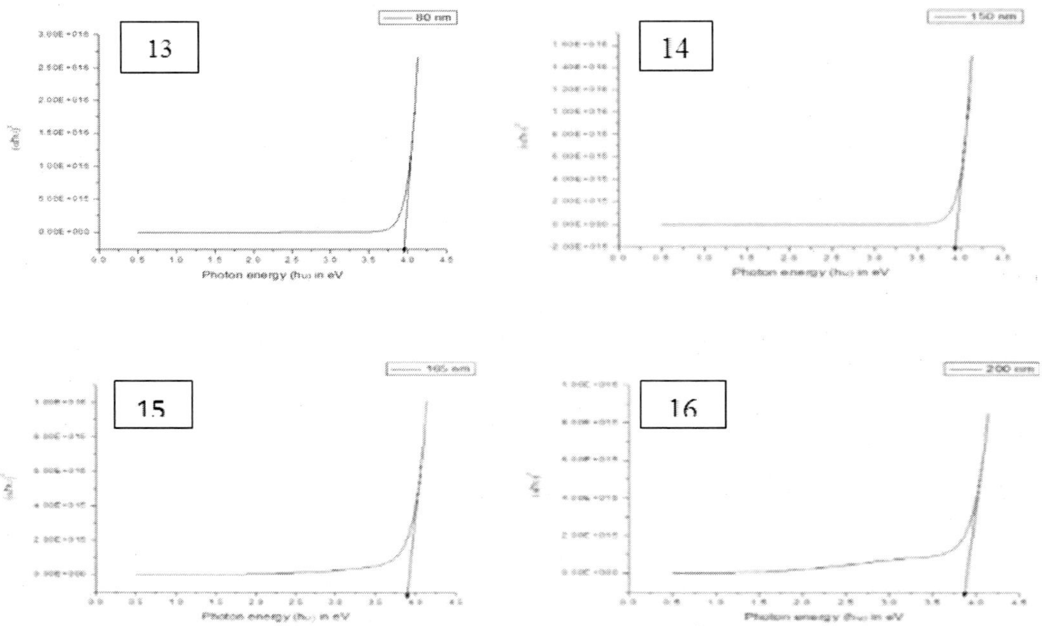

Fig. 13, 14, 15 and 16: The Variation of $(\alpha h\nu)^2$ Versus Photon Energy ($h\nu$) for Various Thicknesses of Deposited Nano thin Films

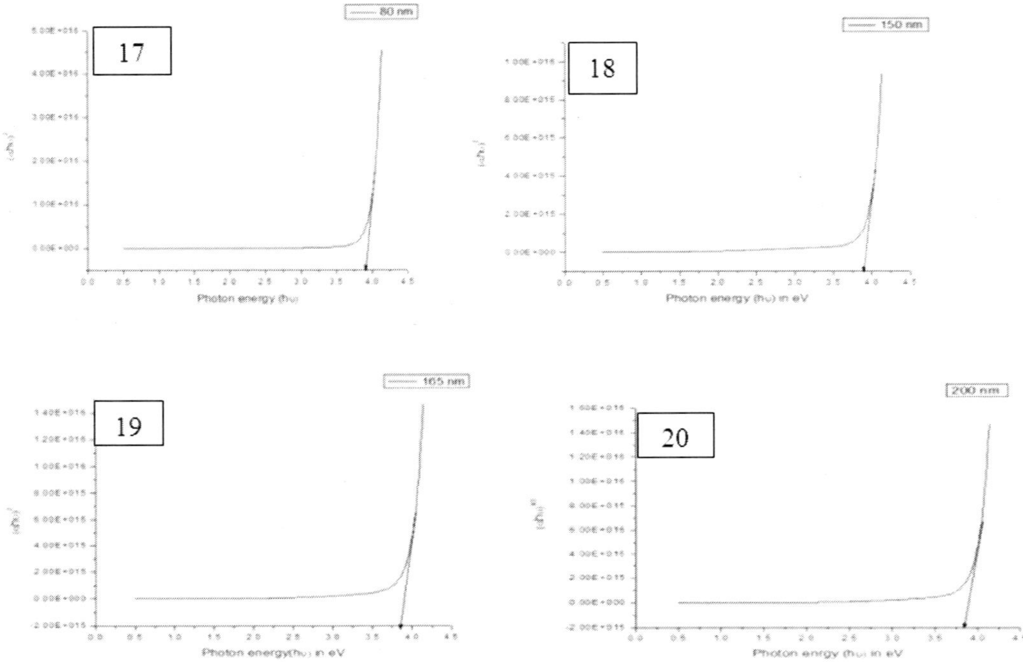

Fig. 17, 18, 19 and 20: The Variation of $(\alpha h\nu)^2$ Versus Photon Energy (hν) for Various Thicknesses of Annealed at 373K for 2 Hour Nano Thin Films

Table. 1: Thickness Dependence on Optical Band Gap

Thickness (nm)	Band gap (eV)	
	As deposited	[illegible]
80	3.95	3.91
150	3.93	3.89
165	3.89	3.88
200	3.87	3.84

CONCLUSION

We have successfully synthesized nao powder of $BaTiO_3$ by low cost wet chemical method using commercially available chemicals such as oxalic acid, TiO_2 and $BaCl_2$. Thin films of few hundred nanometre thickness was prepared on well cleaned glass plate for the first time using thermal evaporation method. The composition of $BaTiO_3$ nanopowder and the $BaTiO_3$ film were confirmed by EDX analysis. XRD indicated the powder has Tetragonal structure and the film predominantly amorphous nature. The presence of small peaks indicated the presence of crystallites of very low dimension in the amorphous region. No pits, pin holes and dendritic features are found on the surface in SEM micrographs. From the transmission spectra, the transmission is found to decrease with increase of film thickness and annealed temperature.The absorption co-efficient and extinction coefficient of the films were determined from optical transmission study which was decreases with increase in thickness and annealed temperature. The optical transition in $BaTiO_3$ films was found to be direct allowed. A decreasing trend in the optical band gap energy was observed with increase in thickness and annealed temperature. The observed smooth surface without any pinholes, pits, impurity - free defect – free microstructure was observed on the SEM micrograph.

REFERENCES

1] N. Setter, D. Damjanovic, L. Eng, G. Fox, S. Gevorgian, S. Hong, A. Kingon, H. Kohlstedt, N.Y. Park, G.B. Stephenson, I. Stolitchnov, A.K. Taganstev, D.V. Taylor and T. Yamada, S. Streiffer, Ferroelectric thin films: Review of materials, properties, and applications *J. Appl. Phys.* 100 (2006) 051606–051652.

2. J. Xu, J. Zhai and X. Yao, Structure and dielectric nonlinear characteristics of $BaTiO_3$ thin films prepared by low temperature process *J. Alloys Compd.* 467 (2009) 567–571.

3. J. Lott, C. Xia, L. Kosnosky, C. Weder J. Shan, Terahertz Photonic Crystals Based on Barium Titanate/Polymer Nanocomposites, *Adv. Mater.* 20 (2008) 3649– 3653.

4. T. Hoshina, S. Wada, Y. Kuroiwa and T. Tsurumi, Composite structure and size effect of barium titanate nanoparticles, *Appl. Phys. Lett.* 93 (2008), 192914–192916.

5. C.-Y. Chang, C.-Y.Huang, Y.-C.Wu, C.-Y.Su, C.-L. Huang, Barium Titanate Nano crystals Prepared under Microwave, *J. Alloys Compd.* 495 (2010) 108–112.

6. S. Yoon, S. Baik, M.G. Kim, N. Shin and I. Kim, Synthesis of Tetragonal Barium Titanate Nanoparticles Via – Hydroxide Sol-Precipitation: Effect of Water Additio, *J. Am. Ceram. Soc.* 90 (2007), 311– 314.

7. K. Sadhana, T. Krishnaveni, K. Praveena, S. Bharadwaj and S. Murthy, Microwave sintering of Nano barium titanate, *Scripta Mater.* 59 (2008), 495–498.

8. H. Hayashi, T. Noguchi, N.M. Islam, Y. Hakuta, Y. Imai and N. Ueno, Hydrothermal synthesis of organic hybrid $BaTiO_3$ nanoparticles using a supercritical continuous flow reaction system, *J. Cryst. Growth* 312 (2010), 1968–1972.

9. C. Fu, W. Cai, L. Zhou, H. Chen and Z. Liu, Appl. Surf. Sci. Synthesis and characterization barium titanate nanoparticles by soft-chemistry routes 255 (2009) 9444–9446.

10. A.K. Jhaa and K. Prasad, Ferroelectric $BaTiO_3$ nanoparticles: biosynthesis and characterization, *Colloids Surf.* B 75 (2010), 330–334.

11. Yoon. S.G, Lee. J. and Safari.A, Sol-gel processing and properties of cerium doped Barium Strontium Titanate thin films Integrated Ferroelectrics, 7 (1995) 329.
12. Sundararajan Tirumalai, Bala Sivanandha Prabu Shanmugavel, Microwave Assisted Synthesis and Characterization of $BaTiO_3$ nanoparticles for Multi Layered Ceramic Capacitor Applications, *Journal of Microwave Power and Electromagnetic Energy*. 45 (3) (2011), 121-127
13. Haizhong Guo, Lifeng Liu, Zhenghao Chen, Shuo Ding, Hubin Lu, Kui- Juan Jin, Yueliangzhou and Bolin Cheng, Structural and Optical Properties of $BaTiO_3$ Ultrathin films, *Europhys. Lett*; 73(1), (2006) 110–115.
14. Z. G. Hu , G.S. Wang, Z.M. Huang, J. H. Chu, Structre–related infrared optical properties of $BaTiO_3$ thin films grown onPt/Ti/SiO_2/Si substrates, *Journal of Physics and Chemistry of Solids* 64 (2003) 2445- 2450.
15. Jinbao Xu .Jiwei Zhai and Xi Yao, .JianqiangXue .Zhiming Huang dielectric and optical properties of $BaTiO_3$ thin films prepared by low-temperature process, *J. Sol-Gel Sci Techn 42* (2007), 209- 212.
16. Liu AY, Meng X.J., Xue J.Q., Sun J.L., Chen J. and Chu J.H., Electrical and optical properties of Pb $(Mg_{1/3}Nb_{2/3})$ O_3–$PbTiO_3$ thin films prepared by chemical solution deposition *Appl. Phys. Lett* 87(2005), 072903.

SYNTHESIS AND CHARACTERIZATION OF POLYMER-SUPPORTED BEADS IMMOBILIZED WITH SILVER AND GOLD NANOPARTICLES: A NEW HETEROGENEOUS NANOPARTICLE CATALYSTS FOR REDUCTION OF CRYSTAL VIOLET DYE

EAGAMBARAM MURUGAN, PARAMASIVAM SHANMUGAM AND S. NATHIYA

Department of Physical Chemistry, School of Chemical Sciences, University of Madras, Maraimalai Campus, Guindy, Chennai, Tamil Nadu
E-mail: dr.e.murugan @gmail.com

ABSTRACT

This study describes the surface grafted 2-acryloxyethyltrimethyl ammonium chloride(PAC) onto the poly(styrene) co (vinylbenzyl chloride) beads (PS-PVBC) through atom transfer radical polymerization (ATRP) and thus obtained insoluble matrix, viz., PS-PVBC-g-PAC beads. This PS-PVBC-g-PAC as a common matrix and to which AgNPs and AuNPs were immobilized individually by simple chemical reduction method and thus obtained the corresponding heterogeneous nanoparticle catalysts, viz., PS-PVBC-g-PAC-AgNPs and PS-PVBC-g-PAC-AuNPs respectively. These two catalysts were characterized by UV-Vis, FTIR, SEM, and EDAX analyses. The catalytic efficiency of these two catalysts was examined through the reduction of Crystal violet (CV) using $NaBH_4$ as a reducing agent and observed that the catalyst effectively accelerate reaction, through kinetics was also performed.

V. Rajendran, P. Manivasakan and K.E. Geckeler (eds.)
Advanced Nano Materials for Industrial Applications, pp. 337-346 (2012).

INTRODUCTION

It is well known that functionalized metal nanoparticles are of great interest in terms of their potential applications in the size-dependent biomedical, electronic, optical materials, magnetic, and catalytic properties [1]. The field of nanocatalysis has under gone an explosive growth during the past decade, both in homogeneous and heterogeneous catalysis. The first report of polymer stabilized metal NPs was appeared in 1998, [2] and was indicated as an effective homogeneous catalyst. However, these homogeneous catalysts were not obtained with high yield during synthesis in aqueous medium. The recovery of such type of homogeneous catalyst has normally been carried out by dialysis, magnetic separation, precipitation, and ultra-centrifugation methods. But, again these methods are practically a tedious one and possess with a lot limitations. More importantly, applications of homogenous nanoparticle catalyst for organic reactions are not considered as economy one owing to their non-recovery. With a view to avoid all the demerits experienced in the homogeneous nanoparticles catalyst, researcher introduced an alternative catalysts, viz., "heterogeneous nanoparticle catalyst". This catalyst was generally prepared by immobilization of metal nanoparticles on to the insoluble polymer-support. Therefore, the research on nanoparticle stabilized onto functionalized polymer beads was started earlier and now it is growing very fast. Today, nanoparticle immobilized/stabilized polymer supported heterogeneous catalyst has variety of applications due to the combination of physical and chemical properties of the encapsulated nanoparticles (NPs) with the tunable solubility and surface reactivity of the polymer templates. The heterogeneous catalyst preparation involves, immobilization of transition metal NPs on various solid matrices is considered to be common method. Many solid support materials, viz., inorganic (silica [5], alumina [6], metal oxides [7]) and organic (carbon [4], polymer [8-9]) have been utilized for the immobilization of metal NPs. However, heterogeneous nanoparticles catalysts developed from an amphiphilic polymer matrixes are received a greater attention owing to its effective catalytic activity. Particularly substitutions of desired functional groups onto the insoluble polymer beads are necessary for stabilization of metal nanoparticles. Even though, many functional modification are performed already in the preparation of cross-linked polymer-supported beads, but still this conventional type of insoluble beads are not popular among the catalytic industries. Alternatively, scientist has been prepared insoluble polymer-supported beads by surface initiator atom transfer radical polymerization technique [10-11]. This techniques provides the controlled chain growth, well defined block, graft copolymer and also formation of homopolymer is avoided [12]. Hence, the SI-ATRP technique is considered to be very efficient one for grafting of surface functional groups on solid surfaces such as nanomaterial, fibres, films, membranes and so on[13]. Among the known nanoparticles, silver (Ag) and gold (Au) are widely studied because of their characteristic optical, spectroscopic and catalytic properties. They have been extensively exploited in the area of photography, photo catalyst, biological labelling, and surface enhanced Raman scattering. Further, it is known that most of dyes are more stable and difficult to biodegrade [14]. Therefore, colour removal from waste-water is a major environmental problem. Reactive dyes are hardly eliminated under aerobic conditions and are probably decomposed into carcinogenic aromatic amines under anaerobic conditions [15]. The conventional methods for treating dyes containing waste-waters are coagulation and flocculation [16], oxidation or ozonation [17], and membrane separation [18]. But these technologies do not show significant efficiency or economic advantage. So, in order to avoid this problem a reusable catalyst which can be effectively utilized for degradation of dyes in waste-water is necessary. Therefore, in this study we reported a new bead shaped heterogeneous nanoparticle catalysts, viz., PS-PVBC-g-PAC-AgNPs and PS-PVBC-g-PAC-AuNPs for effective degradation of crystal violet dyes.

MATERIALS AND METHODS

Materials

Styrene (Aldrich), Vinyl benzyl chloride, divinylbenzene (Lancaster), gelatin (SD fine), boric acid (SRL), poly vinyl alcohol (SRL), sodium hydroxide (SRL), sodium nitrite (SRL), AIBN (SRL), silver nitrate, Auric acid (SRL), ethanol (SRL) and sodium borohydride (SD fine), Crystal violet (SRL), sodium borohydride (SD fine) were used as such without further purification. Azobisisobutyronitrile (AIBN), bipyridine and CuBr were supplied from Sigma Chemical. All other chemicals were of analytical grade and were purchased from Merck AG (Darmstadt, Germany). Double distilled water was used for preparation of solutions.

Synthesis of New Polymer-Supported Poly (Vinyl Benzyl Chloride) (PS-PVBC) Beads Grafted with Poly 2-acryloxyethyltrimethylammonium Chloride (PAC)

The insoluble form of crosslinked poly (styrene)-copoly (vinylbenzylchloride) (PS-PVBC) matrix was prepared as per the earlier methods [19]. The resulting PS-PVBC beads was grafted with 2-acryloxyethyltrimethylammonium chloride (PAC) via atom transfer radical polymerization method (ATRP). Initially, PS-PVBC beads (5 g) was swelled in ethanol for 24 hrs in 100 ml round bottom flask, and then the PAC monomer (20 ml, 0.15 mol), CuBr (0.432 g, 3.0 mmol), 1, 10-phenantracene (0.936 g, 6.0 mmol) and dioxane (10ml)] were added to the flask. The suspension was purged with nitrogen for about 10 min and the RB flask (reactor) was sealed in order to maintain the inert atmosphere, and then polymerization reaction was carried out at 65^{0}C for 18 h. Then the resulting beads were filtered and then transferred into a 100 mL RB flask, to which 10% (w/v) of EDTA solution was added and stirred continuously for 24 hrs, so as to remove copper ions from the beads. The obtained beads were filtered and washed sequentially with water and ethanol and dried under vacuum at 60^{0}C for 24h and thus yielded poly(styrene) supported polyvinyl benzyl chloride grafted 2-acryloxyethyltrimethylammonium chloride (PS-PVBC-g-PAC beads). The PS-PVBC-g- PAC beads were characterized with FTIR and SEM, the observed spectra and images were shown in Fig. 2b, Fig. 3a respectively. The beads were also analyzed with EDAX and the observed result was shown in Table 1.

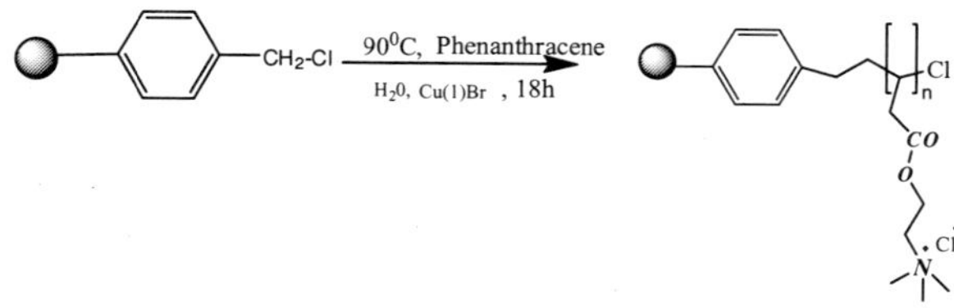

Scheme 1: Synthesis of PS-PVBC-g-PAC

Preparation of Silver and Gold Immobilized Bead-shaped Heterogeneous Nanoparticle Catalysts using PS-VBC-g-PAC AS A Matrix

The immobilization of AgNPs and AuNPs individually onto the PS-VBC-g- PAC polymer matrix was performed using $AgNO_3$ and $HAuCl_4$ as a respective metal precursor (Scheme 2). That is, 500 mg of PS-VBC-g-PAC matrices were taken individually in two different single necked 100 ml round-bottomed flask and swelled in 50 ml of ethanol for overnight. The respective polymer beads in the flask were purged with nitrogen for 1 hr. After 1 hr, $AgNO_3$ (1.5×10^{-4} mol) was added to one flask, then the solution was stirred and refluxed in an oil bath with a magnetic stirrer at 80°C. Care should be taken such a way that the reaction mixture should be protected from light exposure so as to avoid the photochemical reduction of Ag^+. After ten minutes, the colour of the beads was changed in colourless to light yellow. Then, 2 equivalents of $NaBH_4$ were added to the solution, subsequently, after 1 hrs the colour was changed to dark yellow. Again after one hour the flask was refluxed with stirring for 24 hrs and thus produced heterogeneous nanoparticles catalyst, viz., PS-VBC-g-PAC-AgNPs. Similarly, heterogeneous gold nanoparticle catalyst was synthesized by adding $HAuCl_4$ (1.5×10^{-4} mol) solution to the swelled PS-VBC-g-PAC beads, and it was observed that the color of the solution was found to changed from colourless to yellow. Then 2 equivalents of $NaBH_4$ was added to the flask which in turned the yellow colour beads into dark purple and then the beads were refluxed for 24 hrs and finally the reduced forms of AuNPs immobilized PS-VBC-g- PAC heterogeneous nanoparticles catalyst was obtained. The reduced forms of AgNPs and AuNPs immobilized in PS-PVBC-g-PAC-AgNPs and PS-PVBC-g-PAC-AuNPs heterogeneous catalysts were filtered through Buchner funnel, washed three times with ethanol and dried in vacuum oven at 60 °C for 12 hrs. These two catalysts, viz., PS-PVBC-g- PAC-AgNPs and PS-PVBC-g- PAC-AuNPs were characterized with UV-Vis and FTIR the observed spectra were shown in Fig. 1a & 1b and Fig. 2b, 2a & 2c respectively. Similarly, the surface morphology and immobilization of AgNPs and AuNPs onto the respective matrix were studied with SEM and EDAX analyses. The observed SEM images for PS-VBC-g-PAC-AgNPs and PS-VBC-g-PAC-AuNPs were shown in Figs. 3a-3e and observed EDAX result & is given in Table 1.

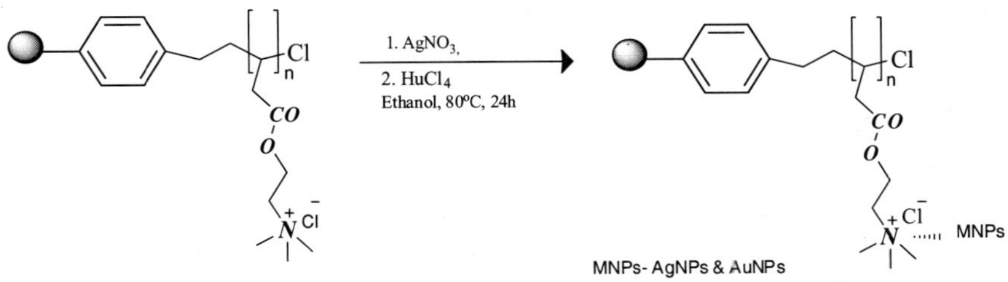

Scheme 2: Synthesis of PS-PVBC-g-PAC-AgNPs and PS-PVBC-g- PAC-AuNPs

Catalysis of PS-PVBC-g-PAC-AgNPs and PS-PVBC-g-PAC-AuNPs in Reduction of Crystal Violet (CV)

The catalytic efficiency of PS-PVBC-g-PAC-AgNPs and PS-PVBC-g-PAC-AuNPs catalysts was examined by the reduction of CV under an identical pseudo-first order condition. The individual kinetic experiments was performed by taking 2 different 25 ml two necked RB flask to which 10 ml of CV (0.2 mM), 10 ml of aqueous $NaBH_4$ (100 mM) and 25 mg of respective catalyst, viz., PS-PVBC-g-PAC-AgNPs and PS-PVBC-g-PAC-AuNPs were added individually. The corresponding reaction medium was agitated with magnetic stirrer at ambient temperature. The progress of the reactions occurred in each container was monitored through UV-Vis spectrophotometer by maintaining the wavelength from 200 to 800 nm. The reaction mixture (3 mL) was withdrawn from the reaction medium at every 5 of minute's interval and the absorption spectrum was recorded quantitatively against the time using UV-Visible Spectrophotometer. The rate constant was calculated using the following formula irrespective of the reaction.

$$kobs = \ln[(A_\infty - A_0)/(A_\infty /0\ A_t)]/t$$

where A_0-initial absorbance, A_t-absorbance at time't', A_∞-absorbance at infinity time.

RESULTS AND DISCUSSION

This study describes the synthesis and characterization of 2 different heterogeneous nanoparticle catalysts, viz., PVBC-g-PAC-AgNPs and PS-PVBC-g-PAC-AuNPs. The catalytic performances of these two catalysts were examined through reduction of CV as a model reaction performed under identical pseudo first order reaction condition.

Characterization of PVBC-g-PAC-AgNPs and PS-PVBC-g-PAC-AuNPs

The reduction/formation of Ag/AuNPs immobilized onto the respective PVBC-g-PAC-AgNPs and PS-PVBC-g-PAC-AuNPs matrixes were inspected through UV-Vis analysis. That is the respective catalyst was swelled with 15 ml of ethanol and then 5 mg of surfactant, viz., cetyltrimethyl ammonium bromide (CTAB) was added to the each catalyst solution. Then the respective catalyst solution was sonicated for 30 minutes and then the obtained filtrate was analyzed with UV. The recorded spectra for PVBC-g-PAC-AgNPs and PS-PVBC-g-PAC-AuNPs were shown in Fig.1a and Fig. 1b respectively. From the Fig. 1a, it is understand that there is clear and sharp peak was noticed at 405 nm and similarly, in Fig. 1b, a sharp peak was abserved at 536 nm. Therefore, the appearance of these two different surface plasma resonance peaks observed at filtrate derived from PVBC-g-PAC-AgNPs and PS-PVBC-g-PAC-AuNPs catalyst has strongly support the formation of AgNPs & AuNPs into the respective catalyst.

The FTIR spectra obtained from PVBC-g-PAC, PVBC-g-PAC-AgNPs and PS-PVBC-g-PAC-AuNPs were shown in Figs. 2b, 2a and 2c respectively. The functionalization PAC on PS-PVBC matrix was established from the appearance of C = 0 (str) and C-N (str) peaks noticed at 1710 cm^{-1} and 1121cm^{-1} (Fig. 2b) respectively and this in turn confirms the functionalization of PAC on PS-PVBC matrixes. Similarly, the formation/immobilization of AgNPs and AuNPs onto the respective matrixes was also ascertained from FT-IR results. That is, in both the catalysts, after individual immobilization of AgNPs and AuNPs, the C-N stretching noticed at 1121 cm^{-1} was shifted to 1126cm^{-1} and 1113cm^{-1} respectively and thus supports the formation/immobilization of AgNPs and AuNPs in the respective matrixes.

SEM analysis was performed both the PS-PVBC-g-PAC-AgNPs and PS-PVBC-g-PAC-AuNPs (Figs. 3b & 3c) catalyst with lower magnification and (Figs. 3d & 3e) with higher magnification and the observed results were compared with the images of PS-PVBC-g-PAC beads (control) (Figs. 3a). From the Figs. 3d & 3e, it is understand that there are white dots noticed on the surface of PS-PVBC-g-PAC matrix which in turn shows the immobilization/ stabilization of AgNPs and AuNPs onto the matrices. Whereas, the control matrix (Figs. 4a) has shown smooth homogeneous surface without any white dots. This observation indicates that the PS-PVBC-g-PAC matrix have shown higher affinity for attraction/immobilizatuin of AuNPs than with AgNPs.

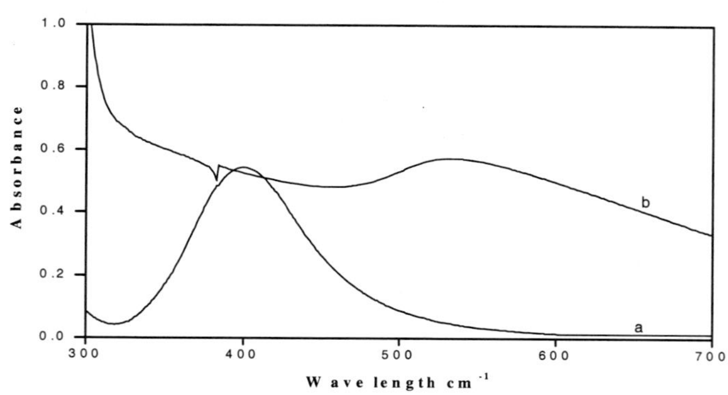

Fig. 1: UV-Vis Spectrum of (a) PS-PVBC-g-PAC-AgNPs & (b) PS-PVBC-g-PAC-AuNPs

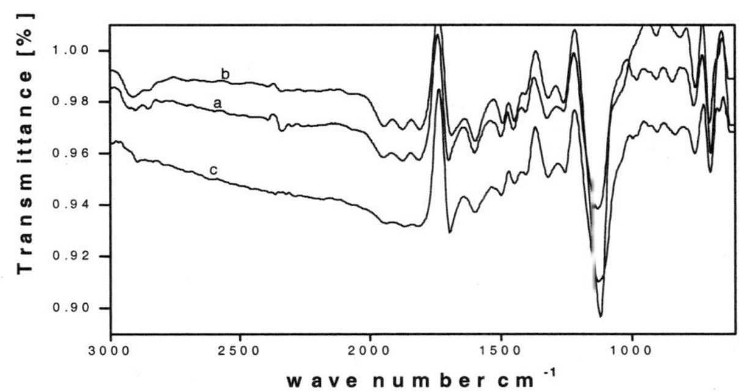

Fig. 2: FTIR Spectrum of (a) PS-PVBC-g-PAC-AgNPs, (b) PS-PVBC-g-PAC (c) PS-PVBC-g-PAC-AuNPs

This is due to the fact, that the standard reduction potential for gold (+1.69) is higher than silver (+0.68) and thus reflecting the presence of rich white dots in the SEM images of PS-PVBC-g-PAC-AuNPs.

The immobilizations of AgNPs and AuNPs onto the PS-PVBC-g-PAC beads were studied extensively through EDAX analysis. The characteristic elements such as C, N, O, Cl, AgNPs, and AuNPs and their quantity (%) available before and after immobilization were given in the Table 1.

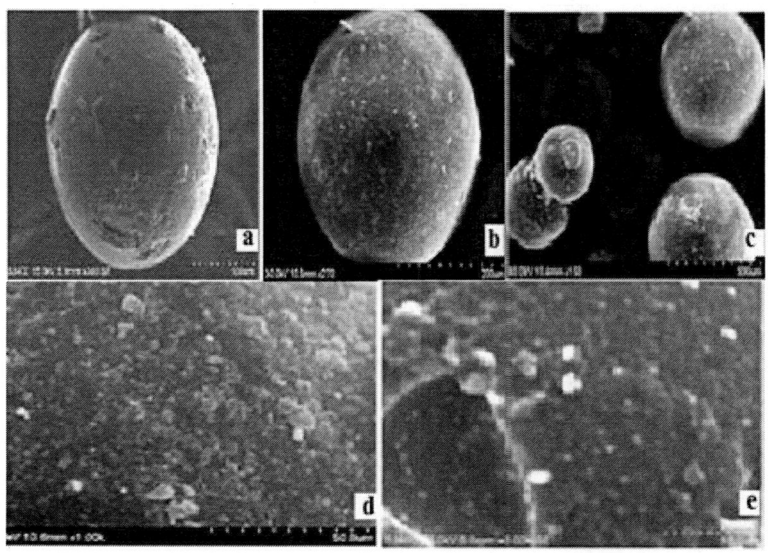

Fig. 3: SEM images of (a) PS-PVS-gPAC (b) PS-PVBC-g-PAC-AgNPs (c) PS-PVBC-g-PAC-AuNPs (d) PS-PVBC-g-PAC-AgNPs (e) PS-PVBC-g-PAC-AuNPs

The weight percentage of AuNPs was higher than that of the AgNPs, which in turn confirms the more amount of formation/immobilization of AuNPs onto the PS-PVBC-g-PAC bead matrix than with AgNPs.

Table 1: Percentage of Elements Determined from EDAX Analysis using Polymer Supported AgNPs and AuNPs

Name of the Compound	*C%*	*N%*	*Cl%*	*O%*	*AgNPs%*	*AuNPs%*
PS-PVBC-g-PAC	65.44	21.09	6.12	7.30	-	-
PS-PVBC-g-PAC-AgNpS	76.53	15.4	5.52	3.20	0.30	-
PS-PVBC-g-PAC-AgNpS	72.14	16.3	4.94	5.86	-	0.34

Comparative Study on the Catalytic Activity of Silver and Gold Immobilized PS-PVBC-g-PAC Beads for the Reduction of Crystal Violet (CV)

The catalytic activity of the PS-PVBC-g-PAC-AgNPs and PS-PVBC-g-PAC-AuNPs were examined through reduction of CV under pseudo first order reaction condition. From the observed UV-Vis spectrum, the characteristic peak observed at 586 nm was found to be gradually decreased after adding the respective catalyst, viz., PS-PVBC-g-PAC-AgNPs and PS-PVBC-g-PAC-AuNPs, i.e., complete reduction of CV was takes place in 90 min and the corresponding UV-Visible spectrum was shown in Fig. 4. The pseudo first order rate constant of the reaction catalyzed by PS-PVBC-g-PAC-AgNPs and PS-PVBC-g-PAC AuNPs were determined to be 16.742×10^4 sec^{-1} and 25.457×10^4 sec^{-1} respectively. The observed rate constants reveal that PS–PVBC-g-PAC–AuNPs catalyst are more superior than PS–PVBC-g-PAC–AgNPs catalyst. Further, the thorough kinetic studies with superior catalyst viz., PS–PVBC-g-PAC–AuNPs was performed by varying the parameters such as [substrate], [catalyst] and [$NaBH_4$] under pseudo first order reaction condition. It was observed that, on increasing the amount of catalyst PS–PVBC-g-PAC–AuNPs, the rate constant was also increases parallelly. This may be attributed to the increase in the number and availability of catalyst nanoparticles with higher catalyst loading. On increasing the [$NaBH_4$], the rate constant was also increased which is due to more adsorption of $NaBH_4$ onto the nanoparticle surfaces, which in turn increases the relay of electrons from BH_4 (donor) to the CV (receptor). On higher [CV], the intensities of penetration of light inside the solution was found to be decreased which in turn decreases the rate constant. The pseudo-first order rate constants were evaluated and plotted against the [substrate], [catalyst] and [$NaBH_4$] and the corresponding graphs are shown in Fig. 5, Fig. 6 and Fig. 7 respectively.

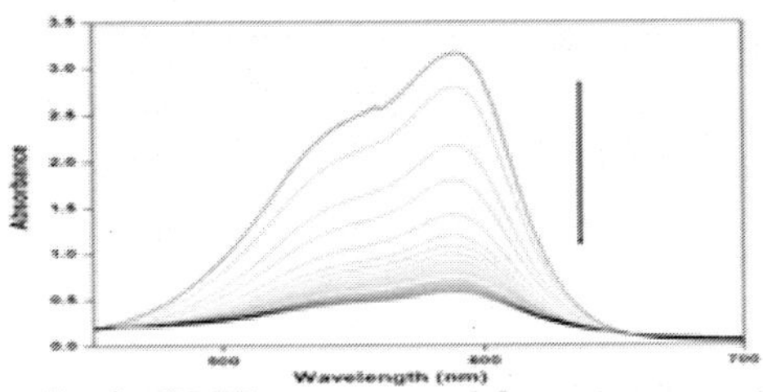

Fig. 4: UV-Vis spectrum of Crystal Violet Day using PS-PVBC-g-PAC-AuNPs

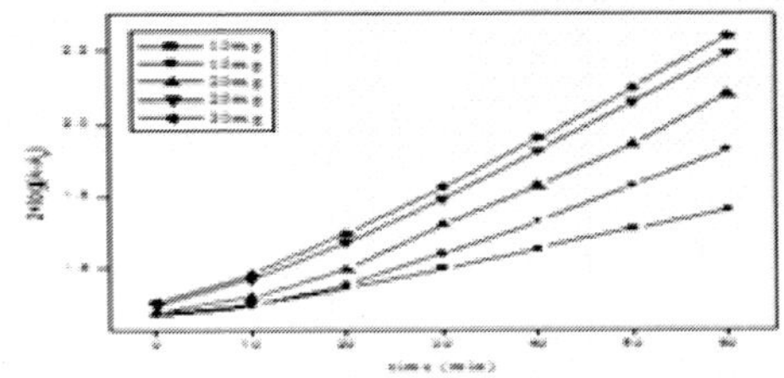

Fig. 5: Effect of Catalyst

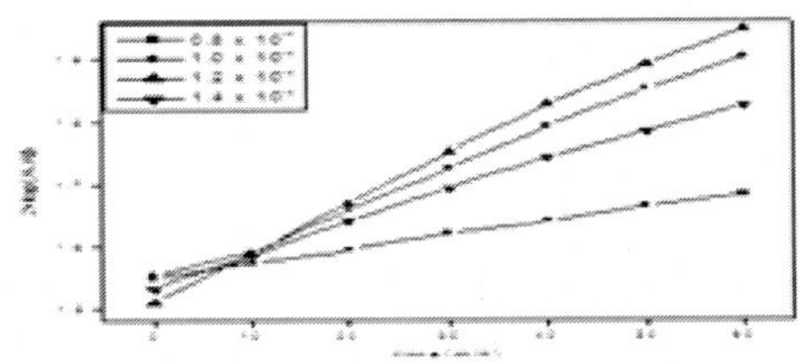

Fig. 6: Effect of $NaBH_4$

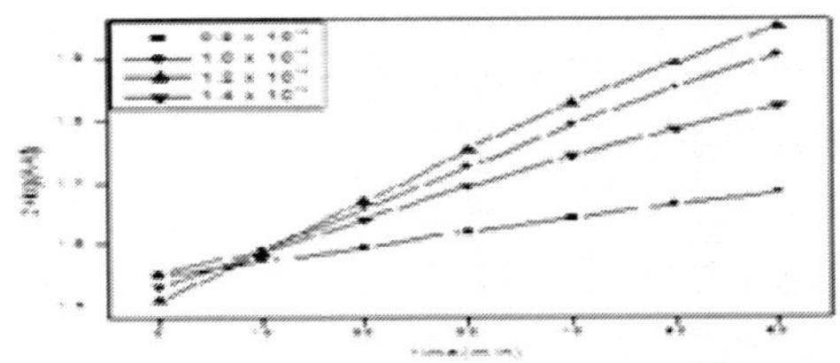

Fig. 7: Effect of Substance

Table 2: Comparative Catalytic Study of PS-PVBC-g-PAC Stabilized NPs Reduction of CV

Through kinetics of CV using Superior catalyst PS-PVBC-g-PACAuNPs

Recycling Efficiency of PS–PVBC-g-PAC–AuNPs through Reduction of CV

The PS–PVBC-g-PAC–AuNPs catalyst are found to effective for in catalysis reduction of CV. That is, at the end of the reaction, the catalyst particles remained active and were separated from the completion of CV reduction. After one course of reaction, the particles were washed thoroughly with distilled water, dried at room temperature and could be recycled a number of times (3) and employed the same for II & III cycles and the observed rate constants are given in Table 2. From the observed results, the rate constant was found to remain constant without any appreciable change and thus confirmed the stability of the catalyst.

CONCLUSION

Two types of new beads shaped heterogeneous nanoparticle catalysts, viz., PS-PVBC-g-PAC –AgNPs and PS-PVBC-g-PAC-AuNPs were prepared by simplified procedure. Initially, PS-PVBC beads was functionalized with PAC via ATRP technique. The obtained insoluble PS-PVBC-g-PAC matrix was used for the immobilization of AgNPs and AuNPs by adopting the simple chemical reduction method. These two types of PS–PVBC-g-PAC–AuNPs and PS–PVBC-g-PAC–AgNPs catalysts were characterized with spectroscopy and microscopy techniques. The catalytic efficiency of these two catalysts was inspected with reduction of CV under pseudo-first order condition. The obtained k_{obs} reveals that PS–PVBC-g-PAC–AuNPs are catalytically superior then with PS–PVBC-g-PAC–AgNPs catalyst. The studies proves that the activity was maintained even upto 3rd cycle irrespective of catalyst.

ACKNOWLEDGEMENTS

The authors thank to the University of Madras for financial support under the scheme URF and NCNSNT, University of Madras for providing characterization facilities.

REFERENCES

[1] J.J. Storhoff, R. Elghanian, R.C. Mucic, C.A. Mirkin and R.L. Letsinger, *J. Am. Chem. Soc.* 120 (1998), 1959.
[2] M. Zhao, L. Sun., R.M. Crooks, *J Am Chem Soc,* 120 (1998), 4877.
[3] Z. Liu, J.Y. Lee, W. Chen, M. Han L. M. Gan, *Langmuir 20 (*2004), 181.
[4] A.G. Boudjahem, S. Monteverdi, M. Mercy, M.M. Bettahar, *J Catal*, 221 (2004), 325.
[5] J. W. Yoo, D.J. Hathcock M.A. El-Sayed, *J Catal*, 214 (2003), 1.
[6] K. Mallick, M.S. Scurrell, *Appl Catal A: Gen* 253 (2003), 527.
[7] K. Suzuki, T. Yumura, M. Mizuguchi, Y. Tanaka, C. W. Chen M.Akashi, *J Appl Polym Sci*, 77 (2000), 2678.
[8] C.W. Chen, M.Q. Chen, T. Serizawa M. Akashi, *Chem. Commun*, (1998), 831.
[9] L. Peng, L. Yushan S. Zhixing, *Ind. Eng. Chem. Res.*45 (2006), 2255.
[10] E. Yavuz, G. Bayramoglu, B.F. Senkal M. Yakup Arica, *Journal of Applied Polymer Science,* 113 (2009), 2661.
[11] D. A. Shipp, J.L. Wang, K. Matjaszewski, *Macromolecules*, 31 (1998), 8005.
[12] R.B. Grubbs, C.J. Hawker, J. Dao, J.M. Frechet, *Angew. Chem., Int. Ed. Engl,* 36 (1997), 270.
[13] P. Liu, *Eur. Polym. J*, 41 (2005), 2693.
[14] S. Seshadri, P.L. Bishop, A.M. Agha, *Waste Management & Research,* 15(1994), 127.
[15] D. Peterson, *In J. Shore*, 1(1990), 32.
[16] T. Panswad S. Wongchaisuwan, *Water Science and Technology,* 18 (1986), 139.
[17] P.K. Malik S.K. Saha, *Separation and Purifÿcation Technology,* 31 (2003), 240.
[18] G. Ciardelli, L. Corsi M. Marucci, *Resources, Conservation and Recycling,* 31 (2000), 189.
[19] T. Balakrishnan Warren T. Ford, *J. of App. Polymer Scienc,* 27 (1982), 133.

DENSITY FUNCTIONAL STUDY OF OZONE ADSORPTION ON SILICON DOPED (8, 0) CARBON NANOTUBES

M.R. SONAWANE, DARSHAN HABALE[1], B.J. NAGARE[2], RITA GHARDE[1] AND R.G. ATRAM

Department of Physics, Ismail Yusuf College, Jogeswari (E), Mumbai
[1]Department of physics, University of Mumbai, Kalina, Santacruz (E), Mumbai
[2]13/002 Om Laxmi narayan C.H.S. Nebula Darshan, Wayle Nagar, Kalyan, Thane
E-mail: smahadev123@gmail.com, bjnagare@gmail.com, gharde.rita@gmail.com

ABSTRACT

We report adsorption of ozone on (i) (8, 0) Pure Carbon Nanotube, and (ii) Silicon doped (8, 0) Carbon Nanotube within the density functional theory calculations based on linear combination of atomic orbital. Particular emphasis is placed on the effects of silicon doping on the adsorption of ozone. The results show that silicon doping keeps the semiconducting character of pure carbon nanotubes with a small reduction in band gap from 0.61 eV to 0.54 eV. The ozone gas molecule is most likely physisorbed with binding energy 0.431eV and minute charge transfer of 0.013 e from pure carbon nanotube, consistent with the strong oxidizing nature of ozone. The binding energy and charge transfer indicate that the ozone adsorption on pure carbon nanotube is gentle and reversible. However, when ozone adsorbs on silicon doped carbon nanotube, a strong chemisorption occurs, leading to relevant structural relaxations and to the formation of a Si-O σ bond with binding energy 3.902 eV and charge transfer of 0.252 e. The band structure and density of states shows that the occupancy of valance band at Fermi energy is completely altered due to ozone adsorption. The charge density analysis also shows formation of sigma bond between silicon and oxygen atoms.

V. Rajendran, P. Manivasakan and K.E. Geckeler (eds.)
Advanced Nano Materials for Industrial Applications, pp. 347-356 (2012).

INTRODUCTION

Over the past 20 years, carbon nanotubes (CNT) [1] have been intensively studied due to their unique geometrical, mechanical [2], chemical [3] and electrical [4] properties. Their electronic properties can vary from semiconducting to metallic depending on its molecular structure. The latest research efforts have aimed to improve both efficiency and selectivity of nanotube. More attention has been given, experimentally as well as theoretically to study electronic, thermal properties and their chemical reactivity [5–7]. These outstanding properties exhibited by carbon nanotubes have been made it suitable for application in nanoelectronics [8], catalysis [9, 10], sensors [11, 12], energy capacitors and hydrogen storage devices [13].

To achieve the enviable results, many research works have been done on the characterization of nanotube, for example, the modiûcation for the properties of nanotube by adsorption of atoms, molecules or clusters on the exterior walls of carbon nanotubes or by insertion of these species into their hollow interior. This offers a promising way in the design of new nanomaterials [14–16]. Furthermore carbon nanotubes were exposed to several different gas molecules such as NO_2, O_2, and NH_3 [17, 18] and observed the change in the electrical resistance. They have also shown that these molecules are weakly adsorbed on pure CNT due to van der Waal interactions. To improve the functionalization, many research groups are now using chemical doping, local defects, radial deformation and metal assistance [19, 20]. The metal assistance is an effective way to functionalize the carbon nanotubes which significantly improves the structural and electronic properties.

Ozone interaction with carbon nanotubes has been studied experimentally [21-23] and theoretically [24, 25]. More recently, it has been observed that ozone can react with the surface defect sites [26]. It can also attack the more reactive end cap, opening it up and making it possible for the physical adsorption of other gases [27]. Ozonolysis can oxidize carbon - carbon double bonds of CNT. It is also observed that ozone exposure induces p-type doping and correspondingly reduces resistance in CNTs. [22, 23]. Theoretical studies have shown that the binding energy of ozone on CNT is near about 0.3 eV depending on geometrical variation [22-25]. Since the binding energy was too small to explain ozone exposure effect on the CNT resistance. It was shown that some defects sites would be responsible for the ozone adsorption and the corresponding resistance change [22]. However, side wall defects are generally hard to detect experimentally. It was noticed that carbon nanotubes becoming gradually thicker could be attributed to a large number of defects accidentally, thus resulting in a diameter increase [28]. All above literature survey indicates that the understanding on the ozone interaction with CNT is not well established till today.

Despite these important contributions, the detailed adsorption mechanism of ozone on the CNTs is still less known and to the best of our knowledge; there is no report on the interaction of ozone molecules on Silicon doped carbon nanotube (Si-CNT). Here we performed extensive DFT calculations and studied the interactions of ozone with the semiconducting CNT and Si-

CNT using electronic properties like binding energy, band structure, density of states and population analysis.

COMPUTATIONAL DETAILS

Calculations were performed using an all-electron linear combination of atomic orbital density functional theory (DFT) approach, implemented in DMol[3] [29, 30] code within the generalized gradient approximation (GGA) and with the Perdew, Burke and Ernzerhof (PBE) correlation functional [31]. The chosen basis set was double numerical plus polarization (DNP), where each basis function was restricted to a global cut-off radius of 4.0 Å. The chosen cut-off value leads to atomic energies with an accuracy of 0.1 eV/atom, allowing calculations without signicant loss of accuracy. The Brilloiouin zone k-point sampling was performed in a $1 \times 1 \times 2$ Monkhorst–Pack mesh [32]. The all calculations were performed using periodic boundary conditions with 64–atoms within the supercell. The tetragonal unit cell of $20 \times 20 \times 8.4$ Å dimension with lateral separation of 13.7 Å between tubes was used. This separation is sufficiently large to avoid interaction between the neighboring cells. The Mulliken population analysis was carried out to predict the charge transfer between molecule and nanotubes.

In this study, we used an (8, 0) zigzag CNT and Si-CNT of diameter 6.26 Å as a model to study the adsorption of ozone molecule. We have examined four different possible adsorption sites such as 'target C atom\Si atom', 'chiral bond', 'axial bond' and 'centriod of hexagon' of CNTs. In all calculations, the nanotubes and isolated ozone molecule were first optimized to minimize their total energies. Further, ozone molecule is kept at four different adsorption sites as mentioned above on each of these individual nanotubes and optimized together to get lowest energy stable structure. The binding energy (E_b) of all ground state structures are calculated

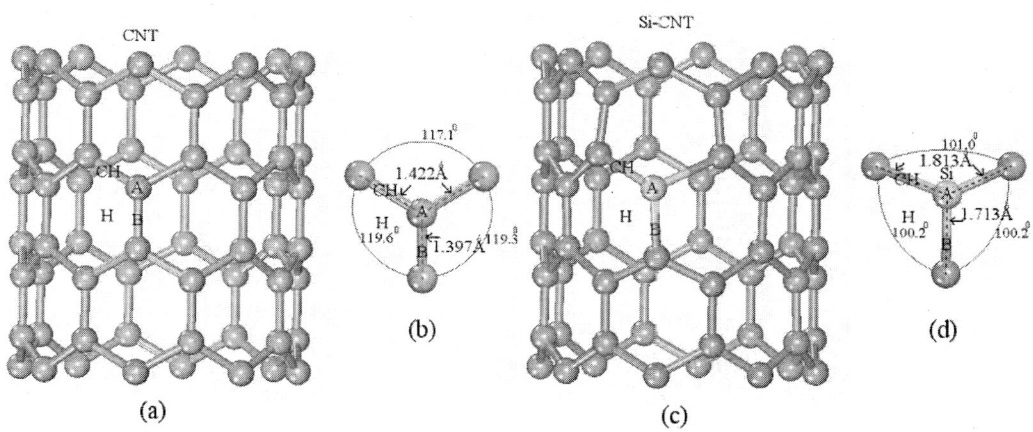

Fig. 1: (a) Ground State Geometry of (8, 0) CNT, (b) Structural Parameters at Target Atom, (c) Ground state geometry of (8, 0) Si-CNT, (d) Structural Parameters at Dopant. The Green and Yellow Color Represent Carbon and Silicon Atom

by $E_b = -[E_{T\,(adsorbent\,+adsorbate)} - E_{T\,(adsorbent)} - E_{T\,(adsorbate)}]$, where $E_{T\,(adsorbent\,+\,adsorbate)}$ denotes the total energy of molecule and CNT system, $E_{T\,(adsorbent)}$ denotes the total energy of CNT and $E_{T\,(adsorbate)}$ is the total energy of isolated gas molecule. To verify the computational accuracy of the model, we have calculated the binding energy of CNTs, density of states and the band gap which are comparable to reported results [33, 34].

RESULTS AND DISCUSSION

Structural Properties of CNT and Si-CNT

The optimized ground state geometry of pure CNT is shown in Fig. 1 (a) along with geometrical parameters in Fig. 1 (b). The bond angles between C–C bonds are fount to be

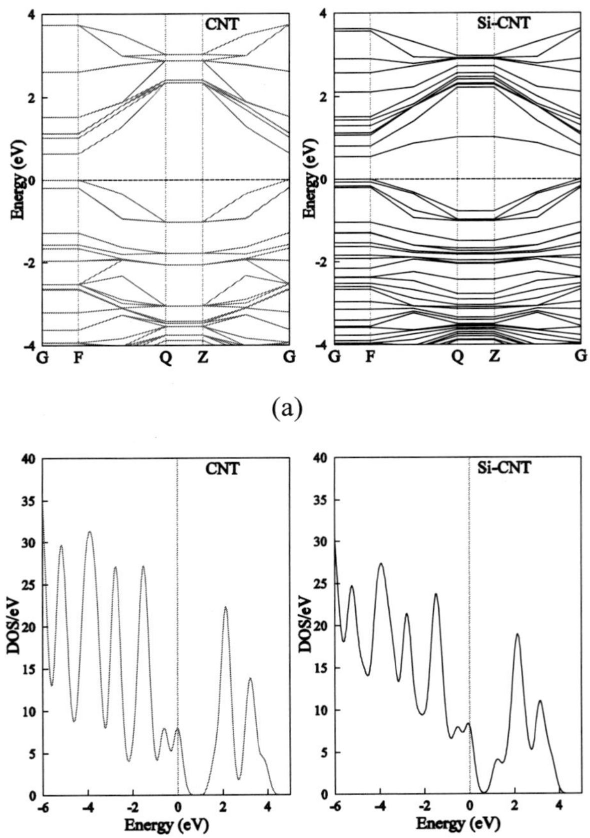

Fig. 2: (a) Band Structure Plot of CNT and Si-CNT, (b) Density of State of CNT and Si-CNT, The Fermi Level is Indicated by Dotted Black Line

less than 120^0 which indicates trigonal structure of bonding and prefers sp^2 hybridization. The stable structure of Si-CNT is shown in Fig. 1(c) where as change structural parameter due to doping of silicon atom is shown in Fig. 1(d). The bond angles between (C-Si-C) bonds for all three combinations are less than 109^0, which indicate tetrahedral structure of bonding and prefer sp^3 hybridization. All structural parameters of our result for CNT and Si-CNT are reported in our previous work found to be in good agreement with reported results [33, 34]. The structural deformation produced by silicon doping results into a signicant change in electronic properties like binding energy, band gap, density of states, charge density and charge transfer.

Electronic Properties of CNT and Si-CNT

The band structure of CNT and Si-CNT are shown in Fig. 2(a). The doping of silicon atom creates an extra unoccupied energy state at 1.26 eV in the conduction band as observed in band-structure plot. The additional state reduces the band gap of Si-CNT to 0.54 eV as compared to band gap of pure CNT. The DOS of CNT and Si-CNT are shown in Fig. 2 (b). The DOS analysis also shows same unoccupied state bellow the conduction band. Such an unoccupied state act as donor like impurity and which plays significant roll of capturing center for atoms or molecules. Further, Mulliken Population analysis shows silicon atom acquires positive charge of magnitude 0.0815 e. This clears that the charge is transferred from the silicon to the vicinity carbon atoms. Such charge polarization at dopant site (silicon atom) acts as an affinity center for chemisorption of the molecules or atom. Thus, the doping of the silicon atom increases the p-type conductivity of the CNT and which improves the reactivity of CNT.

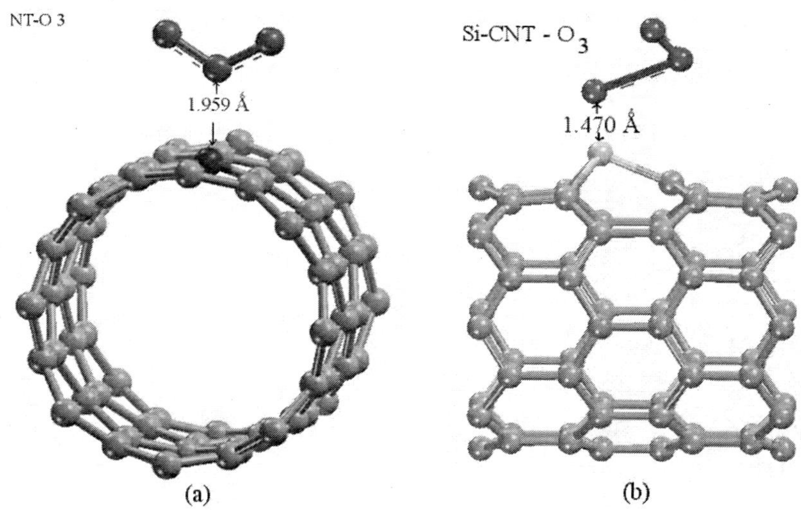

Fig. 3: Ground State Geometries of O_3 Molecule Adsorbed on (a) CNT, and (b) Si-CNT; The Target Carbon Atom is Shcwn by Blue Colour

Structural Properties for Adsorption of Ozone Molecule

To understand the mechanisms of ozone interaction with CNT and Si-CNT, we have calculated structural and electronic properties. Ozone molecule is directly located above the target carbon atom (site-A) of CNT as shown in Fig. 3(a). The ozone molecule is found physisorbed

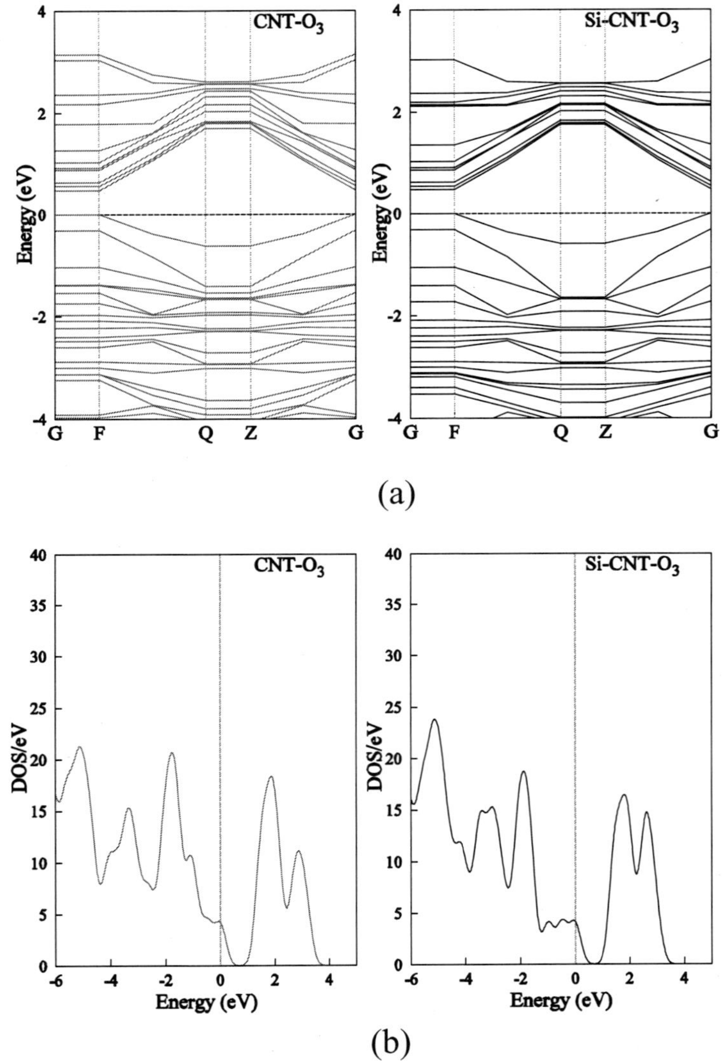

Fig. 4: (a) Band Structure Plot and (b) Density of State of O_3 Molecule Adsorbed on CNT and Si-CNT, The Fermi Level is Indicated by Dotted Black Line

at distance of 1.959 Å. The binding distance also indicates strong physisorption of ozone on CNT. The bond angle between O-O-O oxygen atoms of ozone is decreased from 118^0 to 116^0. The bond distances between O-O are also decreased from 1.279 Å to 1.251 Å. In Si-CNT, ozone molecule is chemisorbed on silicon atom (site A) [see Fig. 3(b)]. The ozone molecule binds to the silicon atom with formation of ó (Si-O) bond and its bond length found to be 1.470 Å. In this case ozone molecule gets dissociated and single oxygen atom gets attached to silicon by ó bond where as oxygen molecule is liberated from ozone. The bond lengths between oxygen atoms are 1.089 Å and 2.871 Å, where as bond angle between O-O-O oxygen atoms of ozone is signicantly decreased to 96.90^0. As shown in Fig. 3(b), the extended bond length between oxygen atoms of ozone confirms dissociation of ozone molecule.

Electronic Properties for Adsorption of Ozone Molecule

The band structure of CNT-O_3 molecule system is shown in Fig. 4(a). The band structure of O_3 adsorption shows that the energy states in valance band near Fermi level are slightly separated as compared to band structure of CNT. Therefore, band gap of CNT-O_3 system is marginally more (0.63 eV) than the band gap (0.607 eV) of pure CNT. In DOS of CNT-O_3 system the occupancy of partially lled states is slightly increased at Fermi level as shown in Fig.4 (b). Further, Mulliken Population analysis shows negligible charge 0.013 e transferred from CNT to O_3 molecule. The isosurface shown in Fig. 5 of CNT did not show overlapping of charge density between the O_3 molecule and CNT. Thus ozone gas molecule is physisorbed with binding energy -0.43 eV and small quantity of charge transfer from carbon nanotube, consistent with the strong oxidizing nature of ozone.

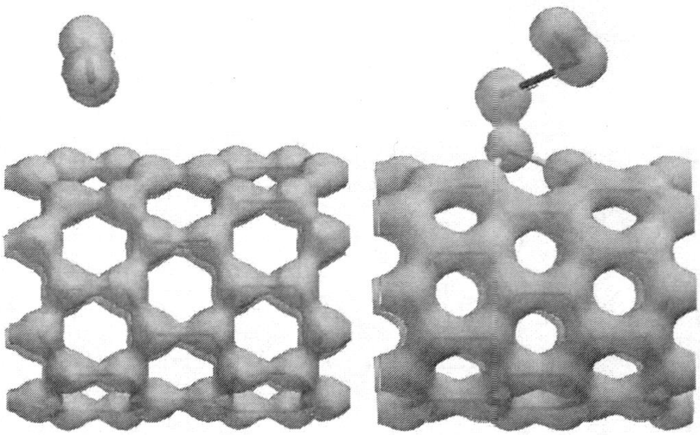

(a) (b)

Fig. 5: Isosurface of the Charge Density of Systems for O_3 Molecule Adsorption on (a) CNT, and (b) Si-CNT

The binding energy, charge transfer and isosurface clearly show the character of the interaction can still be classiûed as physisorption. The adsorption of O_3 on Si-CNT is more pronounced as compared with CNT. However, when ozone adsorbs on Si-CNT, a strong chemisorption occurs, leading to relevant structural relaxations and to the formation of a Si-O ó bond with binding energy - 3.90 eV and charge transfer of 0.252 e. The band structure as shown in Fig. 7 is completely different than band structure of pure Si-CNT. The extra state observed in conduction band of Si-CNT is disappeared and some additional states are observed near Fermi level. These numbers of states are also clearly observed in DOS as shown in Fig. 8. These additional states reduces band gap of Si-CNT-O_3 system to 0.46 eV from 0.54 eV of Si-CNT. The occupancy of valance band at Fermi energy is completely altered due to ozone adsorption. The charge density analysis shows formation of sigma bond between oxygen and silicon atom. Contradictorily the isosurface of charge density shown in Fig. 5 visualize the dissociation of O_3 molecule and formation of silicon-oxide with Si-CNT, where oxygen molecule is liberated from ozone molecule.

CONCLUSIONS

Density functional simulations study has been performed for adsorption of ozone gas molecule on pure carbon nanotube and silicon doped carbon nanotube. Our study has been focused on effect of silicon doping for adsorption of ozone molecule on carbon nanotube. The ozone adsorption on pure CNTs shows a small binding energy and charge transfer indicates the process is most likely of physisorption kind. The extra states in band structure and DOS of Si-CNT plays significant roll in ozone adsorption. The large values of binding energy and charge transfer shows ozone adsorbs on the Si-CNT is more pronounced as compared to CNT. The bond length between two oxygen atoms in ozone molecule has been increased due to adsorption. It is observed that ozone molecule gets dissociated and one oxygen atom gets attached to silicon by ó bond where as oxygen molecule is liberated from ozone. Contradictorily the isosurface of charge density visualize the clear dissociation of O_3 molecule and formation of silicon-oxide with Si-CNT and liberation of oxygen molecule from ozone molecule.

REFERENCES

[1] Ijima S., *Nature* 354, 56, 1991.

[2] Salvetat J.P., Bonard J.M., Thomson N.H., Kulik A.J., Forró L., Benoit W., Zuppiroli L., *Appl. Phys.* A 69, 255, 1999.

[3] Kong J., Franklin N.R., Zhou C., Chapline M.G., Peng S., Cho K., Dai H., *Science*, 287, 622, 2000.

[4] Colbert D.T., Smalley R.E., *Trends Biotechnol.* 17, 46, 1999.

[5] Ajayan P.M., "Nanotubes from carbon," *Chemical Reviews*, 99, 7, 1787–1799, 1999.

[6] Tasis D., Tagmatarchis N., Bianco A., and Prato M., *Chemical Reviews*, 106, 3, 1105–1136, 2006.

[7] Karousis N., Tagmatarchis N., and Tasis D., *Chemical Reviews*, 110, 9, 5366–5397, 2010.

[8] Collins P.G., Arnold M.S., Avouris P., *Science*, 292, 706, 2001.

[9] Groves M.N., Chan A.S.W., Jugroot C.M., Jugroot M., *Chem. Phys. Lett.* 481, 214, 2009.

[10] Kong H., Zhou M., Lin G.D., Zhang H.B., *Catal. Lett.,* 135, 83, 2010.

[11] Ghosh S., Sood A.K., Kumar N., *Science*, 299, 1042, 2003.

[12] Peng S., Cho K., *Nano Lett.*, 3, 5889, 2003.

[13] Fruodakis G.E., *Nano Lett.* 1, 179, 2001.

[14] Harris P.J.F., Carbon Nanotubes and Related Structures: New Materials for the Twenty-First Century, University of Ontario Press, Cambridge, 1999.

[15] Ciraci S., Dag S., Yildirim T., Gulseren O., Senger R.T., J. Phys. *Condens. Matter.* 16, R 901, 2004.

[16] Zakharova G.S., Volkov V.L., Ivanovskaya V.V., Ivanovskii A.L., Usp. Khim. 74, 651, 2005. *Russ. Chem. Rev.* 74, 587, 2005.

[17] Collins P.G., Bradley K., Ishigami M., and Zettl A., *Science*, 287, 1801, 2000.

[18] Kong J., Franklin N.R., Zhou C., Chapline M.G., Peng S., Cho K., and Dai H., *Science*, 287, 601, 2000.

[19] Kong J., *Adv. Mater.* 13, 1384, 2001.

[20] Peng S., Cho K., *Nano. Lett.*, 3, 513, 2003.

[21] Mawhinney D.B., Naumenko V., Kuznetsova A., Yates J.T., Liu J., Smalley R.E., *J. Am. Chem. Soc.*, 122, 2383, 2000.

[22] Picozzi, S., Santucci, S., Lozzi L., Cantalini C., Baratto C., Sberveglieri G., Armentano I., Kenny J.M., Valentini L., Delley B.J., *Vac. Sci. Technol*, A, 22, 1466, 2004.

[23] Wongwiriyapan W., Honda S., Konishi H., Mizuta T., Ikuno T., Ohmori T., Ito T., Shimazaki R., Maekawa T., Suzuki K., Ishikawa H., Oura K., Katayama M., *Jpn. J. Appl. Phys.*, 45, 3669, 2006.

[24] Yim W.L., Liu Z.F., *Chem. Phys. Lett.*, 398, 297, 2004.

[25] Akdim B., Kar T., Duan X., Pachter R., *Chem. Phys. Lett.*, 445, 281, 2007.

[26] Mawhinney D.B., Naumenko V., Kuznetsova A., Yates Jr J.T., Liu J., Smalley R.E., *Chem. Phys. Lett.*, 324, 213, 2000.

[27] Kuznetsova A., Yates Jr J.T., Simonyan V.V., Johnson J.K., Huûman C.B., Smalley R.E., *J. Chem. Phys.* 115, 6691, 2001.

[28] Charlier J.C., Ebbesen T.W., and Lambin *Ph., Phys.* Rev. B. 53, 11108, 1996.

[29] Delley B., An all-electron numerical method for solving the local density functional for polyatomic molecules, *J. Chem. Phys.* 92 (1990), pp. 508–517.

[30] Delley B., From molecules to solids with the $DMol^3$ approach, *J. Chem. Phys.* 113, 7756, 2000.

[31] Perdew J.P., Burke K., Ernzerhof M., *Phys. Rev. Lett.*, 77, 3865, 1996.

[32] Monkhorst H.J., Pack J.D., Special points for Brillouin-zone integrations, *Phys. Rev. B* 13, (12), 5188, 1976.

[33] Guanlun G., Fang W., *I.J. Quantum Chemistry*. 108, 203, 2008.

[34] Haihui J., *Nanotechnology*, 20,145501, 2009.

AUTHOR INDEX